MW00835238

Electron Spin Interactions in Chemistry and Biology

BIOLOGICAL AND MEDICAL PHYSICS, BIOMEDICAL ENGINEERING

The fields of biological and medical physics and biomedical engineering are broad, multidisciplinary and dynamic. They lie at the crossroads of frontier research in physics, biology, chemistry, and medicine. The Biological and Medical Physics, Biomedical Engineering Series is intended to be comprehensive, covering a broad range of topics important to the study of the physical, chemical and biological sciences. Its goal is to provide scientists and engineers with textbooks, monographs, and reference works to address the growing need for information.

Books in the series emphasize established and emergent areas of science including molecular, membrane, and mathematical biophysics; photosynthetic energy harvesting and conversion; information processing; physical principles of genetics; sensory communications; automata networks, neural networks, and cellular automata. Equally important will be coverage of applied aspects of biological and medical physics and biomedical engineering such as molecular electronic components and devices, biosensors, medicine, imaging, physical principles of renewable energy production, advanced prostheses, and environmental control and engineering.

More information about this series at http://www.springer.com/series/3740

Gertz Likhtenshtein

Electron Spin Interactions in Chemistry and Biology

Fundamentals, Methods, Reactions
Mechanisms, Magnetic Phenomena,
Structure Investigation

 Springer

Gertz Likhtenshtein
Department of Chemistry
Ben-Gurion University of the Negev
Beersheba
Israel

and

Institute of Problems of Chemical Physics
Russian Academy of Science
Chernogolovka
Russia

ISSN 1618-7210   ISSN 2197-5647   (electronic)
Biological and Medical Physics, Biomedical Engineering
ISBN 978-3-319-33926-9   ISBN 978-3-319-33927-6   (eBook)
DOI 10.1007/978-3-319-33927-6

Library of Congress Control Number: 2016942036

Printed on acid-free paper

This Springer imprint is published by Springer Nature
The registered company is Springer International Publishing AG Switzerland

Preface

Pivotal role of electron spin interactions in Nature cannot be overestimated. In many processes from energy transduction in biology to specificity of biorecognition these spin interactions appear as defining contributors. Another role involves determining the properties of molecular and functional materials' systems that drive today's technology by finding uses in electronic, spintronic, and magnetic devices.

Electron spin effects are manifesting in the fundamental quantum phenomenon of spin exchange. This interaction enables a number of important elementary processes including electron transfer, triplet energy transfer, and interspin crossing. One should consider interactions of both electronic and nuclear spins including electron–electron, electron–nuclear dipolar, and electron–nuclear contact interactions among the others. Such interactions could be precisely studied by electron magnetic resonance, nuclear magnetic resonance, and related hyphenated resonance techniques providing researchers with unique spectroscopic tools to investigate detailed molecular structure and dynamics of both small and large chemical and biological molecules.

For rather long time the scientists were stuck with paradigm that chemical and biochemical reactions are only ruled by interactions that energetically prevail over the thermal motion. However, contrary to such a strong thermodynamic argument, the last decades of intense research resulted in conclusive evidence of many essential chemical and biological processes being governed by very weak interactions originating from electronic spin systems instead. Advancing the knowledge of molecular mechanisms responsible for photosynthesis in plants and model compounds, radical reactions and influence of magnetic field on these and other processes, as well as rational design of advanced molecular magnets, spintronic devices, catalysts, *etc.*, would not be possible without understanding the spin effects in these systems.

This book represents a collective perspective from physical chemist with long and broad expertise in spin phenomena and related fields. The main intention was not to provide the reader with an exhaustive survey of each topic of vast literature, but rather to discuss the key theoretical and experimental background and focus on recent developments. Thus, chemists and biologists would find the fundamentals of spin phenomena, instrumentation and data interpretation, and a review of the major milestones. This gained knowledge is expected to promote some critical thinking to solve new emerging problem in their fields. Physicists and experts, for example, in magnetic resonance and photoluminescence methods and instrumentation may know already about the above-mentioned technical and quantum mechanical aspects, but would benefit from overview of current problems and achievements in various areas of chemistry and molecular biology, including rapidly evolving fields of natural and artificial photosynthesis, photochemistry, material science, *etc.*

The Chap. 1 of the monograph provides a brief outline of fundamental theories of spin exchange and electron transfer. Non-radiative spin exchange processes involving excited triplet state is the subject of Chap. 2. Electron spin dipolar and electron-nuclear spin contact interactions are described in Chap. 3 as the basis for investigation of molecular structures. A general survey of fundamentals and recent results on spin-selective processes of electron and nuclear spins is presented in the Chap. 4. Three subsequent Chaps. 5–7 are dedicated to experimental methods of investigation of electron spin interactions based on measurements by continuous wave and pulse EPR and by other physical methods. These spectroscopic methods form an experimental basis for investigation of electron spin effects in chosen chemical and physical processes (Chap. 8), effects of magnetic and electromagnetic fields on chemical and biological processes (Chap. 9), establishing structure and spin state of organic and metalloorganic compounds (Chap. 10), and electron transfer in biological systems focusing on the light energy conversion (Chap. 11). Chapter 12 is a brief review of the fundamentals and main results obtained by the methods of spin and triplet (phosphorescence) labels. Chapter 11, Preface and Conclusion have been written in collaboration with professor Alex I. Smirnov

This monograph is intended for scientists working in basic areas related to spin interactions such as spin chemistry and biology, electron transfer, light energy conversion, photochemistry, radical reactions and magneto-chemistry and magneto-biology. The book will be also useful for engineers designing advance magnetic materials, optical and spintronic devices, and photocatalysts. This text as a whole or as separate chapters can also be employed as subsidiary manuals for instructors and graduate and undergraduate students of university physics, biophysics, chemistry, and chemistry engineering departments.

Beersheba, Israel   Gertz Likhtenshtein

Contents

About the Author

 **Gertz Likhtenshtein** received his Ph.D. (1963) and Doctor of Chemical Science (1972) from the Institute of Chemical Physics at the Russian Academy of Science in Moscow, where he was appointed to the position of Head of Laboratory of Chemical Physics of Enzyme Catalysis in 1965, becoming a full professor in 1976.

In 1992 he moved to the Department of Chemistry at the Ben-Gurion University of the Negev, Israel, as a full professor in charge of the Laboratory of Chemical Biophysics and has been an emeritus since 2003. In 2015 Likhteshtein was appointed on a position of the Adviser to the director of Institute of Problem of Chemical Physics of Russian Academy of Science in Chernogolovka, Moscow region. He has authored ten scientific books and around 390 papers, and his many awards include the Medal of the Exhibition of Economic Achievement of USSR (1968), the USSR State Price (1977), the Diploma of Discovery of USSR (1979), Diploma of the Israel Chemical Society (2000). The academician V. V. Voevodsky Award "For Achievements in Investigation of Biological Systems by ESR" (2007), Medal from Academy of Science of Tajikistan Republic (2010), the Gold Medal "Academician N. M. Emanuel, Centenary of Birthday" from Russian Academy of Science and Moscow University (2015). Professor Likhtenshtein was a member of the International ESR Society, the American Biophysical Society, the Israel Chemical Society and Israel ESR Society.

His main scientific interests focus on mechanisms of light energy conversion and novel methods of immunoassay, nitroxide, and antioxidant analysis.

Abbreviation

2D ESTN	Two-dimensional electron spin transient nutation
2D FT ESR	Two-dimensional Fourier transform ESR
2D SECSY	Two-dimensional electron spin echo correlation spectroscopy
BRC	Bacterial reaction center
CE	Cross-effect
CIDEP	Chemical-induced dynamic nuclear polarization
CIDNP	Chemical-induced dynamic nuclear polarization
COSY	Correlated spectroscopy
CP-ENDOR	Circularly polarized radiofrequency fields
CRPM	Pair mechanism
CT	Charge transfer
CW	ESR
DAET	Donor-acceptor energy transfer
DDET	Donor-donor energy transfer
DEER	Double electron-electron resonance
DFT	Density functional theory
DNP	Dynamic nuclear polarization,
DONUT HYSCORE	Double nuclear coherence transfer hyperfine sublevel correlation electron spin echo
DQC ESR	Double quantum coherence pulsed ESR
DSSET	Dexter singlet-singlet energy transfer
DTTET	Dexter triplet-triplet energy transfer
ED ESR	Electrically detected ESR
ELDOR	Electron–electron double resonance
ENDOR	Electron–nuclear double resonance
ESE	Electron spin echo
ESEEM	Electron spin echo envelope modulation
ESR	Electron spin resonance
ESTN	Electron spin transient nutation
ET	Electron transfer

FC	Frank-Condon factor
FID	Free induction decay
FNRO	Fluorophore-nitroxide
FOMO	Fully occupied molecular orbitals
FT ESR	Fourier transform ESR
HFC	Hyperfine couplings
hfi	Hyperfine interaction
HYSCORE	Hyperfine sublevel correlation spectroscopy
ISC	Intersystem crossing
MARY	Magnetically affected reaction yield
MECP	Minimum Energy Crossing Point
MEF	Magnetic isotope effect
MQC	Multiple quantum coherence
MRI	Magnetic Resonance Imaging
MTSL	(1-Oxy-2,2,5,5-tetramethyl-3-pyrroline-3 methyl) methanethiosulfonate
mw	Microwave radiation
NMR	Nuclear Magnsetic Resonance
NOE	Nuclear Overhauser effect
NRO	Nitroxide radical
OEC	Oxygen evolution complex
OLED	Organic light-emitting diode
PDT	PhotoDynamic Therapy
PELDOR	Pulse electron-electron double resonance
PES	Potential energy surface
PIET	Photoinduced electron transfer
PS I	Photosystem I
PS II	Photosystem II
QBE	Quantum beats effect
RIDME	Relaxation-induced dipolar modulation enhancement
RPM	Radical pair mechanism
RTPM	Radical triplet pair mechanism
rw	Radiowave
RYDMR	Reaction yield detection magnetic resonance
SCRP	Spin-correlated radical pair
SE	Spin-exchange
SECSY	Spin echo correlation spectroscopy
SEM	Spin-echo measurements
SMF	Static magnetic fields
SO	Spin-orbit
SOMO	Singly occupied molecular orbital
SR ESR	Saturation recovery ESR
SRM	Saturation recovery measurements
SS	Spin–spin
SSET	Singlet-singlet energy transfer

TM	Triplet mechanism
TNP	Transient nuclear polarization
TR-ESR	Time-resolved ESR
TRIPLE	Electron-nuclear-nuclear triple resonance
TTA	Triplet-triplet annihilation
TTET	Triplet-triplet energy transfer
ZFS	Zero-field splitting

Abstract

The book fills the need to converge understanding of spin effects in chemistry and biology by adopting a chemist view. Electron spin effects in fundamental processes of spin exchange and dipole interactions, triplet triplet energy transfer and annihilation, intersystem crossing, electron and nuclear spins polarization effects are reviewed from both theoretical and experimental perspectives. The physical interpretation of these phenomena forms a basis for understanding molecular mechanisms responsible for essential chemical and biological processes including photosynthesis and influence magnetic field on chemical and biochemical reactions. This knowledge guides the design of advanced molecular magnets, spintronic devices, and photocatalysts, as well as developments of new methods for studying the molecular structure and dynamics of chemical and biological systems.

Chapter 1
Basic Conceptions: Spin Exchange and Electron Transfer

Abstract Spin electron interactions play crucial roles in diverse topics related to electron transfer, environmental and biological issues, respiration, photosynthesis, radical reactions and electron transfer. Biological electron transfer (ET) reactions are essential for the conversion of energy from food or sunlight into the universal energy currency, adenosine triphosphate. Interest to radical processes including oxidative stress, antioxidant activity, some pathology, aging, etc. is fuelled by the many promising applications in biomedicine and clinical medicine. This Chapter is organized as follows. The first two Sections provide a brief preliminary overview on basic conceptions and definitions and the primary quantum mechanical grounds regarding for readers who may be less familiar with the spin physics. In next two Sections, we have presented basic information about different types of spin exchange processes, namely, static direct exchange and super exchange in an encounter complex in solution. Classic theories of electron transfer by Marcus, Levich, Jortner and Zusman and recent progress in the area have been considered. The close interrelations between spin exchange and electron transfer have been emphasized.

1.1 Basic Conceptions and Definitions

Spin of electron or nucleus is an intrinsic form of **angular momentum**, which is related to the rotation or revolution along its axes. Physical behaviour of spin can be described in the frame of two fundamental models. According to a "classical" model; spin, as any charge particle, possesses a **magnetic dipole moment** created by its rotation (Figs. 1.1 and 1.2). From another hand, spin is quantum mechanical species (elementary standing wave) and the spin angular momentum, S, is quantized. The S values can be 0, 1/2, 1, 3/2, 2, etc.

In some cases the spin magnetic moment may be considered as **a spin vector** having a magnitude and direction, and interaction between spins may follow the vector algebra rules.

© Springer International Publishing Switzerland 2016
G. Likhtenshtein, *Electron Spin Interactions in Chemistry and Biology*,
Biological and Medical Physics, Biomedical Engineering,
DOI 10.1007/978-3-319-33927-6_1

Fig. 1.1 Schematic diagram depicting the spin of the electron as the bold arrow and magnetic field lines associated with the spin magnetic moment. https://en.wikipedia.org/wiki/Spin_(physics)

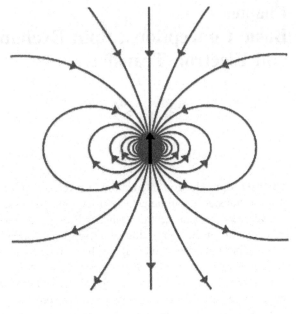

Fig. 1.2 Quantization of the spin angular momentum. https://en.wikipedia.org/wiki/Spin_quantum_number

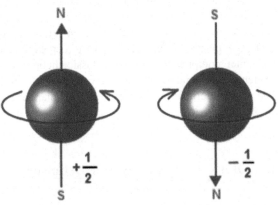

The **spin Larmor precession** (elementary gyroscope motion) is the precession of the spin magnetic moment about an external magnetic field (Fig. 1.3). This model is widely employed in magnetic resonance, in pulse methods in particular.

The **Spin nutation** is a rocking, swaying, or nodding motion in the axis of rotation of a axially symmetric elementary particle bearing spin.

Spin echo is the refocusing (recovery) of spin magnetisation in magnetically non-homogenic media by certain pulse consequence of resonant electromagnetic radiation.

Spin spin dipole–dipole interaction (dipolar coupling) refers to the direct interaction between two magnetic spin dipoles.

Fig. 1.3 Direction of precession for a negatively-charged particle. The *large arrow* indicates the external magnetic field, the *small arrow* the spin angular momentum of the particle. https://en.wikipedia.org/wiki/Larmor_precession

Spin exchange is phenomenon in which the orientation of the oppositely polarized spins may change. The spin-exchange interaction preserves total angular momentum and energy of the system.

Fermi contact interaction is the quantum mechanical interaction between a spin electron and spin nucleus caused by space distribution of electronic cloud resulting in presence of electron spin density on the nucleus.

Spin polarization is the degree to which the spin is aligned with a given direction. In magnetic resonance, a system is called to be polarized if the population of spins oriented along and opposite magnetic field deviates from the Boltzmann distribution.

Spin relaxation is the mechanism by which a spin system, primary deviated from the Boltzmann distribution, exponentially decays towards its equilibrium value.

Electron Magnetic (spin) resonance (EMR) and **Nuclear Magnetic Resonance (NMR)** are physical phenomena in which electrons or nuclei in a magnetic field absorb and re-emit electromagnetic radiation at a specific resonance frequency.

Spin Conservation Rule (Wigner rule) is the fundamental law according to which for any radiative and radiationless transitions, one observes that transitions between terms of the same multiplicity are spin-allowed, while transitions between terms of different multiplicity are spin-forbidden.

Spin-Selective Reactions are processes strongly depending on the spin conservation rule.

Spin chemistry is the interdisciplinary area covering kinetics and mechanisms of spin-selective reactions and physical methods of its investigation. Spin chemistry methods can provide valuable information on spin dynamics, magnetic properties and lifetimes of short-lived radical pairs.

Spin label is a molecule bearing spin, stable radical or metal complex, tethering covalently to an molecular object of interest.

Spin probe is a molecule bearing spin freely diffusing in solution.

1.2   First Elements of Quantum Mechanics

The starting point for the interpretation of quantum mechanical phenomena related to the spin electron interactions is the Shrodinger equation [1, 2] for a model hydrogen molecule-like system with two independent electrons. According to this equation, the antisymmetric combination of product of the single electron wavefunctions, i.e. $\Phi_a(r_1)$ for the first electron and $\Phi_b(r_2)$ for the second electron in position space is:

$$\Psi_A(r_1, r_2) = \frac{1}{\sqrt{2}}[\Phi_a(r_1)\Phi_b(r_2) - \Phi_b(r_1)\Phi_a(r_2)] \qquad (1.1)$$

and the symmetric combination of the product wave functions in position space is:

$$\Psi_S(r_1, r_2) = \frac{1}{\sqrt{2}}[\Phi_a(r_1)\Phi_b(r_2) + \Phi_b(r_1)\Phi_a(r_2)] \qquad (1.2)$$

The energies E_+ for the spatially symmetric solution and E_- for the spatially antisymmetric solution are given as

$$E_{+/-} = E_{(0)} + \frac{C \pm J_{ex}}{1 \pm B^2} \qquad (1.3)$$

where

$$C = \int \Phi_a(r_1)^2 \left(\frac{1}{R_{ab}} + \frac{1}{r_{12}} - \frac{1}{r_{a1}} - \frac{1}{r_{a2}}\right) \Phi_b(r_2)^2 dr_1\, dr_2 \qquad (1.4)$$

is the **Coulomb integral**,

$$B = \int \Phi_b(r_2)\Phi_a(r_2)\, dr_2 \qquad (1.5)$$

is the **overlap integral**, and

$$J_{ex} = \int \Phi_a^*(r_1)\Phi_b^*(r_2)\left(\frac{1}{R_{ab}} + \frac{1}{r_{12}} - \frac{1}{r_{a1}} - \frac{1}{r_{b2}}\right)\Phi_b(r_1)\Phi_a(r_2)\, dr_1\, dr_2 \qquad (1.6)$$

is the **exchange integral**.

The terms in parentheses in (1.4–1.6) correspond to proton–proton repulsion (R_{ab}), electron–electron repulsion (r_{12}), and electron–proton attraction ($r_{a1/a2/b1/b2}$). The strength of the exchange interaction is related to the wavefunction overlap which decreases approximately exponentially when the proton–proton distance increases.

For two-electron systems singlet and triplet states are characterized by four wave functions [3]

$$
\begin{aligned}
{}^{1}\Psi_0 &= \frac{1}{\sqrt{2}}(|\phi_a\alpha\phi_b\beta| - |\phi_a\beta\phi_b\alpha|) \\
{}^{3}\Psi_1 &= |\phi_a\alpha\phi_b\alpha| \\
{}^{3}\Psi_0 &= \frac{1}{\sqrt{2}}(|\phi_a\alpha\phi_b\beta| + |\phi_a\beta\phi_b\alpha|) \\
{}^{3}\Psi_{-1} &= |\phi_a\beta\phi_b\beta|
\end{aligned}
\tag{1.7}
$$

where ϕa and ϕb are orbital functions and α and β the spin function indicating $m_s = \frac{1}{2}$ and $-1/2$, respectively.

The singlet-triplet (S-T) splitting, E_{S-T}, is $2J$, the magnitude of the exchange interaction (Fig. 1.4) which contributes to the Heisenberg spin Hamiltonian:

$$
H_{ex} = -2JS_1S_2
\tag{1.8}
$$

where S_1 and S_2 are spin operators. When J is positive, the triplet states have lower energy than the singlet states, that is the ferromagnetic (spin parallel state) is favored compared to the antiparallel, antiferromagnetic state

In quantum mechanics, the **Bra-ket notation** which was introduced in 1939 by Paul Dirac [4] and is also known as **Dirac notation**, is used as a standard notation for describing quantum states. The inner product of two states is denoted by a <bra| c|ket>; $\langle\phi|\psi\rangle$ consisting of a quantity, $\langle\phi|$, called the **bra** and a quantity, $|\psi\rangle$, called the **ket**. "The **Bra-ket notation** describes quantum leap from the initial state $|\psi\rangle$ to the final state $|\phi\rangle$ with probability of $|\langle\psi|\phi\rangle|^2$. They can be put into one-to-one

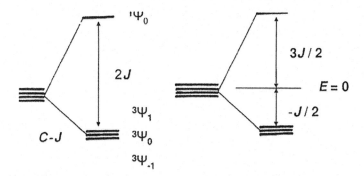

Fig. 1.4 Energy shift due to Coloumb (C) and exchange (I) integrals which attributed to the Heisenberg spin Hamiltonian H_{ex} [3]

correspondence by the operation of complex conjugation which gives a precise meaning to the symbols $\langle\phi|$ and $|\phi\rangle$. For the two-electron integrals, which are defined over spatial orbitals as follows:

$$(ij|kl) = \int \phi_i^*(\mathbf{r}_1)\phi_j(\mathbf{r}_1)\frac{1}{\mathbf{r}_{12}}\phi_k^*(\mathbf{r}_2)\phi_1(\mathbf{r}_2)\,d\mathbf{r}_1\,d\mathbf{r}_2 \qquad (1.9)$$

In many physical systems, electron exchange transitions are accompanied by vibrational, rotational and translational nuclear processes, which form (constitute) a continuum of energy. **Fermi's golden rule** [5] is a way to calculate the transition rate (probability of transition per unit time) from one energy eigenstate of a quantum system into a continuum of energy eigenstates, due to a perturbation. The one-to-many transition probability per unit of time from the state $|i\rangle$ to a set of final states $|f\rangle$ is given, to first order in the perturbation, by

$$T_{i\to f} = \frac{2\pi}{h}|\langle f|H'|i\rangle|^2\rho, \qquad (1.10)$$

where ρ is the density of final states (number of states per unit energy) and $\langle f|H'|i\rangle$ is the matrix element (in bra-ket notation) of the perturbation H' between the final and initial states. This transition probability is also called the decay probability and is related to the mean lifetime of the initial state. Another form of the Fermi's golden rule for a transition with rate constant k_{tr} is

$$k_{tr} = \frac{J_{ex}^2}{h}FC \qquad (1.11)$$

which includes the exchange integral (an electronic coupling term) J_{ex} depending on the overlap of electronic wave functions in the initial and final state of the process; FC is the Franck-Condon factor related to the probability of vibrational and translational transition states accounting for the effects of nuclear motion. FC factor reports on the much slower changes in the nuclear configuration, whereas the exchange integral depends on the instantaneous electronic configuration in the Born-Oppenheimer approximation.

This chapter briefly describes fundamental concepts and current trends in work related to spin exchange and electron transfer.

1.3 Spin Exchange

The following four types of spin exchange interactions are commonly identified:6 (1) direct exchange of pairs with fixed spin-spin distance (r), (2) super exchange through a bridge (3) dynamic exchange in low viscosity solution during an encounter of two radicals (or paramagnetic complexes) bearing spins and (4) exchange in pairs with flexible bridges.

1.3.1 Direct Exchange

The direct spin exchange process with spin Hamiltonian H_{ex} (1.8) occurs via direct overlap of the orbitals of the interacting species (Scheme 1.1).
According to the available data [6] and references therein

$$J = J_0 \exp\left(-\beta r\right). \tag{1.12}$$

where J_0 is the exchange integral at a certain (for example Van der Waals) distance, r is the spin-spin distance and β—is the distance decay (attenuation) parameter.
The Hamiltonian for a trimeric spin system can be written as [7]

$$H = -2(J_1 s_2 s_3 + J_2 s_3 s_1 + J_3 s_1 s_2) \tag{1.13}$$

where all three exchange integrals are in principle different [7]. For a system consisting of a number of spin pairs with different J_{ex}, the appropriate simplified expression for the exchange interaction is

$$H_{ex} = -2\sum_{ij} J_{ij} \tag{1.14}$$

The magnetic interaction between the spins S_1 and S_2 for paramagnetic centers 1 and 2 of many electrons atoms is written in the form suggested by Heisenberg, Dirac, and Van Vleck Hamilnonian (H_{HDVV}) is written as [7]

$$H_{HDVV} = J s_a s_b \tag{1.15}$$

where s_a and s_b are the total spins of the many-electron atoms:

$$
\begin{aligned}
s_b &= \sum_i s_{ib} \\
s_a &= \sum_i s_{ia}
\end{aligned}
\tag{1.16}
$$

H_{HDVV} operates in the space of wavefunctions associated with total spins s_a and s_b.
In comprehensive review phenomenological aspects of spin Hamiltonians (SHs), was presented [8]. Interactions involving spin such as electronic (EZ), and nuclear Zeeman (NZ), terms, electron-exchange interaction (EE), zero-field splitting (ZFS), spin–spin (SS), spin-orbit (SO), nuclear quadrupole (NQ), and hyperfine couplings (HFCs) were briefly described.

Scheme 1.1 Direct spin exchange

A(↑)B(↓) → A(↓)(B(↑)

1.3.2   Superexchange

In some physical systems, spin exchange takes place by superexchange through a bridge or a through space interaction mediated by the molecules of the intervening medium (Scheme 1.2).

In 1959, Anderson [9] in his pioneering work used a perturbational approach to relate the singlet-triplet splitting of a two-spin system to the magnitude of the electron-transfer superexchange coupling between the radical ion pair (RP) state and surrounding states n and that state to which it is coupled at the nuclear coordinate of the relaxed RP state (see details in Sect. 8.4.1). Then McConnel [10] has shown that the electronic coupling matrix element (H_{AB}) between two redox centers separated by a covalent bridge composed of n identical repeat units depends on the coupling strength between the redox sites and the bridge (h_{Ab}, h_{bB}), the coupling between adjacent bridge elements (h_{bb}), and the tunneling energy gap ($\Delta\varepsilon$). The latter was defined as the virtual energy required to remove an electron from the donor, or a hole from the acceptor, and place it on the bridge.

The superexchange model is illustrated in Fig. 1.5

Scheme 1.2 Spin exchange $A(\uparrow)X(\uparrow\downarrow)B(\downarrow) \rightarrow A(\downarrow)X(\downarrow\uparrow)B$
via a bridge

Fig. 1.5 Illustration of the individual parameters involved in the superexchange model. A: acceptor, D: donor, b: bridging unit, h_{ij} coupling strength [11]

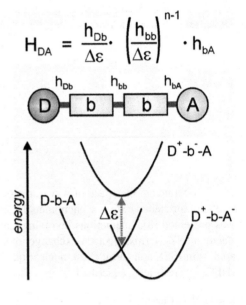

The electronic coupling h_{XY} and H_{DA} was shown to decrease exponentially with distance (d), via the distance decay constant

$$\beta = \frac{2}{\delta} \ln\left(\frac{\Delta\varepsilon}{h_{bb}}\right) \tag{1.17}$$

In references [12–17] the interaction of the donor/acceptor with the bridge was also treated as a perturbation. The electronic coupling matrix element V_{DA} represents the interaction of the electron donor (D) and acceptor (A) orbitals with the bridge (X) and ΔE_X is the energy of the bridge acceptor) orbital relative to the energy of the donor orbital. It was suggested that the donor and the acceptor orbitals have the same energy in the transition state. According to this theory, the electronic coupling (V_{DA}) is given by the expression:

$$V_{DA} = \frac{\sum D_{AX}D_{XA}}{\Delta E_X} \tag{1.18}$$

for spins connected by a bridge X. In more general cases, one finds:

$$J_{AB} = \frac{\sum J_{AX}J_{XA}}{\Delta E_X} \tag{1.19}$$

Pathways tunneling model suggested in [15] treats superexchange as a sequence of steps, each of which is mediated by a covalent bond, a hydrogen bond, or vacuum, and the electronic coupling is proportional to a product of penalties for each step:

$$V_{AB} = V_0 \prod_{1}^{N} \varepsilon_i, \tag{1.20}$$

where V_0 is the coupling between the donor and the first bond of the path way and ε_i is the attenuation factor associated with the exponential decay of electron density from one bond to another and the distance decay parameter (β), taken as $\beta_1 = 1.7\ \text{Å}^{-1}$.

Values of decay factors for longdistance superexchang processes and spin exchange attenuation parameters for specific bridge groups are presented in Section X.

Theoretical aspects of exponential decay of conductance/electron transport and of exchange spin coupling with increasing bridge length, or the increased "conductivity" through the diarylethene and dithiolated bridges in antiferromagnetically coupled diradicals were considered [17]. It was conclude that that common property trends result from either different pairs of orbitals being involved, or from orbital energies. A general survey on the bridged nitroxyl radicals is presented in Section X,

1.3.3   *Dynamic Spin Exchange in Solution*

In solutions of low viscosity, electron spin-spin exchange occurs during an encounter of radicals or paramagnetic complexes [18] (Scheme 1.3).

Under these conditions, electron spin-spin dipolar interactions between paramagnetic species are averaged by fast rotation of the molecules and the short lifetime of the encounter complex and are therefore very weak. A general theory has been developed for exchange relaxation during encounters between paramagnetic particles in solution [18]. According to the theory the probability of spin relaxation of paramagnetic species during encounters depend in the general case on a number of parameters such as the spin-spin $(1/T_{2s})$ and spin-lattice $(1/T_{1s})$ relaxation rates in the presence of a paramagnetic species, the exchange integral (J), the duration of the encounters, and the differences between the resonance frequencies of the spins (δ).

The rate constant of the exchange relaxation (k_{ex}) can be determined by an analysis of the broadening of the Lorentzian line of the radical or from the change in the rates of the spin-spin $(1/T_{2s})$ and spin-lattice $(1/T_{1s})$ relaxation in the presence of a paramagnetic species:

$$\Delta\left(\frac{1}{T_{1s}}\right) = k_{ex}C \tag{1.21}$$

Since $1/T_{1s}$ is usually more sensitive to spin-spin interaction than $1/T_{2s}$, especially in the region of slow rotation and in solids, this method widens the accessible range of k_{ex} values [19–21].

It has been shown [18] for a radical of spin $S = \frac{1}{2}$:

$$k_{ex} = P_{ex}k_d = \frac{f_g f_{ns} k_d J^2 \tau_c^2}{\left(1 + J^2 \tau_c^2\right)} \tag{1.22}$$

where k_d is the rate constant of encounters in solution, P_{ex} is the probability of spin exchange during the life time (τ_c) of the encounter complex, f_g is a geometric steric factor, f_{ns} is a nuclear statistical factor and J is the exchange integral of interaction in the encounter complex arising from direct contact between the particles.

If $J^2\tau_c^2 \gg 1$, k_{ex} is independent of J (strong exchange) and

$$k_{ex} = f_g f_{ns} k_d \tag{1.23}$$

Scheme 1.3 Dynamic spin exchange

$$\mathbf{A(\uparrow) + B(\downarrow) \rightarrow A(\uparrow)B(\downarrow) \rightarrow A(\downarrow)(B(\uparrow) \rightarrow A(\downarrow) + (B(\uparrow)}$$

In the case of weak exchange $J^2\tau_c^2 \ll 1$ and

$$k_{ex} = f_g f_{ns} k_d J^2 \tau_c^2 \tag{1.24}$$

The theory of dynamic exchange appeared to be a basis for development of several technologies [19–23] such as spin label—spin probe methods (Section X), spin label oximetry [20] and measurement of the depth of immersion of paramagnetic centers [21].

Substituting the definition of J from (1.8), with $\beta = 1.3$ Å^{-1} into (1.12) gives

$$k_{ex} = f_g f_{ns} k_d \tau_c^2 10^{28} \exp 2[-1.3(R_0 - r_v)] \tag{1.25}$$

This equation can be used for estimation of the distance of closest approach (or depth of the paramagnetic center immersion), R_0, if other factors in the equation are known[33–34]. For example, in water solution a typical value of f_g f_{ns} $k_d = 2 \times 10^{-9}$ M^{-1} s^{-1} for interaction between a nitroxide radical and a paramagnetic species and $\tau_c = 5 \times 10^{-11}$ s. For an object with a buried paramagnetic center, if the experimental $k_{ex} = 10^{-7}$ M^{-1} s^{-1}, the value of $(R_0 - r_v) = 10$ Å.

1.3.4   Dynamic Spin Exchange in Flexible Biradicals in Solution

In solutions of low viscosity, electron spin-spin exchange in flexible biradicals, in which two stable radicals linked via a flexible bridge, occurs during an encounter of radicals or paramagnetic complexes [22, 23] (Scheme 1.4).

The exchange coupling of two unpaired electrons provides information about dynamics of molecular environment in various systems. In a system of two identical radical fragments with electron spins operators S_1 and S_2 and nuclear spin operator I_1 and I_2 each has the interactions in the systems in the presence of magnetic field H_0 are described by the following Hamiltonian:

$$H = g_e \beta_e H_0 (S_z^1 + S_z^2) + a(S_z^1 I_z^1 + S_z^2 I_z^2) + J S_1 S_2 \tag{1.26}$$

where index z notes projections of the electron and nuclear spins on the Z-axis; g_e is the isotropic g-factor of the radical fragments; β_e is the Bohr magneton; a denotes isotropic hyperfine splitting constant; and J is the exchange integral. According to theory developed in [22] the ESR spectra of a hypothetical biradical depend on the difference in the resonance frequencies of biradical fragments 1 and 2 ($\Delta\omega$) the

A(↑)⁓⁓B(↓) → [A(↑)B(↓) → A(↓)(B(↑)] → A(↓)⁓⁓ (B(↑)

Scheme 1.4   Dynamic spin exchange in flexible biradicals

residence times τ_1 and τ_2 of the contact and remote states of the radical fragments, respectively, the contact (J_{con}) and average (J_{av}) integrals and fractions of the contact (P_1) and remote (P_2) conformations. Examples of thorough investigations various flexible biradical of different structure at wide range of temperature, viscosity and solvent polarity may be found in and references therein [23].

1.4 Electron Transfer

1.4.1 Prerequisite

Electron transfer (ET) is one of the most ubiquitous and fundamental phenomena in Nature. ET is found to be a key elementary step in many important processes in chemistry, biology and physics involving isolated molecules and super molecules, ions and excess electrons in solution, condensed phase, surfaces and interfaces, electrochemical systems, etc. The field has an extensive literature and new studies are constantly being undertaken. This section will give a brief description of the fundamentals of modern electron transfer theory and provide an overview of recent developments in the field.

Electron transfer occurs when an electron moves from one chemical species to another chemical species. The general approach to describe the process of electron transfer (ET) is based on Fermi's golden rule (1.11) and the Born-Oppenheimer approximation, which assumes that nuclei behave quasi-statically and that electrons move on a time scale that is much faster than characteristic nuclear time scales, owing to the greater inertia of the nuclei compared to the electrons. In (1.11), the factors V^2/h and FC are related to the motion of electrons and nuclei respectively.

Electron transfer (ET), as in any other chemical reaction, is accompanied by a change in the nuclear and electronic configurations. According to the commonly accepted model one can visualize ET by constructing a graph (energy E or Gibbs free energy ΔG versus reaction coordinate (Fig. 1.6). Thus, ET is considered as the transition from an initial electronic term (i) to a final electronic term (f) via an interaction that can be modeled as a term crossing.

There are three main effects that influence the ET rate constant (k_{ET}): (1) the process driving force (ΔG_0); (2) the likelihood of sampling the crossover region due to availability of thermal fluctuations; (3) the values of the Bolzman factor exp ($-E_A/RT$), where E_A is the activation energy; and (4) the probability of quantum mechanical jumping (tunneling) of the electronic cloud from its initial to the final configuration (a_{tr}) in the region around the reaction coordinate Q_{tr}:

Thus, the rate constant of the transition (k_{tr}) in the crossing area is dependent on the height of the energetic barrier (the free energy of activation $\Delta G^{\#}$), the frequency of sampling the crossing area (v) and the transition coefficient (a_{tr}):

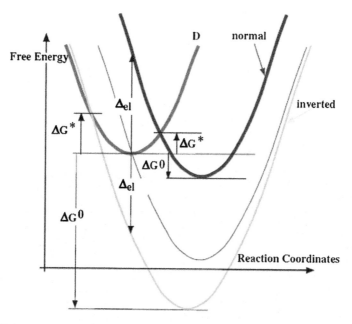

Fig. 1.6 Energy versus reaction coordinates for the reactants (curve D) and the products for normal reaction (curve An), inverted reaction (curve Ai) and at the inversion curve (curve Ac) and standard Gibbs free energy ΔG_0 The electron in the initial state requires a positive excitation energy Δ_{el} for the normal reaction, and a negative excitation energy Δ_{el} for the inverted reaction which could be directly emitted as light [24]

$$k_{tr} = \alpha_{tr} v \exp\left(-\Delta G^{\#}\right) \tag{1.27}$$

All quantitative approaches to electron transfer has been emerged from the fundamental Landau-Zener equation [25, 26] The probability of a transition (a_{tr}) around the reaction coordinate Q_{tr} is given as follows:

$$\alpha_{tr} = \frac{2P}{(1+P)} \tag{1.28}$$

where

$$P = 1 - \exp\frac{-4\pi^2 V^2}{hv\left(S_i - S_f\right)} \tag{1.29}$$

Hear V is the electronic coupling factor (the resonance integral), v is the velocity of nuclear motion, and S_i and S_f are the slopes of the initial and final terms in the Q_{tr} region. The coupling V is

$$V = H_{ab} = \langle \Psi_a | H | \Psi_b \rangle \tag{1.30}$$

where H is the system Hamiltonian. If the exponent of the exponential function is small, then

$$P = \frac{4\pi^2 V^2}{hv(S_i - S_f)} \tag{1.31}$$

and the process is non-adiabatic. Observe that the smaller the magnitude of the resonance integral V the lower the velocity of nuclear motion and the smaller the difference in the curvature of the terms (the difference $S_i - S_f$). The result of these considerations is a reduction in the probability of non-adiabatic transfer. At $P = 1$ the process is adiabatic and obeys classical Arrhenius or Eyring equations.

The theory stressed the key role played by the transition coefficient, which is quantitatively characterized by the value of electronic coupling (resonance integral) V in forming the energetic barrier.

1.4.2 Electron Transfer Theories

Basic Conceptions

The Marcus theory, which provides the basic theoretical description of electron transfer reactions in polar media, appeared to be remarkable breakthrough achievement. According to the fundamental Marcus two state model [27, 28] the distortion of the reactants, products and solvent from their equilibrium configuration is described by identical parabolas in reaction coordinate space for the initial and final states. The process driving force of the value ΔG_0 causes a shift of the parabolas. Within the adiabatic regime (strong electronic coupling, where the resonance integral $V > 200$ cm^{-1}), The value of the electron transfer rate constant is given by the Eyring theory of the transition state:

$$k_{ET} = \frac{k_B T}{h} \exp - \left[\frac{(\lambda + \Delta G_0)^2}{4\lambda k_B T} \right] \tag{1.32}$$

where λ is the reorganization energy defined as the energy for electron excitation without distortion of the nuclear frame. The following relationships based on log $k_{ET} - \Delta G_0$ graphs and dependent on the relative magnitudes of λ and ΔG_0 were predicted from the Marcus model (1.30) (Fig. 1.6): (1) $\lambda > \Delta G_0$, log k increases if ΔG_0 decreases (normal Marcus region), (2) $\lambda = \Delta G_0$, the reaction becomes barrierless, and (3) $\lambda < \Delta G_0$, log k decreases with increasing driving force.

When the initial state distribution remains in thermal equilibrium throughout the ET process, the driving force of the process is related to the standard Gibbs energy (ΔG_0). If the ET occurs, for account of the fast vibrational modes, before the

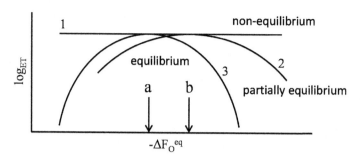

Fig. 1.7 Schematic representation of the dependence of the ET constants logarithm on the equilibrium Gibbs energy $\Gamma 0$: _1_ non-equilibrium conformational and solvational processes; _2_ partial non-equilibrium processes, λ^{neq} and ΔG_0^{neq} are slightly dependent on G_0; and _3_ equilibrium processes. _Arrows a_ and _b_ indicate the conditions for the maximum $\lambda = \Delta G_0$ and $\lambda^{neq} = \Delta G_0^{neq}$, respectively [16]

formation of conformational and solvatational states of the medium, the thermodynamic standard energy for the ET (ΔG_0^{neq}) appears to be less than that which is observed in the case of equilibrium dielectric stabilization of redox centers ΔG_0^{12} (Fig. 1.7) [16, 29].

One powerful approach for studying the effect of solvent dynamics on electron transfer reactions was developed in [30, 31]. According to the Zusman equation (ZE), (1.3) which is based on a description of the solvent complex permittivity in the Debye theory, the dynamics of the electronic degrees of freedom is coupled to a collective nuclear coordinate:

$$\Delta G^{\neq} = (\Delta G_0 + Ep)^2 / 4Ep \qquad (1.33)$$

where

$$Ep = (8\pi)^{-1}(1/\varepsilon_m - 1/\varepsilon_s) Int\, \Delta D^2(r) dr \qquad (1.34)$$

and ε_m is the AC permittivity and ε_s is the DC or static permittivity of the solvent, ΔD (r) is the difference of inductions. In (1.32) the "equilibrium" reorganization energy λ in the corresponding classic Marcus equation (1.35) was replaced with the dynamic reorganization energy (Ep) of the slow degree of freedom of the solvent, while the pre-exponential factor for non-adiabatic reaction includes the coupling factor (V) and effective frequency υ_{ef} of the solvent.

In recent work [32], a brief overview was presented to describe the current nonequilibrium solvation theories and application models. The constrained equilibrium principle in classical thermodynamics and its application to the nonequilibrium solvation were discussed in the framework of the Nonequilibrium Solvation Theory based on Constrained Equilibrium Principle section Solvent Reorganization. The constrained equilibrium was suggested to be a superposition of two equilibrium polarizations; one is the final equilibrium and the other is the equilbrium with constraining external charge and the residual potential. The expression of solvent reorganization energy, with including the electrostatic free energy change of the solvent to the free energy change of the system, was also given.

The theory developed by Levich, Dogonadze [33, 34] was marked advances in understanding non-adiabatic electron transfer between donor (D) and acceptor (A) centers. The theory is based on Fermi's Golden Rule (1.11), the Landau-Zener (1.28). and the Marcus formula (1.32). Using these concepts, the authors proposed a formula for non-adiabatic ET

$$k_{ET} = \frac{2\pi V^2}{h\sqrt{4\pi\lambda k_B T}} \exp\left[-\frac{(\lambda + \Delta G_0)^2}{4\lambda k_B T}\right]$$ (1.35)

where the maximum rate of ET at $\lambda = \Delta G_0$ is given by

$$k_{ET(\text{max})} = \frac{2\pi V^2}{h\sqrt{4\pi\lambda k_B T}}$$ (1.36)

The next principle step in development of the electron transfer theory was taking in account high-frequency vibrational modes involved in the elementary processes. In such a case, a molecular system can be primarily driven by its intramolecular vibrations, which are typically much faster than the solvent reorganization. The involvement of intramolecular high-frequency vibrational modes in electron transfer has been considered for various cases [35–38]. For example, according to [36], when in the low-temperature limit high-frequency modes (hυ) are involved in the electron transfer and the solvent dynamic behavior can be treated classically, the rate constant for non-adiabatic ET in the case of parabolic terms, is given by

$$k_{ET} = \frac{\sum_j 2\pi F_j V^2}{h\lambda k_B T} \exp\left[-\frac{(jhv + \lambda_s + \Delta G_0)^2}{4\lambda k_B T}\right]$$ (1.37)

where j is the number of high-frequency modes, $F_j = e^{-S}/j!$, $S = l_v/h\upsilon$ and l_v and l_s are the reorganization inside the molecule and solvent, respectively.

Anderson pioneered the establishing a principle connection between exchange integral and electron transfer coupling factor [9]. In the Anderson equation for a radical (ion-radical) pair RP: the indicated matrix elements couple the singlet and triplet RP states to states n, E_{RP} and E_n are energies of these states, respectively, and λ is the total nuclear reorganization energy of the charge-transfer reaction [39, 40]:

$$2J = \Delta E_S - \Delta E_T$$
$$= \left[\sum_n \frac{|\langle\Psi_{RP}|V_{RP-n}|\Psi_n\rangle|^2}{E_{RP} - E_n - \lambda}\right]_S - \left[\sum_n \frac{|\langle\Psi_{RP}|V_{RP-n}|\Psi_n\rangle|^2}{E_{RP} - E_n - \lambda}\right]_T$$ (1.38)

The Anderson approach was recently effectively exploited [39, 40]. See also Section X.

Further Developments

This section provides a brief overview of the recent developments and applications of electron transfer theories with primary focus on the various method of calculation of the coupling factor, role vibration modes in reorganization energy, a connection of coupling factor and nuclear motion dynamics (fluctuation-controlled electron tunneling), processes near the inversion point reactions involving a change in spin-state that are formally forbidden, and oscillation phenomena.

Electron Transfer Electronic Coupling

To address a challenging problem of calculation of electronic coupling H_{ab}, a linear correlation between H_{ab} of the two diabatic electronic wave functions for initial and final electron transfer (ET) states, Ψ_a and Ψ_b, respectively ($H_{ab} = \langle \Psi_a | H | \Psi_b \rangle$ and the corresponding overlap $S_{ab} = \langle \Psi_a | \Psi_b \rangle$ has been established [41]. The calculation of H_{ab} and S_{ab} was based on computational scheme in which the overlap of the diabatic state wave function is estimated by the overlap of the highest singly occupied molecular orbital of donor and acceptor. The H_{ab} and S_{ab} correlation was found for a number compounds (Fig. 1.8) [41]:

The transition flux theory was applied to calculation of the coupling matrix element of longdistance electron transfer reactions [42]. It was suggested that in donor (D) acceptor (A) centers located in regions Q_D and Q_A, respectively, interatomic currents J_{ab} connected to the coupling matrix element. The connection is as follows:

$$V_{DA} = -h \sum_{a \varepsilon Q_D, b \varepsilon Q_A} J_{ab} \tag{1.39}$$

Fig. 1.8 Correlation between electronic coupling matrix element (H_{ab}) from the fragment orbital DFT method and exact overlap (S_{ab}) between SOMO orbitals of donor and acceptor [41]

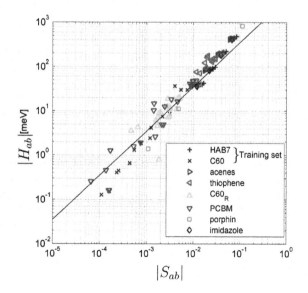

where V_{DA} is the sum of interatomic currents crossing the dividing surface is calculated, interatomic currents Jab, which are currents flowing between atoms. The calculation was performed in framework of the Golden Rule, taking in consideration the multielectronic nature of the transition coupling matrix element, and the electronic Franck—Condon factor.

The Condon approximation, average medium versus fluctuation-controlled electron tunneling, gated and solvent relaxation controlled electron transfer, and the influence of inelastic tunneling on electronic coupling pathway interferences were considered as key factors in modern nonadiabatic theory of electron transfer. It was stressed that nuclear motion also induces fluctuations in the donor-acceptor (DA) electronic coupling For evaluating the electronic Hamiltonian coupling associated with photoinduced electron transfer (PIET) an approach based on constrained real-time time dependent density functional theory (C-RT-TDDFT) using constrained DFT (C-DFT) was developed [44]. In combination with the semi-classical Marcus-Levich-Jourtner theory, the photoexcited ET rate for coherently coupled photoexcitation and electron transfer was determined A generalized description for electronic coupling J, acting between the two tightly coupled states, for geometrical relaxations and in symmetric and antisymmetric modes, and for asymmetry E_0 in the energy of the two states has been provided [45]. In addition, number of important projection and dipole moment operation were presented.

A new non-Condon theory of the rate of electron transfer stressed the role of protein conformational fluctuation [46]. The theory is expressed by a convolution form of the power spectrum for the autocorrelation function of the electronic tunneling matrix element T(DA)(t) with quantum correction and the ordinary Franck-Condon factor and the detailed balance condition for the forward and backward ET rates. The ET rate formula is divided into two terms of elastic and inelastic tunneling mechanisms on the mathematical basis. The Hush vibronic model and the superexchange formalism have been employed to determine the electronic coupling matrix elements (H) [47]. It was shown that the electronic coupling interaction in the mixed-valence complexes can by varied from weakly to moderately strong coupling by modulation of the structural variables. Approaches for electron transfer combining quantum chemical calculations for the reorganization of energy levels and electronic coupling was proposed in [48]. The non-Condon electron transfer rate theory based on Fermi's golden rule and the time-dependent wave-packet method for the description of the coherent motion of the electron. Also was included.

In recent years, the role of dynamics of bridges connecting ET between donor and acceptor centers has received increasing attention [11, 46–50]. Of particular interest in that respect are expressions for the calculation of nonadiabatic electron transfer rates for donor-acceptor systems incorporated fluctuating bridges and their non-Condon electronic couplings presented in [49]. In the frame of developed theory, the non-adiabatic electron transfer rate constant for a donor acceptor pair was given as:

$$k = \frac{1}{\hbar^2} \int dt\, C_{DA}(t) C_B(t), \tag{1.40}$$

where

$$C_{DA}(t) = \frac{1}{Q_D} \mathrm{tr}[e^{-\beta H_D} e^{iH_D t/\hbar} e^{-iH_A t/\hbar}] \tag{1.41}$$

is the usual correlation function of the ET theory and

$$C_B(t) = \frac{1}{Q_B} \mathrm{tr}[e^{-\beta H_B} e^{iH_B t/\hbar} H_{DA}(Q) e^{-iH_B t/\hbar} H_{DA}(Q)]. \tag{1.42}$$

is the non-Condon contribution. The following conclusions have been done: (1) in high temperature limit, a similar property to the Marcus formula was shown, (2) the Marcus parabolic is shifted for the exponential coupling, (3) an overlap of several Gaussian functions for the linear coupling takes place, and (4) the effective couplings are exponentially and linearly dependent on temperature and the squared frequencies of bridge modes for the exponential and linear couplings, respectively. From the theory, it also followed that in the high temperature limit, the Marcus parabola with respect to the driving force is shifted for exponential coupling and becomes an overlap of several Gaussian functions in the linear coupling case. The effective couplings are exponentially and linearly dependent on temperature and the squared frequencies of bridge modes for exponential and linear couplings, respectively.

Electron Transfer Energy Barrier

In light of the importance of the physical processes near the maximum of electron transfer energy barrier several works were devoted this problem [47, 50–53]. The coupling sensitivity to thermal motion, and visualizing pathway fluctuations electron transfer in biomolecules were analyzed using the new Pathways plugin [15]. According to authors, electron transfer can occurr via several pathways and can be involved in quantum interferences. The height and width of effective tunneling barrier are generated by nuclear dynamics. Therefore, the modulation of the nuclear coordinates will lead to a modulation of the tunneling barrier. For systems with flexible bridges and mobile water such proteins and its biomimetic models, this effect can be especially significant.

Formalisms based on density of states (dos) and potential energy curves (pec) were used to model electrochemical electron transfer reactions [51]. The authors stressed that evaluation of an energy barrier is central to the potential energy curve (pec) approach as it is formulated in terms of an exponential of the energy barrier, whereas the dos approach formulates the electrochemical rate constant expression as a product of the reactants' density of states. On the basis of the theory of electronic transitions in the case of localized states, developed in the physics of disorder matter, a mechanism for long distance electron transfer in biological systems was suggested [52]. Authors of this work considered the situation in which the tunneling barrier is highly anisotropic and particular routes, or pathways along which the transmission amplitude propagates can be revealed. It was shown that when more than one such pathway exists, and if the propagation remains coherent, quantum interference among pathways is expected. Chirality and inelastic processes, which are manifested in the tunneling pathway coherence, have been also discussed.

An effect of a driving infrared (IR) field, which can changes the ET kinetics, was considered in [53]. A theoretical analysis showed that IR field may increase the probability of inelastic tunneling and thus enhance the ET rate for a system in which elastic ET is forbidden. IR driving of a nuclear oscillator may promote the oscillator into excited states that couple more strongly to the tunneling electron. The analysis also revealed that interaction with an IR field may enhance the ET rate up to 34 % over the undriven rate for a system in which elastic ET is forbidden.

A theory for processes near the inversion point has been developed [24]. The complex amplitudes of the electronic wavefunctions at different sites were used for describing the quantum tunneling of the electron in the potential generated by its environment which is modeled as a harmonic classical thermal bath. An electron injected on one of the electronic states, a Coherent Electron-Phonon Oscillator (CEPO) appears which generates large amplitude charge oscillations associated with coherent phonon oscillations and electronic level oscillations (Fig. 1.9). Ultrafast Electron Transfer occurs when this fluctuating electronic level may resonate with a third site which captures the electron (catalytic electron transfer).

Article [54] summarized a number of published applications concerning chemical reactions involving a change in spin-state that are formally forbidden. In this work the observed reactivity was determined by two factors: (1) the critical energy required for reaction to occur is defined by the relative energy of the Minimum Energy Crossing Point (MECP) between potential energy surfaces corresponding to the different spin states (Fig. 1.10) and (2) probability of hopping from one surface

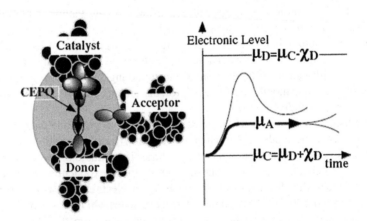

Fig. 1.9 Principle of ET with a Coherent Electron-Phonon Oscillator: Two weakly coupled molecular units donor and catalyst generate a CEPO (see details in [24])

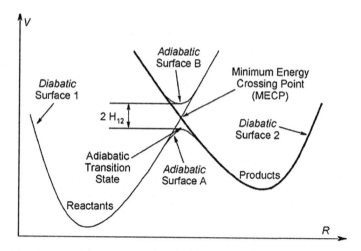

Fig. 1.10 Model system: donor_bridge_acceptor system with *upper* (U) and *upper* (L) bridges. The oscillator couples only to the *upper* bridge electronic state [54]

to the other in the vicinity of the crossing region, which is defined by the spin–orbit coupling matrix element between the two electronic states.

The expression for tunneling at energies below the MECP is given as follows:

$$p_{sh}^{tunnel}(E) = 4\pi^2 H_{12}^2 \left(\frac{2\mu}{\eta^2 F \Delta F}\right)^{2/3} Ai^2 \left[-E\left(\frac{2\mu \Delta F^2}{\eta^2 F^4}\right)^{1/3}\right] \tag{1.43}$$

where ΔF is the relative slope of the two surfaces at the crossing seam, F is the average of the slopes on the two surfaces, Ai denotes the Airy function, H_{12} is the spin–orbit coupling-derived from the off diagonal Hamiltonian matrix element between the two electronic states, E is the kinetic energy of the system as it passes through the crossing region and μ is the reduced mass of the system as it moves along the hopping coordinate.

Reorganization Energy

The values of the driving force (ΔG_0) reorganization energy λ, when solvation is faster than the process of producing photo-initiated charge pairing, can be roughly estimated within the framework of a simplified Marcus model [27] suggesting electrostatic interactions of oxidized donor (D^+), and reduced acceptor (A^-) of radii r_{D+} and r_{A-} separated by the distance R_{DA} with media of dielectric constant e_0 and refraction index n.

The theoretical evaluation of solvent effects on ET has aroused interest and properties of solvents and their impacts on processes was thoroughly analyzed [55–60] (see also Section X). Theory of reorganization energy developed in [55] took in consideration the properties of medium including both size and shape of the solute and solvent molecules, distribution of electron density in reagents and products and the frequency domain appropriate to medium reorganization. More general theory of the reorganization energy was developed in [56]. Energy contributions were divided into those from interactions of donor or acceptor sites with surrounding protein or water environments. In the frame of theory, the difference between energies of the reactant state (U_R) and product state (U_P) with the reaction nuclear coordinate q was suggested to be as:

$$\Delta e(q) = U_P(q) - U_R(q) \tag{1.44}$$

Within this theory, the reorganization energy λ is related to the curvature of the parabolas and equilibrium mean-square fluctuation of the reaction coordinate:

$$\lambda \equiv \frac{1}{2}\beta\left\langle (\Delta e - \langle \Delta e \rangle)^2 \right\rangle \tag{1.45}$$

A novel approach based on constrained real-time time dependent density functional theory (C-RT-TDDFT) in combination with the semiclassical Marcus theory was introduced [57]. In the frame of this theory, the reorganization energy was given as:

$$\lambda_1 = (\sum_{i=1}^{N} U_{A,i} Q_{D,i} - \sum_{i=1}^{N} U_{D,i} Q_{D,i}) \tag{1.46}$$

where $U_{D,i}$ and $U_{A,i}$ are the solvation potentials acting on the ith atom.

In recent years, the specific role of vibrational modes in electron transfer has received increasing attention thanks to the rapid development of quantum mechanics methods. As an example, it was found that atomistic molecular dynamics simulations of the membrane-bound bc 1 bacterial complex revealed that electron transfer is affected by a broad spectrum of nuclear modes, with the slowest dynamics in the range of time-scales ~ 0.1–1.6 μs contributing half of the reaction reorganization energy [58]. A principle role vibrational modes was clearly illustrated by investigating the electron transfer of a betaine-30 molecule from its first excited state to its ground state when being solvated in glycerol triacetate [59]. The total reorganization energy for all vibrational modes under a linear approximation mode was given as:

$$\lambda = \sum_{i=1}^{N_a} \lambda_i = \sum_{i=1}^{N_b} c_i^2 \lambda \tag{1.47}$$

where the coefficient c_i^2 indicates the relative contribution of the ith vibrational mode. As it seen from Fig. 1.11 that for of betaine-30 more than 60 % of the total reorganization energy is related to only seven vibrational modes among 210 available ones.

In work [60] the real-time electron dynamics of two-electron transfer reactions induced by nuclear motion was calculated by three methods: the numerically exact propagation method, the time-dependent Hartree (TDH) method and the Ehrenfest method. It was found that as long as the nuclei move as localized wave packets, the TDH and Ehrenfest methods can reproduce the exact electron dynamics of a simple charge transfer reaction model containing two electrons.

The results of recent theoretical works stressed the role of fluctuation of nuclear frame on coupling factor and reorganization energy and therefore on electron transfer [15, 43, 47, 49–61].

The role of fluctuating media, highlighting issues concerning the validity of the Condon approximation, average medium versus fluctuation-controlled electron tunneling, gated and solvent relaxation controlled electron transfer, and the influence of inelastic tunneling on electronic coupling pathway interference have been discussed in detail [43]. A time-dependent version of Fermi's golden rule with a semiclassical approximation was used. An ET system was described by the Hamiltonian $H = H^D + H^A + H^B + V$, where H^D and H^A are electronic-vibrational Hamiltonians for the donor (D) and acceptor (A) adiabatic potential surfaces, H^B is the vibronic bridge Hamiltonian, and V couples donor (acceptor) and bridge electronic states. In this model, the nonadiabatic ET rate constant is

$$k_{ET}(\omega_{DA}) = \frac{1}{\hbar^2} \int\limits_{-\infty}^{+\infty} dt e^{i\omega_{DA}t} C_{T_{DA}}(t) C_{FC}(t), \qquad (1.48)$$

where ω_{DA} is the frequency corresponding to the energy gap,

$$C_{FC}(t) = \left\langle e^{i\hat{H}^{D(vi)}t/\hbar} e^{-i\hat{H}^{A(vi)}t/\hbar} \right\rangle_D, \qquad (1.49)$$

is the time-dependent Franck-Condon factor, and $C_{TDA}(t)$ is the correlation function of the vibronic DA electronic coupling. The authors concluded that the magnitudes and timescales of molecular fluctuations, which enable access to the activated complex and cause DA electronic interactions to fluctuate can determine the reaction mechanism and that the interplay among electronic coupling and nuclear fluctuations is particularly important in biological electron transfer.

The new non-Marcus theory of electron transfer developed in [62] considered donor and acceptor "supermolecules", which consist of conventional donor and acceptor species plus their associated ionic atmospheres. The theory took into account the fact that charge fluctuations contribute to the activation of electron transfer, besides dielectric fluctuations and revealed that (1) in the inverted region, donor supermolecules are positively charged both before and after the electron transfer event. (2) in the normal region, donor supermolecules change polarity from negative to positive during the electron transfer event. (3) In the superverted region, donor supermolecules are negatively charged both before and after the electron transfer event. Thus. it possible for polar solvents to catalyze electron transfer in the inverted and superverted regions.

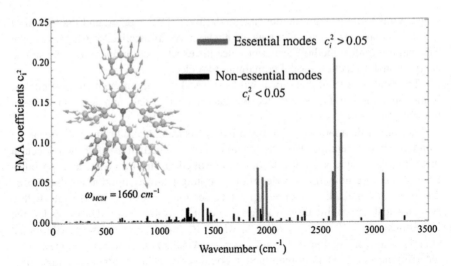

Fig. 1.11 Profile of FMA 9 coefficients, ci 2, for the vibrational modes of the B30 molecule. The seven essential driving modes with ci 2 > 0.05 are colored *grey*, while the others are colored *black* [59]

A situation when the electron transfer rate is comparable or faster relaxation frequencies of the medium has been considered in detail in [63]. When the statistics loses ergodicity (thermal equilibrium for all vibrational and other modes), the ET transition state can not be effectively stabilized as compare with media equilibrium process. And the reaction activation barrier becomes dependent on the medium dynamics. This scenario can be realized in higher viscous systems, for example at temperatures near the point of solvent crystallization. When approaching crystallization, anti-Arrhenius temperature dependence takes place, that is, the reaction rate increases nonlinearly in Arrhenius coordinates (Fig. 1.12).

Various aspects of theories of reorganization energy and their promising applications were discussed in recent publications [58, 64–71]. Long distance electron transfer and the close interrelations between spin exchange and electron transfer are discussed in Sections X and Y.

Fig. 1.12 Arrhenius plots for ergodic and non-ergodic systems [63]

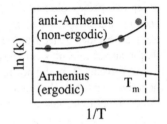

References

1. E. Schrödinger, Phys. Rev. **28**, 1049 (1926)
2. L. Susskind, A. Friedman, *Quantum Mechanics: The Theoretical Minimum* (Basic Books, New York, 2015)
3. J. Yamauchi, in *Nitroxides: Application in Chemistry, Biomedicine, and Materials Science*, ed. dy G.I. Likhtenshtein, J. Yamauchi, S. Nakatsuji, A Smirnov, R. Tamura (WILEY-VCH, Weinhem, 2008)
4. P. Dirac, Math. Proc. Cambridge Philos. Soc. **35**, 416 (1939)
5. E. Fermi, *Nuclear Physics*. (University of Chicago Press, 1950)
6. K.I. Zamaraev, Y.N. Molin, K.M. Salikhov, *Spin Exchange* Theory and Physicochemical application (Springer-Verlag, Heidelberg, 1981)
7. B. Tsukerlat, A. Tarantul, in *Molecular Cluster Magnets,* ed. by R. Winpenny (World Scientific, New Jersy, 2012)
8. M. Mostafanejad, Int. J. Quant Chem. **114**, 1495 (2014)
9. P.W. Anderson, Phys. Rev. **2**, 115 (1959)
10. H.M. McConnell, J. Chem. Phys. **35**, 508 (1961)
11. O.S. Wenger, Acc. Chem. Res. **44**, 25 (2011)
12. R.A. Marcus, in *Understanding Molecular Properties*, ed. by A.E. Hansen, J. Avery, J. P. Dahl, (Reidel, Boston, 1987)
13. Y. Hu, S. Mukamel, Chem. Phys. Lett. **160**, 410 (1989)
14. W.Z. Liang, J. Chem. Phys. **130**, 034111/1 (2009). D.N. Beratan, J.N. Onuchic, J. Chem. Phys. **86**, (1987)
15. A. Balabin, H. Xiangqian, D.N. Beratan, J. Comput. Chem. **33**, 906 (2012)
16. G.I. Likhtenshtein, J. Photochem. Photobiol. A: Chem. **96**, 79 (1996)
17. J. Proppe, C. Herrmann, J. Comput. Chem. **36**, 201 (2015)
18. K.M. Salikhov, A.B. Doctorov, Yu. N. Molin, K.I. Zamaraev, J. Magnet. Reson. (1971)
19. V. Kulikov, G.I. Likhtenstein, Adv. Molecul. Relax. Proc. **10**, 47 (1977)
20. J.S. Hyde, W.K. Subczynski, W. Froncisz, C.S. Lai, Bull. Magn. Reson. **5**, 180 (1983)
21. G.I. Likhtenshtein, in *Magnetic Resonance in Biology*, ed. by L. Berliner, S. Eaton, G. Eaton, vol 18, (Kluwer Academic Publishers, Dordrecht, 2000), p. 309
22. V.N. Parmon, A.I. Kokorin, G.M. Zhidomirov, *Stable Biradicals* (Nauka, Moscow, 1980)
23. G. Ionita, G.A. Vorobieva, V. Chechik, A.I. Kokorin, App. Mag. Reson. **46**, 251 (2015)
24. S. Aubry S, J. Phys. Conden. Matter **19**, 255204/1 (2007)
25. L. Landau, Phys. Zur Sovietunion **2**, 46 (1932)
26. C. Zener, Proc. Royal. Soc. A1 **40**, 660 (1933)
27. R.A. Marcus, N. Sutin, Biochim. Biophys. Acta **811**, 265 (1985)
28. R.A. Marcus, Mol. Phys. **110**, 513 (2012)
29. G.I. Likhtenshtein, *Solar Energy Conversion. Chemical Aspects*, (WILEY-VCH, Weinhem, 2012)
30. L.D. Zusman, Outer-sphere electron transfer in polar solvents. Solvent dynamical effects of polarization diffusion in the rate constant of electron transfer. Electrochim. Acta **36**, 395 (1991)
31. Q. Shi, L. Chen, G. Nan, R. Xu, Y. Yan, J. Chem. Phys. **130**, 164518/1 (2009)
32. X.-Y. Li, **115**, 700 (2015)
33. V.G. Levich, R. Dogonadze, Quantum mechanical theory of electron transfer in polar media. Dokl. Acad. Nauk **78**, 2148–2153 (1959)
34. V.G. Letich, R.R. Dogottadze, Dokl. Akad. Nauk SSSR, **133**, 158 (1960)
35. S. Efrima, M. Bixon, Chem. Phys. Lett. **25**, 34 (1974)
36. R. Neil, N.R. Kestner, J. Logan, J. Jortner, J. Phys. Chem. **78**, 2148 (1974)
37. J.J. Hopfield, Proc. Natl. Acad. Sci. U.S.A. **71**, 3640 (1974)
38. Grigorov LN, Chernavsky DS, Quantum-mechanical model of electron transfer from cytochrome to chlorophyll in photosynthesis, *Biofizika* **17**, 195–102 (1972)

39. E.A. Weiss, M.J. Ahrens, L.E. Sinks, A.V. Gusev, M.A. Ratner, M.R. Wasielewski, J. Am Chem. Soc. **126**, 557(2004)
40. R. Calvo, E.C. Abresch, R. Bittl, G. Feher, W. Hofbauer, R.A. Isaacson, W. Lubitz, M.Y. Okamura, M.L. Paddock, *Am. Chem. Soc.* 122(30), 7327–7341 (2000)
41. F. Gajdos, S. Valner, F. Hoffmann, J. Spencer, M. Breuer, A. Kubas, M. Dupuis, J. Blumberger, J. Chem. Theory Comput. **10** 4653 (2014)
42. M.A. Hagras, A.A. Stuchebrukhov, J. Phys. Chem. B **119**, 7712 (2015)
43. S.S. Skourtis, D.H. Waldeck, D.N. Beratan, Annu. Rev. Phys. Chem. **61**, 461–485 (2010)
44. H. Chen, M.A. Ratner, G.C. Schatz, J. Phys. Chem. C **115**, 18810 (2011)
45. J.R. Reimers, N.S. Hush, Chem. Phys. **299**, 79 (2004)
46. H. Nishioka, A. Kimura, T. Yamato, T. Kawatsu,T. Kakitani, J. Phys. Chem. B, **109**, 15621 (2005)
47. H. Lei, X. Xiao, M. Meng, T. Cheng, Y. Shu, Y.N. Tan, C.Y. Liu, Inorg. Chim. Acta **424**, 63 (2015)
48. Z. Weiwei, Z. Xinxin, S. Yubing, Z. Yi, Prog. Chem. **24**, 1166 (2012)
49. Liang, Wan Zhen, J. Chem. Phys. **130**, 034111/1 (2009)
50. S. Iuchi, N. Koga, J. Chem. Phys. **140**, 024309 (2014)
51. K. Mishra, D.H. Waldeck, J. Phys. Chem. C **115**, 20662 (2011)
52. S.S. Skourtis, D.N. Beratan, D.H Waldeck, Procedia Chem. **3**, 99 (2011)
53. H. Carias, D.N Beratan, S.S Skourtis, J. Phys. Chem. B **115**, 5510 (2011)
54. J.N. Harvey, Phys. Chem. Chem. Phys. **9**, 331 (2007)
55. M.D Newton, in *Advances in Chemical Physics*, ed. by J. Jortner, M. Bixon, vol. 107, Part 1, (John Wiley & Sons. NY, 1999), p. 303
56. O. Miyashita, N. Go, J. Phys. Chem. **104**, 7516 (2000)
57. H. Chen, M.A. Ratner, G.C. Schatz, *J. Phys. Chem. C* **115**, 18810 (2011)
58. D.R. Martin, D.V. Matyushov, J. Chem. Phys. **142**, 161101/1 (2015)
59. H. Chen, J. Phys. Chem. B **118**, 7586 (2014)
60. Y. Suzuki, K. Yamashita, Chem. Phys. Lett. **531**, 216–222 (2012)
61. P.M. Krasil'nikov, Biofizika **56**, 787 (2011)
62. S. Fletcher, J. Solid State Electrochem. **12**, 765 (2008)
63. D.V. Matyushov, J. Phys. Chem. Lett. **3**, 1644 (2012)
64. M. Schröder, M. Schreiber, U.R. Kleinekathöfer, J. Chem. Phys. **126**, 114102 (2007)
65. X.-Y. Li, Int. J. Quant. Chem. **115**, 700 (2015)
66. R. Venkatramani, K.L. Davis, E. Wierzbinski, S. Bezer, A. Balaeff, S. Keinan, A. Paul, L. Kocsis, D.N. Beratan, C. Achim, D.N. Waldeck, J. Am. Chem. Soc. **133**, 62 (2011)
67. D.R Martin, D.N. LeBard, D.V. Matyushov, J. Phys. Chem. Lett. **4**, 3602 (2013)
68. O, Farver, N.M. Marshall, S. Wherland, Y. Lu, I. Pecht, Proc. Nat. Acad. Sci. USA **110**, 10536, S10536/1-S10536/7 (2013)
69. B. Priyanka, V. Anusha, K. Bhanuprakash, J. Phys. Chem. C (2015) (Ahead of Print)
70. T. Xiao, X. Song, J. Chem. Phys. **141**, 134104 (2014)
71. M. Gilbert, B. Albinsson, Chem. Soc. Rev. **44**, 845 (2015)

Chapter 2
Spin Exchange Processes Involving Exited Triplet States

Abstract A general survey of the fundamentals and recent advances in the theory of spin exchange processes such as triplet-triplet and singlet-singlet energy transfer, triplet-triplet annihilation, up conversion and intersystem crossing is presented. These data have formed the theoretical basis for a deep understanding of the details of the mechanisms of numerous chemical and biological reactions, for the development of new methods of structural investigation and analysis, and for the construction of efficient devices for phototherapy, photoelectric processes, photoemission molecular logic systems, light-emitting diodes, light energy conversion, etc.

2.1 General

Triplet excited states ply a key role in any important processes such as the triplet triplet energy transfer and triplet–triplet annihilation [1, 2], upconversion [3], and intersystem crossing [4], which in turn appear to be decisive studies of such important as luminescent oxygen sensing [5], molecular logical devices operation [6], photocatalytic organic reactions [7], and photodynamic therapy [8]. Modern fluorescence and phosphoresces techniques and advance quantum mechanical theories provide remarkable achievements in this area [9–12].

A consistent theoretical description for non-radiative spin exchange processes, that is, Dexter triplet-triplet energy transfer (DTTET), Dexter singlet-singlet energy transfer (DSSET), triplet-triplet annihilation (TTA), singlet-singlet annihilation (SSA) and intersystem crossing (ISC) are based on Fermi's golden rule (1.10, 1.11). The Dexter energy transfer processes involve exchange of donor (D) and acceptor (A) electrons. According to the Fermi Golden rule, the probability of this process occurring is more favorable when the vibrational, rotational or translational levels of the two excited states overlap, since little or no energy is gained or lost in the transition. Thus, for this mechanism to be operative, there must be an overlap of the emission spectra of D and the absorption spectra of A.

© Springer International Publishing Switzerland 2016
G. Likhtenshtein, *Electron Spin Interactions in Chemistry and Biology*,
Biological and Medical Physics, Biomedical Engineering,
DOI 10.1007/978-3-319-33927-6_2

2.2 Dexter Triplet-Triplet Energy Transfer

Dexter triplet-triplet TT energy transfer [13–21] is a process of exchanging both spin and energy between a pair of molecules or molecular fragments (Fig. 2.1).

In their pioneering work Terenin and Ermolaev [14, 15] demonstrated sensitized phosphorescence in solutions of aromatic molecules by the mechanism of triplet-triplet energy transfer.

The general expression for the rate constant of triplet-triplet energy transfer between a donor (D) and acceptor (A) that are tethered by a rigid bridge $^3D-^3A$ is [15].

$$k_{TT} = 2\pi J_{TT} FC \qquad (2.1)$$

This expression includes the exchange integral J_{TT} which quantitatively characterizes the exchange interaction, the Frank-Condon factor, FC, and the energy gap between the donor and acceptor triplet state, $\Delta_{(D-A)}$. In the one electron approximation [15]

$$\begin{aligned}
J_{TT} &= \langle \varphi_D^1(r_1)\varphi_A^0(r_2)|H_{ex}|\varphi_D^0(r_2)\varphi_A^1(r_1)\rangle \\
&\times \langle \alpha_D^1(1)|\alpha_A^1(1)\rangle\langle \alpha_D^0(2)|\alpha_A^0(2)\rangle,
\end{aligned} \qquad (2.2)$$

where (r_1) and (r_2) are the coordinate wave functions corresponding to the donor (D) in the excited states and to the acceptor (A) in the ground state, (r_2) and (r_1) are the wave functions related to the state after the triplet-triplet energy, $\langle \alpha_D^1(1)$ and $\alpha_A^1(1)$ are the spin wave functions in the state before transfer, $\langle \alpha_D^0(2)$ and $\alpha_A^0(2)$ are the same function after the transfer, φ_y^x are corresponding wave functions and Hex is the Hamiltonian of the exchange interaction, and

$$J_{TT} \sim exp(-\beta R) \qquad (2.3)$$

where R is the donor-acceptor distance and β is the attenuation coefficient depending on chemical nature of bridge tethered the donor and acceptor.

According to the Franck–Condon rule, at the moment of energy transition, the energies of the donor and acceptor should be equal and the Franck-Condon factor becomes

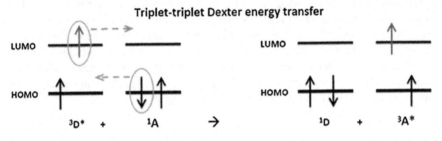

Fig. 2.1 Schematic diagram Dexter triplet-triplet energy transfer chemwiki.ucdavis.edu/ Theoretical.../Dexter_Energy_Transf

$$FC = \int I_{ph}^D(v)\varepsilon_A(v)dv, \tag{2.4}$$

where $I_{ph}^D(v)$ and $\varepsilon_A(v)$ are, respectively, the intensity of the donor phosphorescence and the extinction coefficient of the acceptor.

To summarize, the Fermi Golden equation 2.1 and the Marcus-Levich formula for nonadiabatic electron transfer were used as a the basis for the theory of triplet triplet energy transfer and other related processes [20]. In the endothermic case the transfer rate can still prevail over other triplet quenching channels if the energy gap between donor and acceptor (ΔE_T) is several kcal/mol, and the triplet photosensitization rate constant $k_{TT} = k_o \exp(-\Delta E_T/RT)$. This type of kinetics in solution was first demonstrated by Sandros using a plot of the triplet–triplet energy constant versus the gap between donor and acceptor energies [19].

The importance of triplet excited species in photochemical reactions was emphasized by Zimmerman et al. [16]. As was shown by the Hammond and Saltiel groups [17, 18], *cis-trans* isomerization of the stilbenes by excitation transfer from low energy sensitizers, triplets may arise from the conversion of initially formed singlets or by interaction with a higher energy triplet by sensitization, which can quench a higher energy triplet in the presence of lower energy triplets. In such a case the TTET rate may be considerably faster than that predicted by the Sandros equation [19], an anomaly associated with acceptor molecules with significant conformational flexibility. This mechanism was experimentally revealed by a decrease in the triplet– triplet energy transfer rate constants, which remain slower than expected in the apparent endothermic region, as calculated from the Sandros plot.

Importance of the triplet-triplet energy transfer processes stimulated further developments of theory in this area [20–25]. Two approaches, the direct coupling method and the energy-gap-based configuration-interaction-singlet (CIS) scheme, have been developed to calculate TT energy-transfer couplings between a pair of molecules for both asymmetric and symmetric test systems [21]. The direct triplet-triplet transfer coupling between spin-localized reactant and product adiabatic states was evaluated as

$$T_{rp} = \frac{H_{rp} - S_{rp}(H_{rr} + H_{pp})/2}{1 - S_{rp}^2}, \tag{2.5}$$

here

$$H_{rp} = \langle \Psi_r | \hat{H} | \Psi_p \rangle = \int dx \Psi_r^*(x) \hat{H} \Psi_p(x) \tag{2.6}$$

and

$$S_{rp} = \langle \Psi_r | \Psi_p \rangle = \int dx \Psi_r^*(x) \Psi_p(x), \tag{2.7}$$

where S_{rp} is the overlap integral. where $\mathbf{x}i$ is the spin and spatial coordinates of electron Hi is the Hamiltonian for the system, and Trp is the transfer integral, or transfer-matrix element defined in scattering problems, Ψ_r and Ψ_p are the spin-localized wave functions before and after energy transfer, respectively and S_{rp} is overlap integral. In the frame of the energy-gap-based method, for a two-state model, a CIS calculation for the lowest two triplet states was performed. It was shown that a simple ab initio CIS scheme gave the corresponding coupling value. Furthermore, it was found that for both asymmetric and symmetric test systems, the direct coupling method and the energy-gap-based CIS scheme yielded similar results.

To investigate the dynamic disorder effect on the TTET a time dependent wavepacket diffusion method was used [22]. The partial quantum effect of the nuclear motions was taken into account. The following for the Hamiltonian of a dimer was written:

$$
\begin{aligned}
H(t) &= \sum_{i=1}^{2} \left[\varepsilon_{ii} + V_{ii}(t) \right] |i\rangle \langle i| + \\
&= \sum_{i \neq i}^{2} \left[\varepsilon_{ij} + V_{ij}(t) \right] |i\rangle \langle j|
\end{aligned}
\tag{2.8}
$$

where ε_{ii} and ε_{ij} show the static site energies and electronic couplings between sites respectively, and $Vii(t)$ and $Vij(t)$ parameterize the fluctuations of the site energies and electronic couplings respectively.

A novel theoretical method was developed for establishing the triplet-triplet energy transfer reaction coordinate within the very weak electronic coupling limit (the $0.1–10$ cm^{-1} range) [1]. The proposed method permits the determination of the reaction coordinate for triplet-triplet excitation transfer in both the classical and nonvertical cases for a donor/acceptor pair, that is, (i) the separation of donor and acceptor contributions to the reaction coordinate, (ii) the identification of the intrinsic role of donor and acceptor in the triplet energy transfer process, and (iii) the quantification of the effect of every internal coordinate on the transfer process. This formalism based on the Fermi golden rule formalism allowed to complete the adiabatic potential energy surfaces of the model supermolecules [^{3}D...^{1}A] and [^{1}D...^{3}A] in a TET process as a function of acceptor donor internal coordinates q_A and q_D.

An effective triplet–triplet energy transfer of Dexter type was found to be a cause of low probability of electron transfer from the sacrificial reagent (SR) to the photoexcited Ir complex (bis-(2-phenylpyridinato-)(2,2'-bipyridine)iridium(III) hexafluorophosphate [Ir(ppy)$_2$(bpy)]PF$_6$ (Fig. 2.2) [23].

Interesting photophysical properties related to TTET were revealed in dithienylethene (DTE)-2,6-diiodoBodipy triads (DTE-o-BDT. Figure 2.3 illustrates competition processes, namely, intersystem crossing, singlet-singlet resonance energy transfer (RET) and cyclization in (DTE-o-BDT after excitation at $\lambda = 535$ nm and $\lambda = 268$ nm. The ISC followed by triplet-triplet upconversion [24].

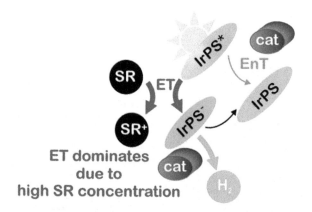

Fig. 2.2 Schematic representation of processes in the photocatalytic system of Ir complex (bis-(2-phenylpyridinato-)(2,2′-bipyridine)iridium(III)hexafluorophosphate used in [23] SR is sacrificial reagent, ET is electron transfer and EnT is the triple-triplet energy transfer

The potential energy surfaces (PESs) crossings, the most likely place for the transition to occur, play important roles in understanding chemical reactivity when a process involves a change in the total electronic spin. A model of nonvertical triplet energy transfer based on transition state theory for nonadiabatic processes was proposed [25] model used PES of reactants and products, as computed from high-level quantum mechanical methods, and a nonadiabatic transfer rate constant (Fig. 2.4). The authors approach was based on triplet photosensitization, first characterized in condensed phases by Terenin and Ermolaev [14, 15].

The electron exchange between donor and acceptor was expressed by the coupling matrix term;

$$H_{if} = \left\langle \Psi_{[*D\cdots A]} \left| \hat{H}' \right| \Psi_{[D\cdots A*]} \right\rangle \tag{2.9}$$

where the indices i, j denote the donor and acceptor electrons, respectively and

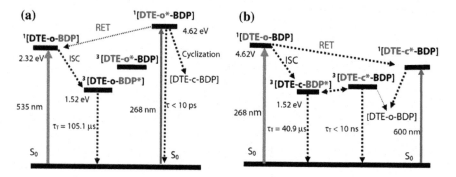

Fig. 2.3 Simplified Jablonski Diagram Illustrating the Photophysical Processes Involved in DTE-o-BDT [24]

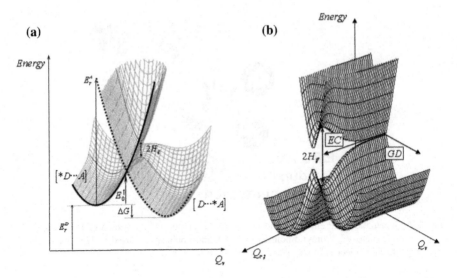

Fig. 2.4 a The triplet energy transfer reaction as pictured by the crossing between the potential energy surfaces (PES) of the donor–acceptor [D⋯A] collision complex. The potential energy surface as a function of the molecular vibrational coordinates (Q_v). E_T^D and E_T^A are the optical excitation energies of the donor and acceptor triplet states, respectively; $E_0^{\ddagger}$ is the energy at the crossing point of the adiabatic PES; H_{if} represents the amplitude of the electron exchange interaction and ΔG the free energy of the process. **b** Another perspective of the PES, now as a function of the donor–acceptor relative position, given by the Qr,t coordinates (See Details in [25])

$$\hat{H}' = \sum_{i,j} \frac{1}{r_{ij}} \tag{2.10}$$

is the electronic coupling Hamiltonian, where the indices $i,\,j$ denote the donor and acceptor electrons, respectively [25].

Recent advances in the theory of triplet-triplet energy transfer and its application have been reflected in [26–30].

2.3 Triplet-Triplet Annihilation

Triplet-triplet annihilation (TTA) discovered by Merrifield [31] is a version of photon upconversion (UC) which is a process in which the sequential absorption of two or more photons leads to the anti-Stokes emission of light at a shorter wavelength than the excitation wavelength. In the TTA, two triplet chemical groups, D* and A*, react to produce two singlet states (Fig. 2.5). A general theoretical approach to a quantitative description of triplet-triplet annihilation on the basis of

Triplet-triplet annihilation (TTA)

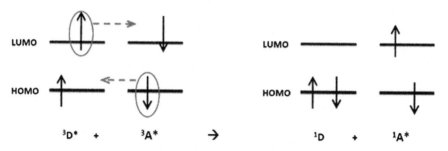

Fig. 2.5 Schematic diagram for triplet-triplet annihilation. chemwiki.ucdavis.edu/Theoretical.../Dexter_Energy_Transf

the Fermi Golden Rule is similar to the requirements for triplet-triplet energy transfer.

The TTA process is important in numerous systems such as homogeneous liquid solutions [15], molecular crystals [31], polymers [32], Langmuir–Blodgett films [33], biomembranes [34–38], etc.

If the energy gap between $^3D^*$ and ^1D is bigger than the energy gap between $^3A^*$ and $^1A^*$, the TTA process might have energy to excite one of them to higher singlet states. This process, in which the $^1A^*$. is populated by T-T annihilation, produces one molecule in the state $^1A^*$ It is principally important, that because characteristic time of the processes is related to life time of the excited triplet state, the life time of delayed fluorescence is much longer than typical for spontaneous fluorescence. whose lifetime is much longer than that of the spontaneous fluorescence.

The triplet-triplet annihilation includes several processes: (1) photoexcitation $S_0 \rightarrow S_1$, (2) unimolecular process $S_1 \rightarrow S_0$, (3) $S_1 \rightarrow T_1$ (intersystem crossing) and final the triplet singlet energy transformation $T_1 + T_1 \rightarrow S_1 + S_0$ being accompanied by delayed fluorescence. The TTA can also occur via a bimolecular with formation of the singlet excimer formation $S_1 + S_0 \rightarrow {''}S_1$. The excimer dissociation being accompanied by The delayed fluorescence and internal conversion ${''}S_1 \rightarrow S_0 + S_0$ or intersystem crossing $S_1 \rightarrow T_1 + S_0$ take place in the process of the excimer dissociation.

Theoretical approaches to T-T annihilation processes may be illustrated by two examples. Results of theoretical and experimental studies of dynamic triplet-triplet annihilation in an organic one-dimensional motion system were reported [39]. Pair-density-matrix theory has been adapted to the triplet-pair exciton annihilation under microwave excitation by including the dimensionality of exciton motion and applied to the fluorescence detected magnetic-resonance spectrum observed for a one-dimensional molecular crystal. This theory gives a satisfactory fit of the observed effects on a 1,4 dibromonaphtalene crystal. The dynamics of the upconversion-induced delayed fluorescence for a model multicomponent organic system, in which high concentrations of triplet states can be sustained in steady-state conditions, have been analyzed [40]. At different excitation powers, two regimes have been identified, depending on the main deactivation channel for the triplets, e.g., the spontaneous decay and the bimolecular annihilation. To have efficient upconversion generation, the threshold (I_{th}), the excitation power density at which triplet bimolecular annihilation becomes dominant, was established. The simple equation obtained for I_{th} allowed predictions of the theoretical efficiency of a generic system on the basis of a few parameters of the constituent molecules.

2.4   Dexter Singlet-Singlet Energy Transfer

Singlet-singlet energy transfer (SSET) may occur for exchange at short distances by the Dexter mechanism (Fig. 2.6) and for exchange at long distances by electric dipole-dipole interactions via the Förster mechanism 2.11 [41].

The SSET rate constant K_{ET} data on photo-processes in cofacial bisporphyrin were analyzed using the following formulas [42].

$$K_{ET}^{Forster} = k_D R_F^6 \left(\frac{1}{R}\right)^2$$

$$K_{ET}^{Dexter} = \frac{2\pi}{h} K J' \exp\left(\frac{-2R}{L}\right), \tag{2.11}$$

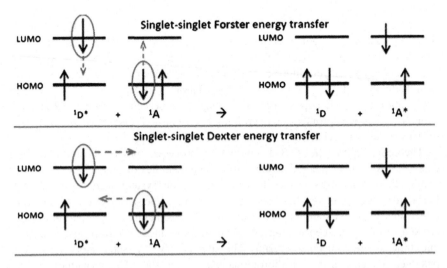

Fig. 2.6 Schematic representation of the Förster and Dexter energy transfer mechanisms. chemwiki.ucdavis.edu/Theoretical…/Dexter_Energy_Transf

where k_D is the emission rate constant for the donor, R_F is the Förster radius, that is, the distance at which transfer and spontaneous decay of excited donors are equally probable, R is the distance between the two macrocycles, J' is the integral overlap, and L is the average Bohr radius. The graphs shown in Fig. 2.7 demonstrate the possibility of a crossover region for the Förster and Dexter mechanisms for suitably chosen parameter values. The straight lines represent what is expected for an energy transfer rate operating exactly according to the Förster mechanism (A) and the Dexter mechanism (B). The Förster mechanism dominates at long distances (more 5–6 Å) while the Dexter mechansim dominates at short distances.

In work [43], it was demonstrated that the mechanism for nonradiative energy transfer can switch from a longer-range dipole-dipole mechanism to a short-range exchange mechanism through energy migration between donors. The migration and trapping rates, W, was estimated using a mechanism for donor-donor energy transfer (DDET) and donor-acceptor energy transfer (DAET):

$$W = y \sum_{i=1}^{\infty} \Gamma_{DA}(R_i) n(R_i) \tag{2.12}$$

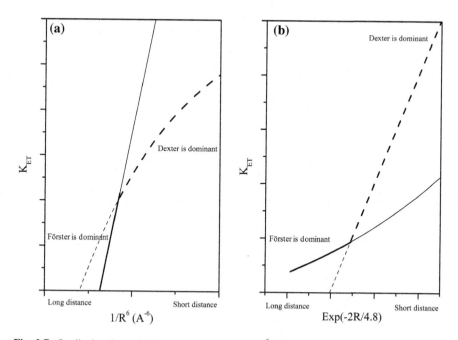

Fig. 2.7 Qualitative theoretical plots for K_{ET} versus $1/R^6$ (Förster, graph A) and K_{ET} versus exp $(-2R/4.8)$ (Dexter, graph B). The *solid lines* correspond to hypothetical situations where the Förster mechanism operates, while the *dotted lines* are hypothetical situations for the Dexter mechanism [42]

where y is the relative acceptor concentration, $\Gamma_{DA}(R_i)$ is the DA transfer rate at a given donor-acceptor distance, R_i, and $n(R_i)$ is the steady-state excitation density, which was approximated by

$$n(R_i) = (1 + \Gamma_{DA}(R_i)\tau_1)^{-1} \tag{2.13}$$

where τ_1 is the decay time for the donor excitation the absence of acceptors and Γ_{DD}, was given by

$$\Gamma_{DD}(R_j) = C_{DD}\exp(-2R_j/L_{DD}) \tag{2.14}$$

where C_{DD} is an interaction coefficient, R_j the donor-donor distance, and L_{DD} an effective Bohr radius. Figure 2.8 shows the dependence of the energy transfer rate on donor concentration. In summary, the crossover from the Dexter mechanism to the Förster mechanism shifts as the donor concentration increases. This theory was applied to an analysis of experimental data on the $Tb^{3+}(^5D_4) \rightarrow Ce^{3+}(5d^1)$ energy transfer as a function of Tb^{3+} concentration in $(Lu,Tb)_3Al_5O_{12}:Ce^{3+}$ garnet phosphors.

Energy pooling (EP) is defined as a three-electron process in which the energy of two electrons in excited states is transferred to the third electron (Fig. 2.9). Two mechanisms of EP were considered: consequence energy transfer in which one donor transfers its energy to a second donor followed by transfers second donor energy to the acceptor and direct energy translation from two donors to acceptors [44]. A combination of molecular quantum electrodynamics, perturbation theory, and ab initio calculation was used for estimating the rate of 3-body singlet upconversion pooling in test systems fluorescein -stilbene-fluorescein. The process was expressed in terms of a coupling matrix element, W(n), and an energy density of states, δ.

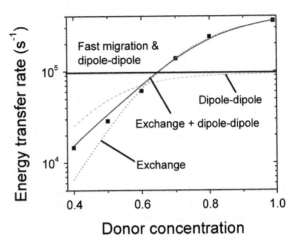

Fig. 2.8 Dependence energy transfer rate on the donor concentration $Tb^{3+}(^5D_4)$ for different mechanisms of singlet-singlet energy transfer [43]

Fig. 2.9 Schematic illustration of a pooling process [44]

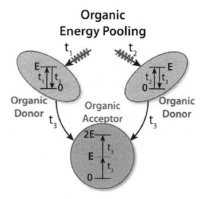

Organic semiconductors was proved to be effective system for realization the singlet fission (a splitting of something into two or more parts) (SF). Commonly, this process is a multistep phenomenon, with different photophysical processes (relaxation, decoherence, exciton diffusion) occurring on multiple time scales. In this objects, the initially excited singlet can rapidly relax to the bottom of the singlet band by relatively slow SF reactions to produce high triplet yields (Fig. 2.10). Singlet → triplet conversion can be spin-allowed when a four-electron system is considered. The fission from the singlet state to the triplet manifold occurs when the singlet projects onto a superposition of the triplet pair states $|xx\rangle$, $|yy\rangle$, and $|zz\rangle$. Interference between the $|xx\rangle$, $|yy\rangle$, and $|zz\rangle$ terms of the triplet state can lead to time-dependent oscillations in the projection. As an example, the time-resolved photoluminescence of a tetracene single crystal showed quantum beats in the delayed fluorescence [45].

Data on upconversion-induced fluorescence in multicomponent systems have been reviewed [46].

Fig. 2.10 Diagrams of the band structures for **a** inorganic semiconductors and **b** organic semiconductors (See details in [45])

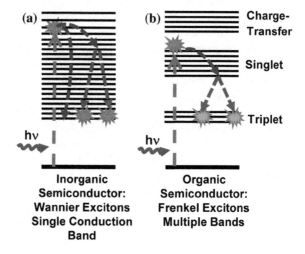

2.5   Intersystem Crossing

2.5.1   Direct Intersystem Crossing

Intersystem crossing (ISC) is a process in which a singlet state nonradiatively transforms into a triplet state, or conversely a triplet transforms into a singlet. The probability of this process occurring is more favorable when the vibrational levels of the two excited states overlap, since little or no energy must be gained or lost in the transition (Fermi Golden Rule, 2.14) [47].

$$k_{ISC} = \frac{2\pi}{h} |v_{SO}|^2 \rho \tag{2.15}$$

where V_{SO} represents the spin-orbit coupling (SOC) strength, consisting of an electronic contribution as well as the Franck-Condon factor, ρ is the density of coupling triplet levels. The SOC is important factor for ISC which is quantitatively characterized by an exchange integral. According to the spin conservation rule for radiative or radiationless transitions, transitions between terms of different multiplicity are spin-forbidden. Nevertheless, coupling of the electron spin degrees of freedom with the non s-state orbital angular momentum (SOC) in paramagnetic metal complexes may allow a system to circumvent the Wigner spin conservation rule.

Specifically according to the Fermi Golden Rule the intersystem crossing rate, k_{isc}, for the non-radiative transition responsible for quenching can be calculated with 2.16:

$$k_{isc} = \left(\frac{4\pi^2}{h}\right) \left({}^1\psi|H_{SO}|^3\psi\right)^2 [FC] \exp - \frac{\Delta_{13}}{RT} \tag{2.16}$$

where $\left({}^1\psi|H_{SO}|^3\psi\right)$ is the value of the spin–orbit coupling term for the non-radiative transition $S_1 \rightarrow T_1$, FC is the Franck–Condon weighted density of states, which characterizes the overlap of the vibrational wave functions, and (Δ_{13}) is the energy difference between the initial and final states.

A typical example of intersystem crossing is characterized by the successive transformations from the exited singlet state (S_1) to excited triplet state (T_1) and from the exited triplet state (T_1) to ground sinlet state (S_0) (Fig. 2.11) [48].

Many chemical reactions involve a change in spin-state that would be spin forbidden in the absence of ISC. A large body of literature exists, both theoretical and experimental, explicating the pivotal role of intersystem crossing in various chemical and physical processes [49–69].

The potential energy surfaces (PESs) crossings, the most likely place for the transition to occur, play important roles in understanding chemical reactivity when a process involves a change in the total electronic spin [55]. Spin-orbit coupling with the approximate one-electron spin–orbit Hamiltonian suggested by Koseki

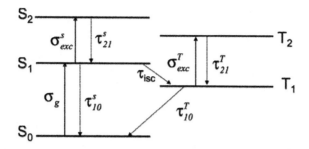

Fig. 2.11 Full Jablonski diagram. S_1 and T_1 represent the singlet triplet states, respectively. $\sigma^s_{exc}, \sigma^T_{exc}$ are singlet and triplet absorption cross-section, respectively. σ_g is the ground state absorption cross-section, t_{ISC} is the intersystem crossing time. en.wikipedia.org/wiki/Jablonski_diagram

et al. [56] was utilized for a theoretical investigation of the reaction of N_2O and CO catalyzed by PtO^+ [55]. In the semiclassical model based on PESs, the probability of intersystem crossing for a molecule passing through a quartet– doublet crossing, in a Landau–Zener type mode was calculated. To localize and characterize the crossing points (CP), in this reaction, the concept of minimum energy crossing point (MECP) on the seam line at which two electronic surfaces of different spin multiplicities intersect were used. The spin–orbit Hamiltonian was given as

$$H_{SO} = \frac{\alpha^2}{2} \sum_i \sum_k \left(\frac{Z^*_k}{r^3_k} \right) Si \cdot L_{ik} \quad \frac{\alpha^2}{2} = \left(\frac{e^2}{2me^2c^2} \right), \tag{2.17}$$

where S_i and L_{ik} are the orbital and spin angular momentum operators for an electron (i) coupled to nearby nuclei. The nuclear charge Z_K was replaced by an effective parameter, Z'_K, which can be taken as the screened nuclear charge.

In article [57] experimental results of Zeeman splitting measurements were summarized. In that work, intersystem crossing from the lowest excited singlet state (S_1) to the triplet state of N-heterocyclic aromatic hydrocarbons and carbonyl compounds with π bonds were considered. The ISC was attributed to spin-orbit coupling with the $^3\pi\pi^*$ state in the same energy region. In the analysis of the experimental data, El-Sayed's rule was utilized [58]. According to this rule, SOC is allowed between states of different symmetry and electronic configurations (Figs. 2.12 and 2.13). ISC is expected to be fast in molecules with the low-lying $n\pi^*$ states, where n represents a nonbonding electron of the nitrogen or oxygen atom. In contrast, ISC is expected to be slow in the S_1 $^1\pi\pi^*$ state of aromatic hydrocarbons with no nonbonding electrons (Figs. 2.12 and 2.13).

In work [60], the results of a theoretical study the reaction mechanism and ISC problem in the reaction of $W + NH_3 \rightarrow N \equiv WH_3$ were presented This reaction ran via four diabatic reaction pathways, corresponding to septet, quintet, triplet and singlet states, respectively and can be predominantly governed by ISC To elucidate the reaction mechanism, the properties of intermediates and transition states along

Fig. 2.12 El-Sayed's rule for spin-orbit coupling [57]

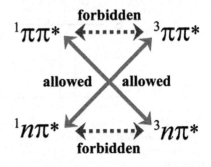

Fig. 2.13 Coupling scheme for ISC in the S1 $^1n\pi^*$ state. SOC and VRC represent spin-orbit coupling and vibronic coupling, respectively [57]

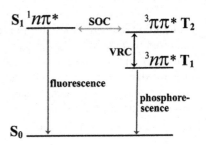

each diabatic pathway, as well as the curve crossing points and the corresponding spin−orbit couplings were calculated. A simplified formula to estimate the following nonadiabtic reaction rate k_{ZN}:

$$Zk_{ZN}\frac{1}{2\pi}\int_0^\infty e^{-E_s/k_B^T}P(E_s)dE_s \tag{2.18}$$

was used, where Z is the partition function and E_s is the translational energy component along the direction normal to the seam surface and $P(E_s)$ represents the ZN transition probability.

Article [59] considered applications of Transition State Theory which can account for the kinetics of spin-forbidden chemical reactions. As an example the reactivity of a typical spin-forbidden reaction $HC + N_2$ determined by two factors was considered. The first factor is the critical energy required for the reaction to occur, which in spin-forbidden reactions is defined by the relative energy of the Minimum Energy Crossing Point between potential energy surfaces corresponding to the different spin states. The second factor is the probability of hopping from one surface to another in the vicinity of the crossing region, which is largely defined by the spin−orbit coupling matrix element between the two electronic wavefunctions.

Cascade decay is a phenomenon associated with excited-state relaxation which follows a complex route of intermediate states with changes of spin and lattice parameters. The cascade decay mechanism for ultrafast intersystem crossing mediated by the spin-orbit coupling in transition-metal complexes was investigated by using a quantum-mechanical description of the cascading process that occurs after photoexcitation [61]. The fastest cascading effects are associated with dephasing of states through the coupling to a continuum, e.g., Fano effects [62]. After the cascade, the system returns to the ground state or to a relatively long-lived metastable state. For a cascading process, in which several levels at energies Ei are involved the following Hamiltonian was used

$$H_s = \sum_i E_i n_i + \hbar \omega a^\dagger a + \lambda_i n_i (a^\dagger + a), \tag{2.19}$$

where $h\omega$ is the level splitting of the vibrational/phonon modes of the surrounding ligands, λ_i is the coupling strength, n_i gives the occupation of state i and $a^\dagger$ is the step operator for the vibrational mode. According the theory, ultrafast cascading occurs when the energy difference between the levels is comparable to the self-energy which is on the order of several hundred meV.

Semiclassical trajectory theories related to intersystem crossing dynamics and a general trajectory surface hopping process were applied to large systems with many degrees of freedom [64]. Substantial electronic spin orbital coupling was took in consideration. Nonstandard situation when triplet state is generated thanks to an three-state crossing involving one singlet and two triplets states was analyzed on example of photoexcited acetophenone using time-dependent functional theory [66]. The singlet-triplet intersystem crossing in the reduced Fenna-Matthews-Olson protein a light-harvesting, bacteriochlorophyll a (BChl a) containing. antenna complex was investigated by e measurement of time-resolved fluorescence and transient absorption [67]. The experiments demonstrated that the apparent quantum yield of singlet-triplet intersystem crossing in the reduced FMO complex is $\sim 11\,\%$ in the most favorable low laser intensities.

The competition between intersystem crossing (ISC) and internal conversion (IC) as nonradiative relaxation pathways is a challenging problem to be solved in many cases. This problem was analyzed on example of the processes of internal conversion and intersystem crossing i in cyclic α,β-unsaturated pentenones excited to their lowest lying $^1\pi\pi^*$ state. These processes were investigated by means of time-resolved photoelectron spectroscopy and ab initio computation [69]. It was found that, both IC and ISC rates decreased with the quantum yields of ISC varying between 0.35 and 0.08. A large spin orbit coupling of 45–60 cm^{-1} over an extended region of near degeneracy between the singlet and triplet state was suggested as cause of the rapid rates of ISC. Sketch of the processes photoinduced dynamics is shown in Fig. 2.14.

The Born-Oppenheimer approximation is used widely in theoretical and computational chemistry and in biomolecular modeling. However, the approximation of separating electronic and nuclear motion can break down and lead to coupling of potential energy surfaces. Relativistic effects must be taken into account, when the

Fig. 2.14 Sketch of the
photoinduced dynamics of
acrolein, 2-cyclopentenone
(CPO) and its methylated
derivatives upon excitation to
the S_2 ($^1\pi\pi^*$) state (See
details in [69])

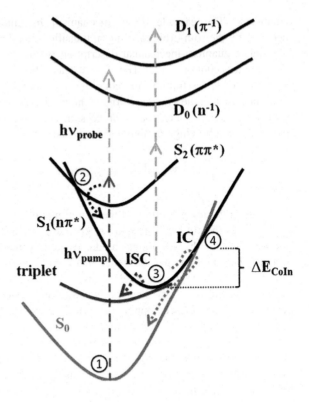

velocity of the electron approaches the speed of light [68]. This is especially relevant in chemical reactions, where even small non–Born-Oppenheimer couplings should be taken in consideration. Various physical effects can contribute to a breakdown of the Born-Oppenheimer approximation; among these, spin-orbit coupling is the most important. Relativistic spin-orbit coupling effects cause spin-forbidden transitions and can play a significant role in various photochemical reactions such as photo- and electroluminescence.

A spin-orbit relativistic long-range corrected (LC) time-dependent density functional theory (LC-TDDFT) for use in investigating spin-forbidden transitions in photochemical reactions has been developed [63]. Relativistic effects including spin-orbit coupling effects were considered by the two-component zeroth-order regular approximation (ZORA) method. In spin-orbit relativistic LC-TDDFT, relativistic effects including spin-orbit coupling effects were incorporated on the basis of the ZORA Hamiltonian,

$$H^{ZORA}\Phi = \left(V + T^{ZORA}\right)\Phi,$$
$$T^{ZORA} = \sigma \cdot p \frac{c^2}{2c^2 - V} \sigma \cdot p, \tag{2.20}$$

where σ, $\mathbf{p}$, and c are a 2×2 Pauli spin-matrix, the momentum operator, and the speed of light, respectively. Potential V contains the nuclear attraction (V_{Ne}), Coulomb (J), and exchange correlation (Vx and Vc) potentials. The spin-orbit couplings were taken into account through T^{ZORA}. This ZORA Hamiltonian was combined with long-range corrected density functional theory (LC-DFT), in which the two-electron operator is divided into short-range and long- parts. In the calculation, the long-range exchange and spin-orbit coupling effects were also taken in consideration.

2.5.2 Induced Intersystem Crossing

Heavy atoms and paramagnetic species can induce intersystem crossing in the processes $S_1 \rightarrow T_1$ and $T_1 \rightarrow S_0$ which is accompanied by quenching of chromophore fluorescence and a reduction of the triplet excited state lifetime, respectively. Induced intersystem crossing (IISC) originates from strong spin–orbit coupling between spins and the inducing heavy atoms, transition complexes or stable radicals. These species can be normal constituents of the chromophores themselves or can be present in the surrounding medium [70–76]. A few typical example are described further on.

In work, commonly was described by The full Breit-Pauli Hamiltonian was used for describing the effect of the charge of nucleus I (Z_I) on intersystem crossing via spin orbital coupling [70]:

$$H_{SO} = \frac{\alpha^2}{2} \left[\overset{\text{electron nucleus}}{\sum_i \sum_A} \frac{Z_A}{r_{iA}^3} L_{iA} S_i - \overset{\text{electron}}{\sum_{ij}} \frac{1}{r_{iA}^3} L_{ij} \left(S_i + 2S_j \right) \right], \qquad (2.21)$$

where α is fine structure constant, L is the orbital momentum operator, S is the spin momentum operator, and Z_A is the nuclear charge. According to 2.21, increasing the nuclear charge, contributes more and more to SOC and SOC is a fairly short-ranged interaction with dropping off $1/r^3$.

In work [71], this equation was also employed. It was found in that two main factors affecting on ISC in metalorganic molecules are dependence on the square of the heaviest atom number and the singlet-triplet energy difference. The authors stressed importance of exciton-spin-orbit-molecular vibration interaction as a perturbation operator for intersystem crossing rate from singlet excited state to triplet excited state of organic molecules.

An ability of the first-row transition metal ions such as Cu^{2+}, Co^{2+}, Ni^{2+}, and Fe^3 and paramagnetic lanthanides to quench fluorescence frequently use for analysis of these metals [72]. The most probable mechanism of the quenching is intersystem crossing. Indeed, the $T_1 \rightarrow S_0$ transition was suggested in work [73]. on the basis of the significant enhancement of the quenching constants in the chelates of benzoyltrifluoroacetone with paramagnetic Gd^{3+} ion as compare with diamagnetic La^{3+}

and Lu^{3+} ions. This conclusion was confirmed in work [74], in which significant decrease of the fluorescence life time τ_{fl} in chelates of methyl salicylate in the presence of Gd^{3+} ions was observed and also by the good linear correlation between the values of magnetic moments, τ_{fl} and the $S_1 \rightarrow T_1$ transition constants for paramagnetic lanthanide(III) texaphyrin complexes [75].

The induced $S_1 \rightarrow T_1$ intersystem crossing at fluorescence quenching of 7-amino-1,3-naphthalenesulfonic acid by paramagnetic ions Cu^{2+}, Ni^{2+}, Co^{2+}, Mn^{2+}, Sm^{3+}, and Ce^{3+} cations has been found in an aqueous medium [76]. This mechanism was strongly sustained by the combination of monitoring fluorescence quantum yield and the triplet state yield of the fluorophore. The constants of static (Kst) and dynamic (kq) fluorescence quenching, and rate constant of amino-G-acid triplet state quenching kTby paramagnetic the metal ions were measured using the Stern-Volmer plots. It is important that a qualitative correlation between triplet state quenching rate constants and values of the efficient paramagnetic susceptibility and spin of the cations has been established.

Fluorescence and phosphorescence quenching by the stable nitroxide free radicals solution were described in references [77, 78]. The fluorescence quenching of phenanthrene (Phen), 9-cyanophenanthrene (CPhen), 9-cyanoanthracene (CA), perylene (Per), 9,10-dicyanoanthracene (DCA), and 9,10-diphenylanthracene (DPA) using stable nitroxide radicals as quenchers has been studied by steady state and flash photolysis measurements [79]. On the basis of flash photolysis measurements as well as the overlap between the emission spectra of hydrocarbons and the absorption spectra of radicals, a resonance energy transfer mechanism was suggested in case of Per, DPA, DCA, and CA, while for Phen and CPhen where the energy gap between the first excited singlet and the nearest lower triplet state is small, an induced intersystem crossing was judged to be more likely.

As was first shown by Likhtenshtein group et al. [80], in a dual fluorophore-nitroxide compound (Dansyl-TEMPO, FNRO•).

Formula 2.1 The irradiation of the chromophore segment of dual fluorophore-nitroxide compound

The irradiation of the chromophore segment of dual fluorophore-nitroxide compound (Dansyl-TEMPO, FNRO•) in a glassy liquid invoked producing the hydroxylamine derivative accompanying a decay of the nitroxide ESR signal and parallel eight-fold increase in fluorescence. Both processes run with the same rate constant k_{red} under identical conditions. In order to establish a mechanism of intramolecular fluorescence quenching (IFQ) and photo-reduction of the nitroxide segment in the dual molecules, a series of dansyl-nitroxides of different structures and flexibility of the spacer group, and different redox potentials of nitroxide were synthesized and investigated (Fig. 2.14) [81].

The rate constants of IFQ (k_q), fluorescence quantum yields of paramagnetic and diamagnetic forms, rate constant of photoreduction (k_{red}) in various solvents for a series of dansyl compound containing various nitroxides were measured [82, 83]. It was shown that, the positive correlation between the rate constant of the nitroxide fragment photoreduction k_{red} and the equilibrium constant K_{eq} for the chemical exchange reaction between different nitroxides takes place. The K_{eq} value strongly depends on the nitroxide redox potential Nevertheless the kq values were found to be independent of K_{eq}. On the basis of these and other available data, two mechanisms of IFQ were proposed: the dominant mechanism, intersystem crossing and a secondary mechanism, irreversible intramolecular electron transfer (ET) from the excited singlet of the fluorophore (donor D) to nitroxide (acceptor A) followed by fluorophore segment regeneration and hydroxyl derivative formation. The second mechanism is responsible for photoreduction (Fig. 2.15).

Electron exchange interactions with paramagnetic species have been known to facilitate the excited-state relaxation of chromophores. It was proves that the quenching mechanism originates from changes in the spin multiplicity of the electronic states [81] and references therein. For example, the singlet (S_0) state and the lowest excited singlet (S_1) state of the chromophore and the unpaired electron spin of the doublet nitroxide radical (NR) become the doublet (D_0 and D_n), respectively (Fig. 2.16).

The lowest excited doublet (D_1) and quartet (Q_{A1}) states are generated by an interaction between the nitroxide radical spin and the T_1 chromophore triplet. Because the spin-forbidden transitions of the chromophore, $S_1 \rightarrow T_1$ and $T_1 \rightarrow S_0$, can be partially transformed into the non-forbidden the Dn $\rightarrow$ D_1 and $D_1 \rightarrow D_0$ transitions, respectively. The doublet states (Dn, D_1 and D_0) have the same spin multiplicity, therefore, the lifetimes of the excited state were expected to be very short compared to those of the $S_1 \rightarrow S_0$ transition. As an example, copper porphyrins form short-lived (a few ten nanoseconds) D_1 and Q_{A1} states.

A series of chromophores bonded to nitroxides have been studied by time-resolved ESR (TRESR) spectroscopy to investigate directly the excited multiplet states consisting of a photoexcited triplet chromophore and doublet nitroxides. The D_1 and Q^{A1} states have been observed in solution and in frozen solutions for several systems. The photophysical properties of silicon phthalocyanin (SiPc) covalently linked to one or two nitroxide radicals have been systematically studied by fluorescence, transient absorption and TRESR spectroscopies [81, 84, 85]. The electronic states of R0, R1 and R2 are depicted in Fig. 2.17.

Pyrene Derivatives (FI = PYR)

Dansyl Derivatives (FI = DAN= NR-Dns)

Spacer
Nitroxide

SO_2
N–R
Spacer
Nitroxide

R = H (a)
    Me (b)
    Et (c)
    Ph (d)

Spacer =

*—CHCHONH—   (A)
      CH$_2$Ph

*—(CH$_2$)$_2$NHCO—   (B)

*—(CH$_2$)$_3$NHCO—   (C)

*—(CH$_2$)$_5$NHCO—   (D)

*—CH$_2$NHCO—   (E)

*—CH$_2$CONH—   (F)

*—(CH$_2$)$_3$CONH-   (G)

* -Linked to FI

*—CONH-   (H)

*—NHCO—   (I)

*—CH$_2$NHC(=O)C=NCO—   (J)
            CH$_2$Ph

*—CH$_2$NHC(=O)CHNHCO—   (K)
            CH$_2$Ph

Nitroxide ˙R =

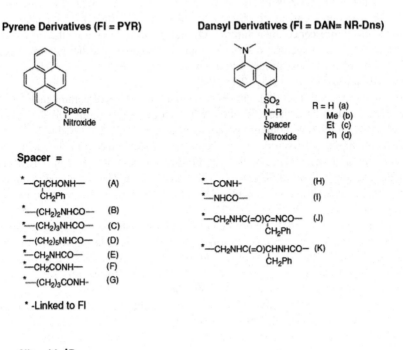

1   2   3   4   5   6   7   8

-Linked to spacer

Fig. 2.15 Structure of pyrene and dansyl nitroxides (private communication from Drs. V.V. Martin and A. Weis, Lipitek International, Inc.) [81]

Electron Exchange   Relaxation

π^* D_n π^* D_1 π^* D_0

π SOMO π SOMO π SOMO

Pc   NR   Pc   NR   Pc   NR

Fig. 2.16 The $D_n \rightarrow D_1$ and $D_1 \rightarrow D_0$ transitions via electron exchange processes [81]

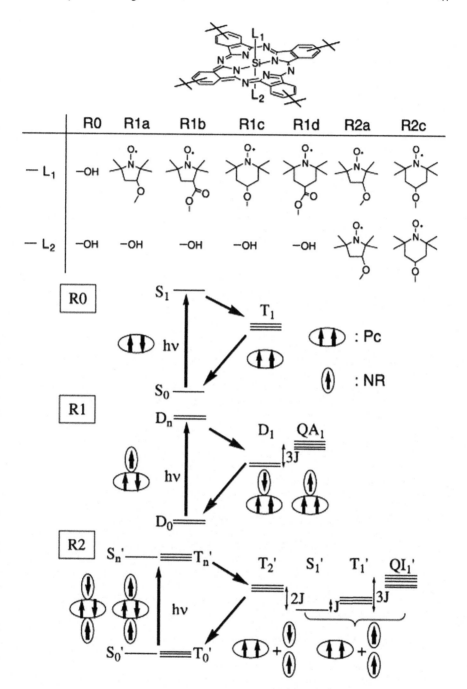

Fig. 2.17 Molecular structures and electronic states of SiPc covalently linked to nitroxide radicals NRs [81]

In the case of the $D_n \rightarrow D_1$ and $D_1 \rightarrow D_0$ transitions, the electron exchange process is dominant, as these transitions occur along with the electron exchange process between the SiPc and nitroxil radical moieties.

Both the theoretical and experimental data presented in clearly demonstrate the significant progress that has been recently made in applying dual chromophore-nitroxide compounds to areas of photophysics and biological analytical chemistry [82–89].

References

1. J. Vura-Weis, S.H. Abdelwahed, R. Shukla, R. Rathore, M.A. Ratner, M.R. Wasielewsk, Science **328**, 1547 (2010)
2. J. Zhao, K. Xu, W. Yang, Z. Wang, F. Zhong, Chem. Soc. Rev. (2015) Ahead of Print
3. B. Zhou, B. Shi, D. Jin, X. Liu, Nat. Nanotechnol. **10**, 924 (2015)
4. J. Vura-Weis, S.H. Abdelwahed, R. Shukla, R. Rathore, M.A. Ratner, R. Wasielewski, Science **328**, 1547 (2010)
5. Y. Feng, J. Cheng, L. Zhou, X. Zhou, H. Xiang, Analyst **137**, 4885 (2012)
6. D. Gust, Faraday discussions (2015). Ahead of Print
7. S. Fukuzumi, K. Ohkubo, Chem. Sci. **4**, 561 (2013)
8. A. Kamkaew, S. Lim, H. Lee, L. Kiew, L. Chung, K. Burgess, Chem. Soc. Rev. **42**, 77 (2013)
9. J.R. Lakowicz, *Principles of Fluorescence Spectroscopy*, 3rd edn. (Springer, 2006)
10. M. Sauer, J. Hofkens, J. Enderlein, *Handbook of Fluorescence Spectroscopy and Imaging: From Ensemble to Single Molecules* (Wiley, 2011)
11. D.W. Ball. Phosphorescence. in *Field Guide to Spectroscopy* (SPIE, Press books, 2015)
12. C.R. Ronda. *Luminescence: From Theory to Applications* (Wiley, 2008)
13. D.L. Dexter, J Chem. Phys. **21**, 8361953)
14. A.N. Terenin, V.L. Ermolaev, Trans. Faraday Soc. **52**, 1042 (1956)
15. V.L. Ermolaev, E.N Bodunov, E.B., Sveshnikova, T.A. Shakhverdov, *Radiationless Transfer of Electronic Excitation Energy* (Nauka, Leningrad, 1977)
16. H.E. Zimmerman, G.L. Grunewald, J. Am. Chem. Soc. **88**, 183 (1966)
17. G.S. Hammond, J. Saltiel, J. Amer. Chem. Soc. **85**, 2516 (1963)
18. J. Saltiel, J.M. Mace, L.P. Watkins, D.A. Gormin, R.J. Clark, O. Dmitrenko, J. Amer. Chem. Soc. **125**, 16158 (2003)
19. K. Sandros, Acta Chem. Scand. **18**, 2355 (1964)
20. J.E. Subotnik, J. Vura-Weis, A.J. Sodt, M.A Ratner, J, Phys. Chem. A **114**, 8665 (2010)
21. Z.Q. You, C.-P. Hsua, G.R. Fleming, J. Chem. Phys. **124**, 044506 (2006)
22. Y.-B. Si, X.-X. Zhong, W.-W. Zhang, Y. Zhao Chin, J. Phys. Chem. C **116**, 12499 (2012)
23. A. Neubauer, G. Grell, A. Friedrich, S.I. Bokarev, P. Schwarzbach, F. Gaertner, A.-E. Surkus, H. Junge, M. Beller, O. Kuehn, S. Lochbrunner, J. Phys. Chem. Lett. **5**, 1355 (2014)
24. J. Ma, X. Cui, F. Wang, W. Xueyan, J. Zhao, X. Li, J. Org. Chem. **79**, 10855 (2014)
25. L.M. Frutos, O. Castano, J.L. Andres, M. Merchan, A.U. Acuna, J. Chem. Phys. **120**, 1208 (2004)
26. F. Zapata, M. Marazzi, O. Castaño, A.U. Acuña, L.M. Frutos, J. Chem. Phys. **140**, 034102 (2014)
27. D. Chantal, Coord. Chem. Rev. **282–283**, 19 (2015)
28. M. Durchan, J. Tichy, R. Litvin, V. Slouf, Z. Gardian, P. Hribek, F. Vacha, T. Polivka, J. Phys. Chem. B **116**, 8880 (2012)
29. F.S. Steinbacher, R. Krause, A. Hunze, A. Winnacker, Phys. Status Solidi A **209**, 340 (2012)
30. Z.-Q. You, C.-P. Hsu, J. Chem. Phys. **133**, 074105 (2010)

31. R.E. Merrifield, J. Chem. Phys. **48**, 4318 (1968)
32. R. Islangulov, J. Lott, C. Weder, F.N. Castellano, J. Am. Chem. Soc. **129**, 12652 (2007)
33. Ibrayev NK, V.A. Latoni. Dynamics of triplet excitations in Langmuir-Blodgett films of aromatic molecules. J. Lumin. **2000**, 87–89, 760–763
34. V.M. Mekler, O.V. Belonogova, L.I. Borodenko, G.I Zhiznevskaya, S.F. Izmailov, in *Spectroscopy of Biological Molecules*, ed. by J.C. Merlin, S. Turrell, J.P. Huvenne, European Conference on the Spectroscopy of Biological Molecules, 6th, Villeneuve d'Ascq, Fr., Sept. 3–8, 1 405–406 (1995)
35. V.M. Mekler, F.T. Umarova, Biofizika **33**, 720 (1988)
36. V.M. Mekler, G.I. Likhtenshtin, Biofizika **31**, 568 (1986)
37. V.M. Mekler, A.I. Kotel'nikov, G.I. Likhtenshtin, M.A. Berkovich, Biofizika **27**, 641 (1982)
38. T. Miteva, V. Yakutkin, G. Nelles, S. Baluschev, New J. Phys. **10**, 103002 (2008)
39. A. Benfredj, F.L. Hachani, S. Romdhane, H. Bouchriha, Phys. Rev. B **71**, 075205 (2005)
40. A.R. Monguzzi Meinardi, Phys. Rev. B **80**, 039904(E) (2009)
41. T. Förster, Naturwissenschaften **33**, 166 (1946)
42. S. Faure, C. Stern, R. Guilard, P.D. Harvey, J. Am. Chem. Soc. **126**, 1253 (2004)
43. A.A. Setlur, J.J. Shiang, C.J. Vess, J. Phys. Chem. C. **2011**(115), 3475–3480
44. M.D. LaCount, D. Weingarten, N. Hu, S.E. Shaheen, J. van de Lagemaat, G. Rumbles, D.M. Walba, M.T. Lusk, J. Phys. Chem. A **119**, 4009 (2015)
45. G.B. Piland, J.J. Burdett, R.J. Dillon, C.J. Bardeen, J. Phys. Chem. Lett. **5**, 2312 (2014)
46. A. Monguzzi, J. Mezyk, F. Scotognella, R. Tubino, F. Meinardi, Phys. Rev. B **78**, 195112 (2008)
47. M. Bixon, J.J. Jortner, Chem. Phys. **48**, 715 (1968)
48. A. Jabłoński, Nature **131**, 839 (1933)
49. J.M. Bowman, Science **319**, 40 (2008)
50. A. Nakata, T. Tsuneda, K. Hirao, J. Chem. Phys. **135**, 224106 (2011)
51. Q. Li, Y.-X. Qiu, X.-Y. Chen, W.H.E. Schwarz, S.G. Wang, Phys. Chem. Chem. Phys. **14**, 6833 (2012)
52. C.M. Marian, Wiley Interdisciplinary Rev: Comput. Mol. Sci. **187**, 203 (2012)
53. Y. Si, W. Zhang, Y. Zhao, J. Phys. Chem. A **116**, 2583 (2012)
54. D. Chantal, Coord. Chem. Rev. **282–283**, 19 (2015)
55. L.-G. Gao, X.-L. Song, Y.-C. Wang, L.-L. Lü, Comput. Theor. Chem. **3**(968), 31–38 (2011)
56. S. Koseki, M.S. Gordon, M.W. Schmidt, N. Matsunaga, J. Phys. Chem. **99**, 12764 (1995)
57. M. Baba, J. Phys. Chem. A **115**, 9514 (2011)
58. M.A. El-Sayed, Acc. Chem. Res. **1**, 8 (1968)
59. J.N. Harvey, Phys. Chem. Chem. Phys. **9**, 331–343 (2007)
60. Y. Si, W. Zhang, Y. Zhao, J. Phys. Chem. A **116**, 2583 (2012)
61. J. Chang, A.J. Fedro, van M. Veenendaal, Phys. Rev. B **82**, 075124 (2010)
62. U. Fano, Phys. Rev. **124**, 1866 (1961)
63. A. Nakata, T. Tsuneda, K. Hirao, J. Chem. Phys. **135**, 224106 (2011)
64. S. Mai, P. Marquetand, L. Gonzalez, Intern. J. Quant. Chem. (2015). Ahead of Print]
65. M. Kleinschmidt, C. van Wüllen, C.M. Marian, J. Chem. Phys. **142**, 094301/1 (2015)
66. M. Huix-Rotllant, N. Ferre, J. Chem. Phys. **140**, 134305/1 (2014)
67. G.S. Orf, D.M. Niedzwiedzki, R.E. Blankenship, J. Phys. Chem. B **118**, 2058 (2014)
68. L.D. Landau, E.M. Lifshitz *Quantum Mechanics*, (Pergamon, Oxford, 1976)
69. O. Schalk, M.S. Schuurman, G. Wu, P. Lang, M. Mucke, R. Feifel, A. Stolow, J. Phys. Chem. A **118**, 2279 (2014)
70. E.F. Healy, S. Manzer, J. Gorman, A. Jones, N. Cristea, Chem. Phys. Lett. **485**, 258 (2010)
71. D. Ompong, J. Singh, Physica Status Solidi C: Current Topics in Solid State Physics (2015). Ahead of Print
72. S. Pal, N. Chatterjee, P.K. Bharadwaj, RSC Adv. **4**, 26585 (2014)
73. S. Tobita, M. Arakawa, I. Tanaka, J. Phys. Chem. **89**, 5649 (1985)
74. M. Wierzchaczewski, A. Sidorowicz, E. Lukowiak, W. Strek, P. Gawryszewska, J. Appl. Spectrosc. **62**, 221 (1995)

75. D.M. Guldi, T.D. Mody, N.N. Gerasimchuk, D. Magda, J.L. Sessler, J. Am. Chem. Soc. **122**, 8289 (2000)
76. V. Volchkov, V. Ivanov, B. Uzhinov, Journal of Fluorescence **20**, 299 (2010)
77. J.A. Green, J.A. Singer, J.H. Parks, J. Chem. Phys. **58**, 2690 (1973)
78. V.A. Kuzmin, A.S. Tatikov, J. Chem. Phys. Let. **53**, 606 (1978)
79. A.F. Tarek, G. Grampp, S. Landgraf, Int. J. Photoenergy **1**, 173 (1999)
80. I.M. Bystryak, G.I. Likhtenshtein, A.I. Kotel'nikov, O. Hankovsky, K. Hideg, Russ. J. Phys. Chem. **60**, 1679 (1986)
81. G.I. Likhtenstein, K. Ishii, S. Nakatsuji, Photochem. Photobiol. **83**, 871 (2007)
82. G.I. Likhtenshtein, Pure Appl. Chem. **80**, 2125 (2008)
83. G.I. Likhtenshtein, D. Pines, E. Pines, V. Khutorsky, Appl. Magn. Reson. **35**, 459–472 (2009)
84. K. Ishii, Y. Hirose, N. Kobayashi, J. Am. Chem. Soc. **123**, 702–708 (2001)
85. E. Lozinsky, V.V. Martin, T.A. Berezina, A. Shames, A.L. Weis, G.I. Likhtenshtein, J. Biochem. Biophys. Meth. **38**, 29 (1999)
86. N. Medvedeva, V.V. Martin, G.I. Likhtenshten, J. Photochem. Photobiol. A Chem **163**, 45 (2004)
87. E.M. Lozinsky, L.V. Martina, A.I. Shames, N. Uzlaner, A. Masarwa, G.I. Likhtenshtein, D. Meyerstein, V.V. Martin, Z. Priel, Detection of NO from pig trachea by a fluorescence method, Analyt. Biochem. **326**, 139 (2004)
88. G.I. Likhtenshtein, Appl. biochem. biotech. **152**, 135 (2009)
89. G .I. Likhtenshtein, J. Yamauchi, S. Nakatsuji, A. Smirnov, Tamura R, *Nitroxides: Application in Chemistry, Biomedicine, and Materials Science*. (WILEY-VCH, Weinhem 2008)

Chapter 3
Spin Electron Dipolar and Contact Interactions

Abstract In solving problems of enzyme catalysis, biophysics of biopolymers, biomembranes, molecular biology and physico-chemistry of polymers it is necessary to know the spatial disposition of individual parts of systems under interest and the system molecular dynamics. One must also know the depth of immersion of paramagnetic centers in a biological matrix, i.e. the availability of enzyme sites to substrates, distance of electron tunneling between a donor and an acceptor group, position of a spin-label in a membrane and in a protein globule, etc. Structural and dynamic studies of many of such systems that cannot be obtained as pure single crystals are beset with serious difficulties. This problem can be solved with the use of methods of Electron Spin Resonance (ESR) and some time in combination with Nuclear Magnsetic Resonance (NMR) techniques. The present chapter briefly described theories of electron spin–electron spin dipolar interactions and electron spin–nuclear spin dipolar and contact interactions. The theories form a basis for methods of measurement of distance between the paramagnetic centers, the study of its motion, and the establishment of spin distribution. In combination, these methods have been proven to be a powerful approach in the investigation of the structure and dynamics of biological and chemical molecular objects, including those which cannot be obtained as single crystals.

3.1 Introduction

The ESR phenomena involves the resonance absorption or dispersion of a microwave frequency 0.3–250 GHz [1–14] of electro magnetic field (ν) by a system of particles with the intrinsic spin moment of an unpaired electron in a constant magnetic field of strength H_0. The absorption leads to magnetization in the exited state of the system. Accordingly, the electron magnetic resonance condition is:

$$h\nu = g_e \beta_B \mathbf{H_0}, \qquad (3.1)$$

© Springer International Publishing Switzerland 2016
G. Likhtenshtein, *Electron Spin Interactions in Chemistry and Biology*,
Biological and Medical Physics, Biomedical Engineering,
DOI 10.1007/978-3-319-33927-6_3

where g_e is a g-factor, characterizing the value of the intrinsic electron spin moment (free electron g-value is 2.002319), β_B is the Bohr magneton ($9.27400968 \times 10^{-24}$ J T^{-1}). The values of g-factors and magnetic field strength H_0 dictate position of resonance frequencies in the NMR and ESR spectra. There are two fundamental mechanisms of spin electron spin electron interactions (SSI), that is, quantum mechanical exchange interaction (Section X) and semiclassical dipole dipole interaction (DDI). The DDI arises because the magnetic dipole of one paramagnetic center induces a local magnetic field at the site of another paramagnetic group. This interaction manifest itself in the dipolar energy splitting or change of spin-lattice (T_1) and spin-spin (T_2) relaxation times. These phenomena form a basis for various methods including distance estimation and determination of parameters of molecular dynamics.

The interaction of an unpaired electron, by way of its magnetic moment, with nearby nuclear spins, results in additional allowed energy states. Two common mechanisms by which electrons and nuclei interact are: the Fermi contact interaction and dipolar interaction. The Fermi contact interaction is the quantum mechanical magnetic interaction between an electron and an atomic nucleus when the electron is inside that nucleus. This interaction leads to appearance of spin density on s-orbits of atoms bearing nucleus and applies to the case of isotropic interactions which are independent of sample orientation in a magnetic field. The dipole-dipole interaction between spin electron and spin of nucleus depends on the sample orientation. The both phenomena serve as a tool for investigation of structure paramagnetic spices and its rotational dynamics.

Dynamic nuclear spin polarization is a third mechanism for interactions between an unpaired electron and a nuclear spin. Spin polarization is the degree to which the spin is aligned with a given direction. In the thermal thermal equilibrium, the alignment of electron and nuclear spins at a given magnetic field and temperature is described by the Boltzmann distribution. Dynamic nuclear polarization (DNP) relates to transferring spin polarization from electrons to nuclei, resulted in additional aligning (polarizing) the nuclear spins to the extent that electron spins are aligned [13, 14]. Those electrons can be aligned to a higher degree of order by chemical reactions leading to Chemical-Induced DNP, CIDNP) (Section X). This phenomenon is widely employed for investigation of radical reaction mechanisms.

3.2   Spin Relaxation

Spin relaxations times are important parameters of electron spin resonance and nuclear magnetic resonance. Two approaches have been used for describing relaxation phenomena. In the frame of the first approach, the characteristic time of recovery of the induced magnetization to the ground state on account a transfer of magnetic energy to energy of media (lattice) is noticed as spin-lattice relaxation ($1/T_1$). Another type of relaxation, spin-spin or spin-phase relaxation ($1/T_2$), is related to time of return of the spin system to equilibrium in the exited magnetic state as a result of spin-spin interaction with environment.

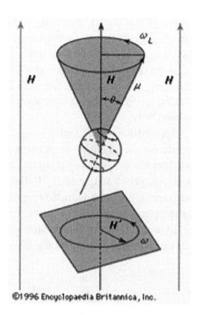

The second approach is based on properties of the electron Larmor precession, which is the precession of the magnetic moments of electrons about external magnetic field (Fig. 3.1) The angular momentum vector $\vec{J}$ precesses about the external field axis with an angular frequency known as the Larmor frequency,

$$\omega = -\gamma B \tag{3.2}$$

where ω is the angular frequency,

$$\gamma = \frac{-eg}{2m} \tag{3.3}$$

is the gyromagnetic ratio, B is the magnitude of the magnetic field and g is the g-factor.

In NMR and ESR the Bloch equations [15] are used to calculate the dynamic of spin magnetization M = (M_x, M_y, M_z), after the microwave resonance absorption, as a function of time when relaxation times T_1 and T_2. For **M**(t) the time dependent spin magnetization $\mathbf{M}(t) = (M_x(t), M_y(t), M_z(t))$, the Bloch equations read:

$$\frac{dM_x(t)}{dt} = \gamma(\mathbf{M}(t)) \times \mathbf{B}(t))_x - \frac{M_x(t)}{T_2}$$

$$\frac{dM_y(t)}{dt} = \gamma(\mathbf{M}(t)) \times \mathbf{B}(t))_y - \frac{M_y(t)}{T_2} \tag{3.4}$$

$$\frac{dM_z(t)}{dt} = \gamma(\mathbf{M}(t)) \times \mathbf{B}(t))_z - \frac{M_z(t) - M_0}{T_1}$$

where $\mathbf{B}(t) = (B_x(t), B_y(t), B_0 + \Delta B_z(t))$ is the magnetic field experienced by the spins. In general, the z component of the magnetic field $\mathbf{B}$ can be bancomposed of two terms: B_0, constant in time, and $\Delta B_z(t)$, time dependent. $\mathbf{M}(t) \times \mathbf{B}(t)$ is the cross product of these two vectors. M_0 is the steady state spin magnetization in the z direction. The relaxation times T_1 and T_2 are called as longitudinal and transverse relaxation times, being synonyms of spin-lattice and spin-spin relaxation time, respectively.

Spin-lattice ($1/T_1$) and spin-spin ($1/T_2$) rates of a paramagnetic depend significantly on the intensity of phonon dynamics in the medium [16, 17]. The phonon oscillations are accompanied by electric field oscillations within distribution of wide range of frequencies. If the distribution partially overlaps the microwave frequency, the paramagnetic magnetization energy will be transferred to the energy of the phonon oscillation through the spin-orbital interaction.

In work [18] the temperature-dependent relaxation rates were modeled as the sum of contributions from the direct, Raman, and local modes:

$$\frac{1}{T_1} = A_{\text{direct}}(T + b) + A_{\text{Ram}}\left(\frac{T}{\theta_D}\right)^9 J_8(\theta_D/T)$$
$$+ A_{\text{loc}}\frac{\exp[\Delta_{\text{loc}}/T]}{(\exp[\Delta_{\text{loc}}/T] - 1)^2} \tag{3.5}$$

where T is temperature in Kelvin, A_{direct} and b are the coefficients for the contribution from the direct process, A_{Ram} is the coefficient for the contribution from the Raman process, Θ_D is the Debye temperature, J_8 is the transport integral,

$$J_8(\theta_D/T) = \int_0^{\theta_D/T} x^8(e^x/(e^x - 1)^2) \, dx, \tag{3.6}$$

A_{loc} is the coefficient for the contribution from local mode, and Δ_{loc} is the energy of the local mode. Methods of measurement of the spin relaxation parameters related to spin-spin interaction will be outlined in Section X. For nitroxide radicals in liquids, Eaton's group [19]. Proposed a spin lattice relaxation process, called the thermally activated process *relared to* relaxation in the slower tumbling regime which dominated by a frequency-dependent process.

Comrehensive review on the spin relaxation phenomena in organic radicals and transition memal ions was published by SS Eaton and Eaton [20] Electron spin relaxation mechanisms in various systems were discussed in recent works [21, 22].

3.3 Electron Spin-Spin Dipole-Dipole Interaction

3.3.1 General

According to theory of the Electron Spin Resonance, factors affecting interaction between two magnetic dipoles in a static magnetic field (H_0) are: its magnetic moments, interspin distance and the angle between the interspin vector and the external field. Other important factors are the spin-spin (T_2) and spin-lattice (T_1) relaxation times, the ESR frequencies (ω) of both interacting paramagnetics [1–6]. When the product of relaxation time T_1 for the second center and the interaction energy ($\Delta\omega$) expressed in frequency units and $\Delta\omega^2 T_1^2 \gg 1$, the effect of the second spin may be regarded as an interaction with a permanent dipole moment (slow relaxing spin). When $\Delta\omega^2 T_1^2 \ll 1$, the interaction may be considered as a weak perturbation because of the fast relaxation (fast relaxing spins).

3.3.2 Interaction Between Slow Relaxing Dipoles

Dipolar splitting. Pake dublet
The first applications of dipole-dipole spin-spin interactions to the investigation of protein surface topography (measurement spin-spin distances) were based on changes in the lineshape of ESR spectra of nitroxide radical or a paramagnetic complex [25–30] or in the spin relaxation parameters [31–33] arising from interaction with a second nitroxide radical or a paramagnetic ion (Sections X, Y). For rough estimation of distance between nitroxide radicals by analysis of its ESR spectra, it is convenient to use the Kokorin parameter [30].

The dipolar coupling between two spin leads to splitting of the ESR and NMR lines with formation of the Pake doublet [23]. The shape of the Pake doublet (Fig. 3.2) is due to the fact that there are only two possible orientations along the external magnetic field (parallel and antiparallel), but many orientations perpendicular to the B_0-field.

A triplet biradical with a total spin of S = 1 is characterized by three magnetic sublevels with the quantum number Ms + 1, 0, and −1 (zero-field splitting, ZFS), is parametrized by a matrix **D** within the phenomenological spin Hamiltonian [22]:

$$\hat{H}_{ZFS} = \hat{S}\mathbf{D}\hat{S} \tag{3.7}$$

The Hamiltonian that describes interaction between two permanent spins in a point-dipole approximation is given by [24]

$$\hat{H}_{SS} = \frac{g_e^2 \alpha^2}{8} \sum_{i \neq j} \left[\frac{\hat{S}(i)\hat{S}(j)}{r_{ij}^3} - 3\frac{(\hat{S}(i)\mathbf{r}_{ij})(\hat{S}(j)\mathbf{r}_{ij})}{r_{ij}^5} \right] \tag{3.8}$$

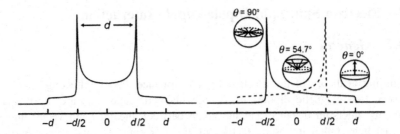

Fig. 3.2 The shape of the Pake doublet at different dipolar coupling d and the angle Θ between the axis connecting the two spins and the external magnetic field vector B_0. www.chemie.uni-hamburg.de/nmr/.../dipolar_coupling.htm

Here, $\mathbf{r}_{ij} = \mathbf{r}_i - \mathbf{r}_j$ for electrons i and j at positions $\mathbf{r}_i$ and $\mathbf{r}_j$, and with spins sˆ(i) and sˆ(j), g_e is the free-electron g-value.

The doublet splitting and line shape depend on orientation of the principal axis of the coupling interaction relative to the magnetic field direction. In high magnetic fields, the orientation dependence is expressed with the secular dipole-dipole coupling constant d_{12} through the angle Θ between the axis connecting the two spins and the external magnetic field vector $\mathbf{B}_0$:

$$d_{12} = b_{12} \frac{1}{2}(3 \cos^2 \Theta_{12} - 1) \tag{3.9}$$

where

$$b_{12} = -\frac{\mu_0}{4\pi} \frac{\gamma_1 \gamma_2 \hbar}{r_{12}^3} \tag{3.10}$$

in units of rad/s.

Spin-spin anisotropic interaction

The general spin Hamiltonian in the high-field approximation for a pair of dipolar coupled paramagnetics without taking into consideration spin-orbital effects is given by [34]:

$$
\begin{aligned}
\hat{\mathcal{H}} = & g_{zz}^1 \beta_e \vec{H}_0 \hat{S}_z^1 + g_{zz}^2 \beta_e \vec{H}_0 \hat{S}_z^2 - \omega_n(\hat{I}_z^1 + \hat{I}_z^2) + \gamma_e \mathbf{A}_{xz}^1 \hat{I}_x^1 \hat{S}_z^1 \\
& + \gamma_e \mathbf{A}_{yz}^1 \hat{I}_y^1 \hat{S}_z^1 + \gamma_e \mathbf{A}_{zz}^1 \hat{I}_z^1 \hat{S}_z^1 + \gamma_e \mathbf{A}_{xz}^2 \hat{I}_x^2 \hat{S}_z^2 + \gamma_e \mathbf{A}_{yz}^2 \hat{I}_y^2 \hat{S}_z^2 \\
& + \gamma_e \mathbf{A}_{zz}^2 \hat{I}_z^2 \hat{S}_z^2 + \frac{D}{4} \hat{S}_+^1 \hat{S}_-^2 + \frac{D}{4} \hat{S}_-^1 \hat{S}_+^2 + D\hat{S}_z^1 \hat{S}_z^2 + J\hat{S}^1 \cdot \hat{S}^2,
\end{aligned}
\tag{3.11}
$$

where β_e is the Bohr magneton, g^1 and g^2 are the tensors defining the interaction of the electron spin of paramagnetic 1 ($\hat{S}^1$) and paramagnetic 2 ($\hat{S}^2$) with the magnetic field (H_0), ϖ_n is the Larmor frequency of the nucleus, γ_e is the gyromagnetic ratio for the electron, A^1 and A^2 are the hyperfine tensors defining the interaction of the nuclear spins ($\hat{I}^1$ or $\hat{I}^2$) with $\hat{S}^1$ or $\hat{S}^2$, D is the element of the dipolar coupling tensor, and J is the scalar exchange interaction. Significant contribution of J coupling in pair of paramagnetics tethered with a nonsaturated bridge (super exchange, **Section X**) can be observed only for distances fewer than 7–8 Å.

The corresponding Hamiltonian can be expressed in a tensor form as [35]

$$\mathscr{H}_{SS} = g_e^2\mu_B^2[S_{1x}S_{1y}S_{1z}] \cdot \begin{bmatrix} \left\langle \frac{r^2-3x^2}{r^5} \right\rangle & \left\langle -\frac{3xy}{r^5} \right\rangle & \left\langle -\frac{3xz}{r^5} \right\rangle \\ \left\langle -\frac{3xy}{r^5} \right\rangle & \left\langle \frac{r^2-3y^2}{r^5} \right\rangle & \left\langle -\frac{3yz}{r^5} \right\rangle \\ \left\langle -\frac{3xz}{r^5} \right\rangle & \left\langle -\frac{3yz}{r^5} \right\rangle & \left\langle \frac{r^2-3z^2}{r^5} \right\rangle \end{bmatrix} \cdot \begin{bmatrix} S_{2x} \\ S_{2y} \\ S_{2z} \end{bmatrix} \tag{3.12}$$

$\langle\,\rangle$ means integration over the spatial coordinats.

The dipolar interaction of paramagnetics with anisotropic g-tensor was described by the Hamiltonian [36]

$$H_d = -J_d S_1 S_2 + D_d(3S_{1z}S_{2z} - S_1 S_2) + E_d(S_{1x} - S_{2x} - S_{1y}S_{2y}) \tag{3.13}$$

For isotropic interaction:

$$J_d = -(\beta/3r^{-3})\left(g_{1x}g_{2x} + g_{1y}g_{2y} - 2g_{1z}g_{2z}\right) \tag{3.14}$$

For the anisotropic interaction with the axial and rhombic symmetry of the g-factor,

$$D_d = -(\beta/3r^{-3})[(2g_{1z}g_{2z} + 0.5(g_{1z}g_{2z} + g_{1y}g_{2y})] \tag{3.15}$$

and

$$E_d = -(\beta/3r^{-3})(g_{1x}g_{2x} + g_{1y}g_{2y}) \tag{3.16}$$

respectively. Thus, the general Hamiltonian should include terms associated with anisotropy hyperfine interaction.

The spin-spin interaction term for spins was calculated as a first order term from perturbation theory [37]. The calculated zero-field splitting parameters include both the spin-spin coupling to first order and spin orbital coupling to second order of perturbation theory. The dipolar spin-spin contribution was presented as

$$D_{kl}^{SS} = \frac{g_e^2\alpha^2}{4S(2S-1)} \left\langle \Psi_0^{SS} \left| \sum_i \sum_{i\neq j} \frac{r_{ij}^2\delta_{kl} - 3r_{ij;k}r_{ij;l}}{r_{ij}^5}\{2\hat{s}_z(i)\hat{s}_z(j) \right.\right.$$
$$\left.\left. - \hat{s}_x(i)\hat{s}_x(j) - \hat{s}_y(i)\hat{s}_y(j)\}\right|\Psi_0^{SS} \right\rangle \tag{8}$$
$$\tag{3.17}$$

Fig. 3.3 Diagram defining
parameters for the
tether-in-a-cone model. The
axes X, Y, and Z are the axes
of the two nitroxides with
X along the N-O bond and
Z perpendicular to the plane
of the nitroxide and with all
subscripts referring to
nitroxide 1 or nitroxide 2.
Other detais see in [40]

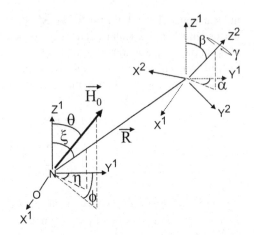

Here, Ψ_0^{SS} denotes the Ms = S component of the wave function for the state
under investigation. The authors stressed than the direct dipolar spin–spin interaction contains an "exchange" contribution that is of different origin than the isotropic Heisenberg type exchange interaction The Heisenberg isotropic exchange
interaction is responsible for the splitting of the energy levels into singlet and triplet
states of the interacting radical pair, while the "exchange" contribution to the spin–
spin interaction in (3.19) represents a quantum mechanical correction to the magnetic dipole–dipole interaction. Equations (3.16)–(3.19), can be used for determination interspin distances up to 25–30 Å.

A theoretical method for *ab initio* calculations of the zero-field splitting
parameters from electron spin–spin interaction contributions was described within
the density-functional theory [38]. For organic biradicals, such as carbene, vinyl-
methylene, phenylcarbene, and diphenylcarbene, the relationship between their spin
orbitals, magnetic axes, and effects of conformational changes and the zero-filed
splitting parameters (D- and E) have been revealed. With the aim to develop tools
for determining both the distance distribution and the relative orientation of the
labels from experimental spectra, a tether-in-a-cone model (Fig. 3.3) was developed. Such an approach allowed the simulation of electron paramagnetic resonance
spectra of dipolar coupled nitroxide spin labels attached to tethers statically disordered within cones of variable halfwidth [39, 40]. In the frame of the model, the
spin Hamiltonian in the high-field approximation for a pair of dipolar coupled
nitroxides was used.

The Hamiltonian that describes the interacting system of a spin $\mathbf{S} = 1$ ($\mathbf{S}^{Fe}$) with
a spin $\mathbf{S} = 1/2$ ($\mathbf{S}^{rad}$) is given by [41]

$$\mathbf{H} = \beta\,\mathbf{S}^{rad} \cdot \mathbf{g}^{rad} \cdot \mathbf{H} + \beta\,\mathbf{S}^{Fe} \cdot \mathbf{g}^{Fe} \cdot \mathbf{H} + \mathbf{S}^{Fe} \cdot \mathbf{D} \cdot \mathbf{S}^{Fe} + \mathbf{S}^{rad} \cdot \Delta \cdot \mathbf{S}^{Fe} \qquad (3.18)$$

where β is the Bohr magneton; $\mathbf{H}$, the applied magnetic field; $\mathbf{g}^{rad}$ and $\mathbf{g}^{Fe}$ are the **g**-
tensors for the radical and the iron center; $\mathbf{D}$ is the axial zerofield splitting tensor for

the iron and is; Δ is the diagonal interaction tensor; and $\mathbf{S}^{\mathrm{rad}}$ and $\mathbf{S}^{\mathrm{Fe}}$, are the spin operators g. In work [41], The Hamiltonian was expanded as

$$
\begin{aligned}
\mathbf{H} = \beta(&g_x^{\mathrm{rad}}\hat{S}_x^{\mathrm{rad}}H_x + g_y^{\mathrm{rad}}\hat{S}_y^{\mathrm{rad}}H_y + g_z^{\mathrm{rad}}\hat{S}_z^{\mathrm{rad}}H_z + g_x^{\mathrm{Fe}}\hat{S}_x^{\mathrm{Fe}}H_x \\
&+ g_y^{\mathrm{Fe}}\hat{S}_y^{\mathrm{Fe}}H_y + g_z^{\mathrm{Fe}}\hat{S}_z^{\mathrm{Fe}}H_z) + \mathrm{D}\left(\hat{S}_z^{\mathrm{Fe}}\hat{S}_z^{\mathrm{Fe}} - \frac{2}{3}\right) \\
&+ \Delta_x(\hat{S}_x^{\mathrm{rad}}\hat{S}_x^{\mathrm{Fe}}) + \Delta_y(\hat{S}_y^{\mathrm{rad}}\hat{S}_y^{\mathrm{Fe}}) + \Delta_z(\hat{S}_z^{\mathrm{rad}}\hat{S}_z^{\mathrm{Fe}})
\end{aligned} \tag{3.19}
$$

The interaction between the iron center and the radical results in six energy levels. Because the values for the coupling are small compared to the positive zero-field splitting parameter D, only transitions between the two lower levels were taken in consideration

3.3.3 Interaction Between a Slow Relaxing Dipole and Fast Relaxing Dipole

Of particular interest for the solution of structural problems is the case involving one slower-relaxing paramagnetic center with T_{1s} and T_{2s} times that are long enough for the ESR line to be observed and for the ESR saturation curve to be obtained, and a second center characterized by short relaxation times, T_{1f}, T_{2f} and i.e., $T_{1f}\,\mu_{f/r3} \ll 1$. For such a pair, assuming that $T_{1f} = T_{2f}$, $(\omega_f - \omega_s)^2 T_{1f}^2 \gg 1$, $(\omega_s)^2 T_{1f}^2 \gg 1$, and the spherical average of orientation-dependent terms for the contribution of the second paramagnetic center to T_{1s} and T_{2s}, [31–33, 42], the Solomon and Bloembergen Equation [43] gives:

$$
\Delta\frac{1}{T_{1s}} = \frac{1}{T_{1s}} - \frac{1}{T_{1s}^0} = \frac{\mu_f^2\gamma^2}{6r^6 T_{1f}}\left[\frac{24}{5(\omega_f + \omega_s)^2} + \frac{12}{5\omega_s^2} + \frac{4}{5(\omega_f - \omega_s)^2}\right] \tag{3.20}
$$

$$
\Delta\left(\frac{1}{T_{2s}}\right) = \frac{1}{T_{2s}} - \frac{1}{T_{2s}^0} = \frac{4\mu_f^2\gamma^2}{15r^6}T_{1f} \tag{3.21}
$$

The relaxation times can be estimated experimentally and by analysis of CW power saturation curves in solids from saturation curves recorded under conditions of rapid passage [31–33, 44] using effect of cross relaxation of free radicals in partially ordered solids [45], and by pulse methods [46–48].

For application (3.20) and (3.21) without knowing T_{1f} two approaches have been proposed [31–33, 44]. In the first approach the T_{1f} factor cancels in the product $\Delta(1/T_{1s})\Delta(1/T_{2s})$. The second approach is based on the fact in solids T_{1f} usually decreases monotonically with increasing temperature (if no other dynamic process occurs that affects T_{1f}, such as a phase transition). In such a condition the T_{1f} can disappear at the temperature at which the dipolar contribution is maximum. At this temperature $T_{1f} = \Delta\omega$ and the dipolar contribution is given by (3.22).

$$\Delta\left(\frac{1}{T_{1s}}\right) = \frac{\mu_f^2\gamma^2}{15r^6\Delta\omega} \tag{3.22}$$

Because μ_f depends on electron spin, the theoretical maximum distances (r_{max}) to a second paramagnetic center with spin S_f that could be determined were estimated as: $r_{max} = 100$ Å for $S_f = 1/2$ and $r_{max} = 150$ Å for $S_f = 5/2$. Such a high sensitivity of dipolar interaction based on the measurement of spin relaxation time as compare to measurement of spin-spin relaxation time is due to the fact that at low temperature for slow relaxing radicals $T_{1s} \gg T_{2s}$. Such an approach allowed to apply (3.22) directly for interspin distances without knowledge of T_{1f} and T_{2f}.

Application of (3.20) and (3.21) based on analysis of the saturation recovery with knowledge T_{1f} and T_{2f} was described in [48]. The analog of this equation for the case of $S = 5/2$ with zero-field splitting D much greater than the Zeeman interaction is (3.23):

$$\begin{aligned}
\frac{1}{T_{1s}} ={}& \frac{1}{T_{1s}^0} + b^2\left[\frac{10(P_5 + P_3)T_{2f}}{1 + (4D - \omega_S)^2T_{2f}^2} + \frac{16(P_3 + P_1)T_{2f}}{1 + (2D - \omega_S)^2T_{1f}^2}\right. \\
&\left. + \frac{18P_1T_{2f}}{1 + (\omega_f - \omega_S)^2T_{2f}^2}\right] + c^2\left[\frac{(25P_5 + 9P_3 + 4P_1)T_{1f}}{1 + \omega_S^2T_{1f}^2}\right. \\
&\left. + \frac{5(P_5 + P_3)T_{1f}}{1 + (4D)^2T_{1f}^2} + \frac{8(P_3 + P_1)T_{1f}}{1 + (2D)^2T_{1f}^2} + \frac{9P_1T_{1f}}{1 + \omega_f^2T_{1f}^2}\right] \\
&+ e^2\left[\frac{10(P_5 + P_3)T_{2f}}{1 + (4D + \omega_S)^2T_{2f}^2} + \frac{16(P_3 + P_1)T_{2f}}{1 + (2D + \omega_S)^2T_{1f}^2}\right. \\
&\left. + \frac{18P_1T_{2f}}{1 + (\omega_f + \omega_S)^2T_{2f}^2}\right].
\end{aligned} \tag{3.23}$$

Here P_1, P_3 and P_5 are the population of the $m_s = \pm1/2$, $\pm3/2$, $\pm5/2$ energy level, respectively; 2D and 6D are the splitting between the ground and excited state, respectively. It was assumed that values of $T_{1f} = T2_f$.

Aforementioned approach based on interaction between a slow relaxing dipole and fast relaxing dipole was applied by to determine interspin distances in various biological systems [44] and references therein. The determination of electron-electron distances in various systems was considered in Chaps. 5 and 6.

3.4 Electron Spin Nuclear Spin Interaction

3.4.1 General

The electron-nuclear spin interaction originates from the coupling of the nuclear magnetic moment to the magnetic field generated by the electron magnetic moment or equivalently from the coupling of the electron magnetic moment to the magnetic field generated by the nuclear magnetic moment. The hyperfine interaction between the electron and the nuclear spins consists of the isotropic Fermi contact interaction and the anisotropic dipole-dipole interaction [49–52].
The contact interaction is given by.

$$\hat{H}_{eN} = \sum_n a_n \hat{S}_e \cdot \hat{J}_n, \tag{3.24}$$

where $\hat{S}_e$ and $\hat{J}_n$ are the spin and nucleus operators, respectively and

$$a_n = \frac{\mu_0}{4\pi} \gamma_e \gamma_n \frac{8\pi}{3} |\Psi(\mathbf{R}_n)|^2, \tag{3.25}$$

Here μ_0 is the vacuum magnetic permeability, $\mathbf{R}$n denotes the coordinates of the nth nucleus, γ_n and γ_e are the nuclear gyromagnetic ratio, and the electron gyromagnetic ratio, respectively; $|\psi(\mathbf{R}_n)|^2$ is the spin density of s-electron on the nucleus.
The total Hamiltonian of the system of an electron and many nuclear spins is given by [51]:

$$\hat{H} = \hat{H}_e + \hat{H}_N + \hat{H}_{eN} + \hat{H}_{NN}, \tag{3.26}$$

and composed of the single spin Zeeman energies, Hamiltonian for interaction in the applied magnetic field along the z axis, the hyperfine interaction and the intrinsic nuclear-nuclear interaction, respectively. Both the spin and orbital angular momentum of the electron contribute to nuclear magnetic moment. The electron-nuclear spin interactions for an electron in an s-symmetry orbital and in a non-s-symmetry orbital are expressed as follows:

$$H_{HF} = \frac{16\pi}{3} \gamma I \mu_B \mu_N \delta(\mathbf{r})[\mathbf{S} \cdot \mathbf{I}], \quad l = 0$$

$$H_{HF} = \frac{2\gamma I \mu_B \mu_N}{r^3} \left[(\mathbf{L} - \mathbf{S}) \cdot \mathbf{I} + 3 \frac{(\mathbf{S} \cdot \mathbf{r})(\mathbf{I} \cdot \mathbf{r})}{r^2} \right], \quad l \neq 0. \tag{3.27}$$

Here, γ_I, μ_B, and μ_N are the gyro-magnetic factor of the nuclear spin, the Bohr magneton, and the nuclear magneton, respectively. S and I are the spin operators for the electron and the nucleus, and L is the angular momentum operator for the electron. The nuclear magnetic moment m_I originating from the nuclear spin quantum number I is $m_I = 2I + 1$.

For anisotropic hyperfine dipolar of anisotropic electron spin with anisotropic nuclear spin Hamiltonian was given by [35]

$$
\mathscr{H}_{dip} = -g\mu_B g_n \mu_n [S_x\, S_y\, S_z] \cdot
\begin{bmatrix}
\left\langle \frac{r^2-3x^2}{r^5} \right\rangle & -\left\langle \frac{3xy}{r^5} \right\rangle & -\left\langle \frac{3xz}{r^5} \right\rangle \\
-\left\langle \frac{3xy}{r^5} \right\rangle & \left\langle \frac{r^2-3y^2}{r^5} \right\rangle & -\left\langle \frac{3yz}{r^5} \right\rangle \\
-\left\langle \frac{3xz}{r^5} \right\rangle & -\left\langle \frac{3yz}{r^5} \right\rangle & \left\langle \frac{r^2-3z^2}{r^5} \right\rangle
\end{bmatrix}
\cdot
\begin{bmatrix} I_x \\ I_y \\ I_z \end{bmatrix}
\tag{3.28}
$$

$$
= \mathbf{S} \cdot \mathbf{A}_{dip} \cdot \mathbf{I}
$$

Here $\langle\ \rangle$ means integration over the spatial coordinates.

Various aspects of the electron spin and nuclear spin interactions can be illustrated by examples presented in [52–59]. For instance, a quantum solution to the electron spin decoherence by a nuclear pair-correlation method for the electron-nuclear spin dynamics under a strong magnetic field and a temperature high for the nuclear spins but low for the electron was presented in [53]. The theory incorporates the hyperfine interaction, the intrinsic both direct and indirect nuclear interactions, and the extrinsic nuclear coupling mediated by the hyperfine interaction with the single electron. A quantum cluster expansion method was developed for the problem of localized electron spin decoherence due to dipolar fluctuations of lattice nuclear spins [54]. Modern powerful pulse methods such as HYperfine Sublevel CORrElation spectroscopy (HYSCORE)54 and Electron Spin Echo Envelope Modulation (ESEEM) [54, 55] are ones of the most widely applied EPR techniques for the investigation of hyperfine couplings in a wide range of spin systems (Sect. 6.7).

3.4.2   Nuclear Magnetic Resonance of Paramagnetic Molecules

- The nuclear magnetic resonance properties are governed by the total Hamiltonian of the system of an electron and many nuclear spins. Effects of electron spin on nuclear spins on position of NMR spectra are revealed in contact and pseudocontact chemical shifts, which are caused by electronic screening external magnetic field, and the relaxation rates enhancement. Self-orientation residual dipolar couplings cross correlations between Curie relaxation (Sect. 3.4.3) and dipolar relaxation also should be taken in consideration [60–65]. All these effects are used for investigation of structure and electron distribution of biological and non-biological molecules.
- The total observed NMR chemical shift (δ_{tot}) includes both a diamagnetic or orbital contribution (δ_{dia}) from paired electrons and a hyperfine contribution (δ_{hf}) from spin electrons–spin nucleas interactions. If it is relevant, spin-orbit coupling effects can be also taken into account [65]. The hyperfine shift, in turn,

can be broken down into Fermi contact (δ_{FC}) and pseudocontact (δpc) terms. δ_{FC} of a given nucleus depends on the spin state (S) of the system, the spin density at the nucleus ($\rho\alpha\beta$), and on temperature (T) [66].

- The Fermi contact shift

$$\delta_{FC} = m(S+1)\rho_{\alpha\beta}/T \tag{3.29}$$

when m is a collection of fundamental physical constants and $\rho\alpha\beta$ is the spin density at the nucleus ($\rho\alpha\beta$).

- The pseudocontact chemical shift (PCS) is a contribution to chemical shift of nuclear Larmor frequency caused by the presence of centers with unpaired electrons, where the probability density of the unpaired electron at the site of nucleus is negligible [62, 67, 68]. Such a shift is related the dipolar coupling between the magnetic moments of the nucleus and of the unpaired electron, when the magnitude of the magnetic moment of the unpaired electron depends on the molecular orientation with respect to the magnetic field vector. Magnitude of the pseudocontact shift is proportional to r^{-3} and, therefore, PCS is more sensitive to relatively long-distance interactions then that the Fermi contact shift.

- Pseudocontact shifts, described as early as 1958 by McConnell and Chesnut [67], is given by [62]:

$$\delta^{pcs} = \frac{1}{12\,\pi r^3}\left[\Delta\chi_{ax}(3\cos^2\theta - 1) + \frac{3}{2}\Delta\chi_{rh}\sin^2\theta\cos 2\varphi\right]$$
$$\Delta\chi_{ax} = \chi_{zz} - \frac{\chi_{xx} + \chi_{yy}}{2} \text{ and } \Delta\chi_{rh} = \chi_{xx} - \chi_{yy} \tag{3.30}$$

where r is the distance between observed nuclei and electron, $\Delta\chi_{ax}$ and $\Delta\chi_{rh}$ are the axial and rhombic anisotropy parameters of the magnetic susceptibility tensor of the system, θ and φ are the polar coordinates of the nucleus in the frame of the electronic magnetic susceptibility tensor, and r is the distance of the atom from the paramagnetic center.

In work [65], based on density functional theory (DFT), scalar-relativistic and spin-orbit coupling effects NMR shifts were taken into account, and a Fermi-contact term was included in the NMR shielding tensor expression. [65]. It was shown that the combination of spin-orbit coupling and magnetic field mixes spin triplet states into the ground state, inducing a spin density. This spin density can then interact with the nuclei of the molecule via the Fermi-contact interaction, and, therefore, effects the NMR spectra. Proposed approach allowed separation of pseudocontact (PCS) from Fermi contact (FC) shifts from NMR data.

In work [64], the combination of pseudocontact shifts induced by a site-specifically bound lanthanide ion and knowledge of three-dimensional (3D) structures of the lanthanide-labeled protein ε186 was used to achieve: (i) rapid assignments of NMR spectra, (ii) structure determinations of protein–protein

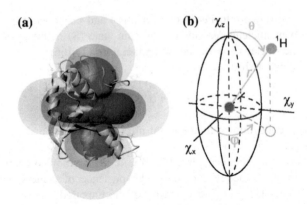

Fig. 3.4 a Isosurfaces depicting the pseudocontact shifts (PCSs) induced by Dy^{3+} plotted on a ribbon representation of the crystal structure of ε186. The PCS isosurfaces constitute a representation of the $\Delta\chi$ tensor. **b** Schematic representation of the χ tensor, illustrating the electron–nucleus distance r (the nuclear spin being exemplified by a ^{1}H) and the angles θ and φ [64]

complexes, and (iii) identification of the binding mode of low-molecular weight compounds in complexes with proteins. The spatial the pseudocontact shifts (PCS) distribution detected in [64] and shown in Fig. 3.4 is described by

$$\Delta\delta^{PCS} = \frac{1}{12\,\pi r^3}\left[\Delta\chi_{ax}(3\cos^2\theta - 1) + \frac{3}{2}\Delta\chi_{rh}\sin^2\theta\cos 2\varphi\right]\ (1) \tag{3.31}$$

where $\Delta\delta^{PCS}$ denotes the difference in chemical shifts measured between diamagnetic and paramagnetic samples, r is the distance between the metal ion and the nuclear spin, $\Delta\chi_{ax}$ and $\Delta\chi_{rh}$ are the axial and rhombic components of the $\Delta\chi$ tensor, and the angles θ and φ describe the position of the nuclear spin with respect to the principal axes of the $\Delta\chi$ tensor. Figure 3.4 illustrates isosurfaces PCSs induced by Dy^{3+} In ε186.

3.4.3   *Nuclear Relaxation Rates Enhancement by Electron Spin*

The unpaired electron–nucleus dipolar interaction causes R_1 (longitudinal) and R_2 (transverse) nuclear relaxation because of its modulation. The nuclear relaxations are enhanced by the presence of a paramagnetic ion. Commonly, the major source of relaxation of nuclei I may be the coupling with the unpaired electrons but not the coupling with nuclei. For the Solomon and Curie contributions, proton relaxation enhancement (PRS) is described by equations [69, 70].

$$R_2^{\text{PRE}} = \frac{k_{\text{Solomon}} + k_{\text{Curie}}}{r^6} \qquad (3)$$

$$k_{\text{Solomon}} = \left(\frac{\mu_0}{4\pi}\right)^2 \frac{\gamma_I^2 g_e^2 \mu_B^2 S(S+1)}{15} \left[4\tau_c + \frac{\tau_c}{1 + (\omega_I - \omega_S)^2 \tau_c^2}\right.$$
$$\left. + \frac{3\tau_c}{1 + \omega_I^2 \tau_c^2} + \frac{6\tau_c}{1 + (\omega_I + \omega_S)^2 \tau_c^2} + \frac{6\tau_c}{1 + \omega_S^2 \tau_c^2}\right] \qquad (3.32)$$

$$k_{\text{Curie}} = \frac{1}{5}\left(\frac{\mu_0}{4\pi}\right)^2 \frac{\omega_I^2 g_e^4 \mu_B^4 S^2(S+1)^2}{(3kT)^2}\left[4\tau_{\text{Curie}} + \frac{3\tau_{\text{Curie}}}{1 + \omega_I^2 \tau_{\text{Curie}}^2}\right]$$

where ω_I is the nuclear Larmor frequency, ω_S is the electron Larmor frequency, r is the electron–nucleus distance, g_e is the electron g factor, μ_B is the electron Bohr magneton, g_I is the nuclear magnetogyric ratio, k is the Boltzmann constant, T is the temperature, S is the electron spin quantum number. The correlation times related to the Solomon and Curie mechanisms responsible for relaxation are:

$$\tau_c^{-1} = \tau_e^{-1} + \tau_r^{-1} + \tau_M^{-1} \qquad (3.33)$$

and

$$\tau_{\text{Curie}}^{-1} = \tau_r^{-1} + \tau_M^{-1} \qquad (3.34)$$

If the proton exchange rate τ_M^{-1} is negligible, τ_c is determined by both the reorientation time of the electron–nucleus vector τ_r and the electron relaxation time τ_e, whereas τ^{Curie} is determined by τ_r only.

Cross correlation rates between Curie spin relaxation and dipole-dipole coupling arising from the interference of relaxation mechanisms due to the presence of unpaired electron spin density were first observed in paramagnetic molecules by the Bertini group [71, 72] The relaxation allowed coherence transfer peaks in 2D correlated spectroscopy (COSY) spectra (Sect. 6.7). For instance, the cross correlation rates and H-N dipole-dipole coupling depend on the metal-to-proton distances and on the M-H-N angle.

The nuclear Overhauser effects (NOEs) [73] cause changes in the intensity of a signal at one frequency when the resonance frequency of a different nucleus is irradiated, due to dipole-dipole interactions between the magnetic moments of the pair of nuclei. Delocalization of the unpaired electron spin density on resonating nuclei through chemical bonds causes a further contribution to the paramagnetic nuclear relaxation enhancement, which is called contact relaxation. The enhancement of the longitudinal R_{1M} and transverse R_{2M} contact relaxation rates are given by the Bloembergen equations [60, 74]

$$R_{1M}^{con} = \frac{2}{3}S(S+1)\left(\frac{A}{\hbar}\right)^2 \frac{\tau_c}{1+\omega_S^2\tau_c^2}$$

$$R_{2M}^{con} = \frac{1}{3}S(S+1)\left(\frac{A}{\hbar}\right)^2 \left(\tau_c + \frac{\tau_c}{1+\omega_S^2\tau_c^2}\right)$$
(3.35)

where A/h is the contact coupling constant (rad s^{-1}). Because the electron relaxation and chemical exchange modulates the coupling, the correlation time was provided by

$$\left(\tau_c^{con}\right)^{-1} = \tau_s^{-1} + \tau_M^{-1}$$
(3.36)

The R_{1M} and R_{2M} are proportional to r^{-6}, where r is the distance between the unpaired electron and the resonating nucleus. This dependence allows to determine the nuclear–metal distances for protons close to the paramagnetic center and to have an upper distance limit of protons far from the paramagnetic center.

Spin dynamic in a system containing with two spins, which are involved in the exchange interaction and hyperfine interaction with two nuclear spins in strong magnetic field (Fig. 3.5) have been considered [75]. In the high magnetic field, the two-spin system can be described by a singlet state (S) and three triplet states (T_0, T_{+1}, T_{-1}). The total Hamiltonian is given as

$$H = \omega_1 S_1^z + \omega_2 S_2^z - JS_1 \cdot S_2 + \sum_i a_1^i S_1 \cdot I_1^i + \sum_j a_2^j S_2 \cdot I_2^j + \frac{J}{4}$$
(3.37)

where J is the isotropic part of the spin exchange, $\omega_1 = g_1\mu_B B$ and $\omega_2 = g_2\mu_B B$ are the isotropic Zeeman energies, and $a_1^{\ i}$ and $a_2^{\ j}$ are isotropic hyperfine couplings. If exchange integral J is close to magnitude to the electron Zeeman interaction, weak isotropic electron–nuclear hyperfine coupling leads to population of $|T_{+1}\rangle$ state. In

Fig. 3.5 Schematic presentation of a system in which radical pair spins S_1 and S_2 are coupled to the nearby nuclei I_1 and I_2 and interact with each other with the exchange coupling J [75]

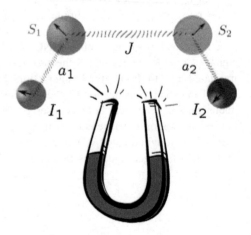

such a condition triplet quantum yield (QY) the nuclear polarization is almost equal to the QY of T_{+1} with opposite sign.

In paper [76], author concentrated on two aspects of NMR of paramagnetic system: composition of high-resolution NMR spectra on the base of detection of ESR superfine interaction in ESR spectra and peculiarities of NMR of multielectron molecules. For paramagnetic centers possessing a small magnetic susceptibility (electron spin S = 1/2) method of very fast magic angle spinning (VFMAS), proposed by Ishii and co-workers in 2003 [77], has proved to be effective experimental tool for detecting high resolution NMR spectra. Spinning the sample (frequency of 1–100 kHz) at the magic angle θ_m (54.74°) with respect to the direction of the magnetic field allowed to average anisotropic dipole dipole interaction and, therefore, to make narrow the NMR spectra lines.

The recent developments and applications of paramagnetic magic-angle spinning NMR focus on the investigations of metalloproteins and natively diamagnetic proteins modified with covalent paramagnetic tags, for example derivative of Dy^{3+}, Tb^{3+}, Tm^{3+}, Yb^{3+} or Y^{3+} [78, 79]. Data on pseudocontact shifts and spin relaxation enhancements derived from solid-state NMR spectra provide information about electron–nucleus distances on the 20 Å length scale [80]. Another area of application of the VFMAS is structure and dynamics of complexes of transition metals, cobalt(II) in particular [81]. ^{13}C and ^2H solid-state fast magic angle spinning revealed Fermi contact and hipper fine coupling in Cu^{2+} complexes with amino acid ligands [82].

3.4.4 Contrast Enhancement of Water Protons in Magnetic Resonance Imaging

Magnetic resonance imaging (MRI), or magnetic resonance tomography (MRT) has been proved to be a powerful medical imaging technique based on a scanning strong magnetic fields, radio waves, and field gradients to form images of the body [83–87]. In many cases this technique appears to be indispensable in medical diagnosis.

Contrast enhancement in Magnetic Resonance Imaging (MRI) is defined as the difference in intensity between the same pixel or the same collection of pixels of region of interest in the pre-contrast (Ipre) and post-contrast (Ipost) MRI. The ability of paramagnetic probes to increase the relaxation rates r of solvent water is a linear function of the concentration of the contrast agent [CA], the slope of which is known as r^{CA} relaxivity.

$$r = r^0 + r^{CA} [CA] \qquad (3.38)$$

where r^0 represent the relaxation rate of the water protons in the absence of contrast agent. As contrast agents, superparamagnetic particles of iron oxide, iron, platinum, manganese, gadolinium chelates, iron salts and, protein-based MRI agents, utilized the abilities of some amino acids residues to bind with gadolinium, can serve.

The electron spin interacts through dipole-dipole coupling with the nuclear spin residing in the paramagnetic complex (the inner-sphere case) as well as outside of it (the outer-sphere case). The Fermi contact mechanism can be involved as well. An example of such an interaction was described in work [86]. Magnetic field-dependent measurements of the paramagnetic relaxation enhancement for water protons in the presence of Mn(II) complexes (S = 5/2), were analyzed using theoretical models which took in account the electronic Zeeman interaction and the zero-field splitting in the complex. It was also suggested that suggested. That the relaxation of the electron spin for S > 1 is multiexponential. The measured ^{1}H relaxation rates in the presence of paramagnetic complexes were given by a sum of two terms:

$$T_{1P}^{-1} = \frac{P_{Mq}}{T_{1I} + \tau_M} + T_{1OS}^{-1} \tag{3.39}$$

The first and the second term of the equation correspond to the inner- and outer-sphere contribution to the PRE, respectively. P_{Mq} is the mole fraction of ligand protons in the bound position and q is the number of bound ligand molecules, T_{1I} is the spin-lattice relaxation time for the nuclear spin I in the complex, τ_M is the chemical exchange lifetime.

In another work [87], NMR Fermi contact and dipolar shifts in the first water hydration shell with nuclea–electron distance r_{IS}^{-3} were considered. The electron–proton dipole–dipole correlation function $C_P^{DD}(\tau)$, writing it as a product of an isotropic reorientational diffusion correlation function and the distance correlation function, $\langle r_{IS}^{-3}(0) r_{IS}^{-3}(\tau) \rangle$, was presented as

$$C_p^{DD}(\tau) \approx \frac{1}{5 \langle r_{IS}^3 \rangle^2} e^{-\tau/\tau_R}.$$

were τ_R is the correlation time in the picosecond regime. In addition Molecular dynamic MD simulations was employed for analyzing transient zero-field splitting (ZFS) in the frame of the pseudo-rotation model. It was shown that, the fast wagging and rocking motions of the water protons is important in the time dependence of the transient ZFS interaction. The model described the fluctuating zero-field interaction as a constant amplitude in the principal frame but reorienting according to a rotational diffusion equation of motion. A three-exponential correlation function which oscillates at short times were predicted. Magnetic Resonance Imaging finds wast application in biomedical research and medical practice [88–91].

References

1. P.C. Poole, *Electron Spin Resonance: A Comprehensive Treatise on Experimental Technique.* (Dover Publcation, 1997)
2. B. Abebe, *Fluorescence and Electron Paramagnetic Resonance (EPR) Spectroscopy: Basic Principles, the Different Techniques and Applications* (LAP LAMBERT Academic Publishing, 2011)
3. A. Lund, M. Shiotani, S. Shimada, *Principles and Applications of ESR Spectroscopy* (Springer 2011)
4. G.R.. Eaton, S.S. Eaton, D.P. Barr, R. TT Weber, *Quantitative EPR* (Springer, 2010)
5. B.C. Gilbert, M.J. Davies, K.A. McLauchlan. (Eds.)., *Electron Paramagnetic Resonance, a Specialist Periodical Report*, vol. 17 (Royal Society of Chemistry, 2010)
6. G.M. Smith, J. Keeble, O. Schiemann (Eds.), *Introduction to Modern Epr Spectroscopy* (CRC, 2010)
7. S. Moebius, A.N Savitsky, *High-Field EPR Spectroscopy of Proteins and their Model Systems: Characterization of Transient Paramagnetic States* (Royal Society of Chemistry, 2009)
8. M.R. Brustolon, *Principles and Applications of Electron Paramagnetic Resonance Spectroscopy* (Blackwell Publishers, 2008)
9. W.R. Hagen, *Biomolecular EPR Spectroscopy* (CRC, 2008)
10. G.I. Likhtenshtein, J. Yamauchi, S. Nakatsuji, A. Smirnov, R. Tamura, *Nitroxides: Application in Chemistry, Biomedicine, and Materials Science* (WILEY-VCH, Weinhem, 2008)
11. G.R. Eaton, S.S. Eaton, K.M. Salikhov, *Foundations of Modern Epr* (World Scientific, 1998)
12. J.H. Freed, Ann. Rev. Phys. Chem. **51**, 655 (2000)
13. A.M. Abragam, M. Goldman, M. Rep. Prog. Phys. **41**, 395 (1976)
14. K.M. Salikhov, *Spin Polarization and Magnetic Effects in Radical Reactions (Studies in Physical and Theoretical Chemistry)* (Elsevier Science Ltd, 1984)
15. F. Bloch, Phys. Rev. **70**, 460 (1946)
16. S. Alexander, O. Entin-Wohlman, R. Orbach, Phys. Rev. **33**, 3935 (1986)
17. G.I. Likhtenshtein, *Biophysical Labeling Methods in Molecular Biology* (Cambridge University Press, Cambridge, N.Y., 1993), p. 36
18. H. Sato, V. Kathirvelu, G. Spagnol, S. Rajca, A. Rajca, S.S. Eaton, G.R. Eaton, J. Phys. Chem. B **112**, 2818 (2008)
19. J.R. Biller, V.M. Meyer, H. Elajaili, G.M. Rosen, S.S Eaton, G. R Eaton. J. Magn. Reson. **225**, 52 (2012)
20. S.S. Eaton, G.R. Eaton, Relaxation times of organic radicals and transition memal ions, in: *Magnetic Resonance in Biology. Distance Measurement in Biological Systems* by ESR, ed. by L. Berliner, S. Eaton, G. Eaton (Kluwer Academic Publishers. Dordrecht)
21. A.A. Kuzhelev, D.V. Trukhin, O.A. Krumkacheva, R.K. Strizhakov, O.Y. Rogozhnikova, T.I. Troitskaya, M.V. Fedin, V.M. Tormyshev, E.G. Bagryanskaya, J. Phys. Chem. B **119**, 13630 (2015)
22. K. Kundu, D.R. Kattnig, B. Mladenova, G. Grampp, R. Das, J. Phys. Chem. A **119**, 3200 (2015)
23. G.E. Pake, J. Chem. Phys. **16**, 327 (1948)
24. S. Sinnecker, F. Neese, J. Phy. Chem. A **110**, 12267 (2006)
25. Molec Likhtenshtein, Biol. (Moscow) **2**, 234 (1968)
26. G.I. Likhtenshtein, PKh Bobodzhanov, Biofizika **13**, 757 (1968)
27. J.C. Taylor, J.S. Leigh, M. Cohn, Proc. Natl. Acad. Sci. U.S.A. **64**, 219 (1969)
28. A.I. Kokorin, K.I. Zamaraev, G.L. Grigoryan, V.P. Ivanov, E.G. Rozantsev, Biofizika **17**, 34 (1972)
29. A.I. Kulikov, G.I. Likhtenshtein, E.G Rozantzev, V. Suskina, A.B. Shapiro, Biofisika **17**, 42 (1972)

30. A.I. Kokorin, in *Nitroxides—Theory, Experiment and Applications* ed. by A.I. Kokorin (InTech, 20120)
31. A.I. Kulikov, G.I. Likhtenstein, Biofizika **19**, 420 (1974)
32. A.V. Kulikov, Determination of distance between the nitroxide label and paramagnetic center in spin-labeled proteins from the parameters of the saturation curve of the ESR spectrum of the label at 77K. Molecul. Biol. (Moscow) **10**, 109–116 (1976)
33. A.I. Kulikov, G.I. Likhtenshtein, Adv. Molecul. Relax. Proc. **10**, 47 (1977)
34. F. Neese, in: *Calculation of NMR and EPR Parameters. Theory and Applications*, ed. by M. Kaupp, M. Buhl, V.G. Malkin, (Wiley Heidelberg, 2004), p. 541
35. J. Yamauchi, *Nitroxides: Application in Chemistry, Biomedicine, and Materials Science*, ed. by G.I. Likhtenshtein, J. Yamauchi, S. Nakatsuji, A. Smirnov, R. Tamura, (WILEY-VCH, Weinhem, 2008)
36. G.R. Eaton, S.S., Biol. Magn. Reson. **8**, 339 (1989)
37. C. Riplinger, J.P.Y. Kao, G.M. Rosen, V. Kathirvel, G.R. Eaton,. S.S., A. Eaton, A. Kutateladze, F. Neese, J. Am. Chem. Soc. **131**, 10092 (2009)
38. M. Shoji*, K. Koizumi, T. Hamamoto, T. Taniguchi, R. Takeda, Y. Kitagawa, T. Kawakami, M. Okumura, S. Yamanaka, K. Yamaguchi, Polyhedron **24**, 2708 (2005)
39. E.J. Hustedt, A.H. Berth, Annu. Rev. Biophys. Biomol. Struc. **28**, 129 (1999)
40. E.J. Hustedt, R.A. Stein, L. Sethaphong, S. Brandon, Z. Zhou, S.C. DeSensiy, Biophys. J. **90**, 340 (2006)
41. A. Ivancich, P. Dorlet, D.B Goodin. S Un. J. Am Chem. Soc. **123**,5050 (2001)
42. G.I. Likhtenshtein, *Spin Labeling Method in Molecular Biology* (Wiley Interscience, N.Y., 1976)
43. J. Solomon, N. Blombergen, *Nuclear magnetic interaction in HF molecule.* J. Chem. Phys. **25**, 261–266/ (1956)
44. G.I. Likhtenshtein, Depth of immersion of paramagnatic centers, in *Magnetic Resonance in Biology*, ed. by L. Berliner, S. Eaton, G. Eaton, (Kluwer Academic Publishers). Dordrecht, V. 18.2000, p. 309
45. M.K. Bowman, J.R. Norris, J. Phys. Chem. **86**, 3385 (1982)
46. J.S. Hyde, W.K., Subczynski, W. Froncisz, C.S. Lai, Bull Magn. Reson. **5**, 180–182 (1983)
47. G.I. Likhtenshtein, *Biophysical Labeling Methods in Molecular Biology* (Cambridge University Press, Cambridge N.Y., 1993)
48. S.S. Eaton, G. Eaton, in: *Magnetic Resonance in Biology*, ed. by L. Berliner, S. Eaton, G. Eaton, vol. 18 (Dordrecht, Kluwer Academic Publishers, 2000), p. 309
49. H.M. McConnell, R.E. Robertson, J. Chem. Phys. **29**, 1361 (1958)
50. M. Karplus M.J. Chem. Phys. **30**, 11 (1959)
51. S. Lee, P. von Allmen, F. Oyafuso, K. Klimeck, K.B. Whaley, J. Appl. Phys. **97**.043706/1 (2005)
52. H.B. Stuhrmann, Acta Cryst. A **63**, 455 (2007)
53. W. Yao, R.-B. Liu, J. Sham, Theory of electron spin decoherence by interacting nuclear spins in a quantum dot. Phys. Rev. B **74**, 195301 (2006)
54. W.M. Witzel, R. Rogerio de Sousa, S. Das, Sarma. Phys. Rev. B **72**, 161306 (2005)
55. L. Cywinski, W.M. Witzel, S. Das Sarma, Phys. Rev. Lett. **102**, 057601 (2009)
56. A.F. Bermudez, F. Jelezko, M.B. Plenio, A. Retzker, Phys. Rev. Lett. **107**, 150503 (2011)
57. L. Cywinski, W.M. Witzel, Electron spin dephasing due to hyperfine interactions with a nuclear spin bath. Phys. Rev. Lett. **102**, 057601 (2009)
58. C.E. Tait, P. Neuhaus, H.L. Anderson, C.R. Timmel, D. Carbonera, M. Di Valentin, App. Magn. Reson. **46**, 389 (2015)
59. R.I. Samoilova, A.T. Taguchi, P.J. O'Malley, S. Dikanov, Appl. Magn. Reson. **45**, 941. (2014)
60. N. Bloembergen, E.M, Purcell, R.V. Pound. Phys. Rev. **73**, 679 (1948)
61. I. Bertini, C. Luchinat, G. Parig, Concepts Magn. Reson. **14**, 259 (2002)
62. I. Bertini, C. Luchinat, P.G. Giacomo, R. Pierattelli, Chem. Bio. Chem. **6**, 1536 (2005)
63. Y. Zhang, J. Am. Chem. Soc. **130**, 3814 (2008)

64. G. Pintacuda, M. John, X.C. Su, G. Otting, Acc. Chem. Res. **40**, 206 (2007)
65. S.K Wolff, T. Ziegler T. J. Chem. Phys. **109**, 895 (1998)
66. Y. Zhang, E. Oldfield E. J. Am. Chem. Soc. **130**, 3814 (2008)
67. H.M. McConnell, D.B. Chesnut, J. Chem. Phys. **28**, 107 (1958)
68. S. Di Pietro, S. Lo Piano, L. Di Bari, Coord. Chem. Rev. **255**, 2810 (2011)
69. I.A. Solomon, Phys. Rev. **99**, 559 (1955)
70. I. Bertini, C. Luchinat, M. Nagulapalli, G. Parigia, E. Ravera, Phys. Chem. Chem. Phys. **14**, 9149 (2012)
71. I. Bertini, G. Cavallaro, M. Cosenza, R. Kümmerlec,, C. Luchinata, M. Piccioli, P. Poggi. J. Biomol. NMR **23**, 115 (2002)
72. I. Bertini, C. Luchinat, D. Tarchi, Chem. Phys. Lett. **203**, 445 (1993)
73. A.W. Overhauser. Phy.Rev. **l92**, 411 (1953)
74. N. Bloembergen, J. Chem. Phys. **27**, 572 (1957)
75. M. Zarea, R. Carmieli, M.A. Ratner, M.R. Wasielewski, J. Phys. Chem. A **118**, 4249 (2014)
76. V.K. Voronov, Magn. Reson. Chem. **53**, 188 (2015)
77. Y. Ishii, N.P. Wickramashinghe, S. Chimon, J. Am. Chem. Soc. **125**, 3438 (2003)
78. P.J. Christopher, P. J. Magn. Reson. **253**, 50 (2015)
79. V. Ladizhansky, Isr. J. Chem. **54**, 86 (2014)
80. D.J. Crick, J.X. Wang, B. Graham, J.D. Swarbrick, H.R. Mott, D. Nietlispach, J. Biomol, NMR **61**, 197 (2015)
81. V.V. Novikov, A.A. Pavlov, A.S. Belov, A.V. Vologzhanina, A. Savitsky, Y.Z. Voloshin, J. Phys. Chem. Lett. **5**, 3799 (2014)
82. G. Szalontai, R. Csonka, G. Speier, J. Kaizer, J. Sabolović, Inorg. Chem. **54**, 4663 (2015)
83. A.E. Merbach, E. Tóth, *The Chemistry of Contrast Agents in Medical Resonance Imaging* (Wiley, Chichester, 2001)
84. W. Krause, *Contrast Agents I: Magnetic Resonance Imaging*, vol. 221 (Springer, Berlin, 2002)
85. M.P. Rohrer, H. Bauer, J. Mintorovitch, M. Requardt, H.-J. Weinmann, Invest. Radiol. **40**, 715 (2005)
86. D. Kruk, J. Kowalewski, J. Biol. Inorg. Chem. **8**, 512 (2003)
87. M. Lindgren, A. Laaksonen, P.-O. Westlund, Phys. Chem. Chem. Phys. **11**, 10368 (2009)
88. S. Xue, J. Qiao, K. Hubbard, N. White, L. Wei, S. Li, Z.-R. Liu, J.J. Yang, Med. Res. Rev. **34**, 1070 (2014)
89. J. Qiao, S. Xue, F. Pu, N. White, Z.-R. Liu, J.J. Yang, Molecular imaging of EGFR/HER2 cancer biomarkers by protein MRI contrast agents. J. Biol. Inorg. Chem. **19**(2), 259–270 (2014)
90. Yi-X.J. Wang, Quant. Imaging Med. Surg. **1**, 35 (2011)
91. C.F.G.C. Geraldes, S. Laurent, Contrast Media Mol. Imaging **4**, 1 (2009)

Chapter 4
Spin-Selective Processes of Electron and Nuclear Spins

Abstract A detailed knowledge of spin-selective phenomena in magnetic field is necessary for the development of molecular magnetic materials and for deep insight to mechanisms of many physical, chemical and biological processes. The spin conservation is an absolute requirement to any selective process. Examples of spin-selective processes are: (1) the quenching of electronically excited molecules in singlet and triplet states by paramagnetic species, (2) the generation and decay of electron spin-polarized states, intersystem crossing (ISC, Sect. 4.2.5), (3) electron, energy, and spin transfer, and magnetic isotope effect. In this chapter, the fundamentals and applications of methods based on the spin selected phenomena of dynamic nuclear polarization DNP, chemically induced spin correlated dynamic electron polarization CIDEP and chemical-induced dynamic nuclear polarization (CIDNP) have been discussed. These methods, with their capacity for monitoring radical and ion-radical intermediates, have been successfully used as effective tools for investigation of spin elective processes in chemistry, physics and biology including native and artificial photosynthesis.

4.1 Introduction

A key step for spin-selective physical and chemical processes, leading to the generation of spin polarized states of electrons and nuclei, is the conversion from the singlet state (S_1) to the triplet state by the intersystem crossing mechanism. The magnetic field B_0 splits the initial triplet state to T_0, T_+, T_- (Fig. 4.1). $T_0 - T_-$ and $T_0 - T_+$ transitions can be induced by electron spin relaxation or an applied resonant microwave magnetic field. Using these basic chemical and physical facts, the following advanced spin selected methods have been developed: chemically induced spin correlated dynamic electron polarization. dynamic nuclear polarization and chemical-induced dynamic nuclear polarization.

The main source of information on the spin selective processes is the modern paramagnetic spin resonance EPSR sometime in combination with nuclear magnetic resonance NMR. These processes are involved in elementary steps in

© Springer International Publishing Switzerland 2016

G. Likhtenshtein, *Electron Spin Interactions in Chemistry and Biology*,
Biological and Medical Physics, Biomedical Engineering,
DOI 10.1007/978-3-319-33927-6_4

Fig. 4.1 Singlet triplet
transition and triplet state
degeneration in magnetic
field. *S* denotes singlet state;
T_0, T_-, T_+ denote triplet states

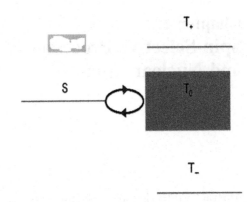

molecular magnetic materials [1, 2], including systems of switching between paramagnetic and ferromagnetic states [2], organic-based spin polarizer, [3], molecular devices designed to convert light energy [4] and probably in molecular compassing (Section X).

4.2 Chemically Induced Spin Correlated Dynamic Electron Polarization

Chemically induced spin correlated dynamic electron polarization (CIDEP) effects take place when the populations of the spin states deviate substantially from thermal equilibrium. Such an effect can arise in the electron spin system of reactive radicals generated as transient intermediates following the sudden initiation of chemical reactions. The CIDEP phenomenon can arise from a variety of mechanisms involving the mixing of singlet and triplet spin states [5–8]. The first CIDEP spectra were first detected in 1963 by Fessenden and Schuler [9]. The main mechanisms causing spin polarization in free radicals are: the triplet mechanism (TM), the radical pair mechanism (RPM), the correlated coupled radical-pair mechanism (CRPM), the radical triplet pair mechanism (RTPM) and triplet-triplet annihilation mechanism (TTAM). These mechanisms have been extensively reviewed and many examples of CIDEP have been reported in the literature [8–12].

4.2.1 Triplet Mechanism

The triplet mechanism (TM) is observed only in irradiated systems of lower than cubic symmetry [5, 6]. In the context of the triplet mechanism, the following processes occur: (1) singlet excited molecules may undergo inter-system crossing (ISC, Sect. 4.2.5) to their triplet state and subsequently form radicals, which can lead to differently populated triplet sub-levels owing to the ISC selection rules (Fig. 4.1), (2) the resulting initial spin polarization of the triplet state can then be transferred to radical reaction products, thus producing triplet molecules with pronounced spin polarization in the molecular frame (3) the zero-field spin sub-levels of a low-symmetry triplet molecule, T_x, T_y and T_z are rendered non-degenerate state due to the dipolar or zero-field coupling between the electrons. The interaction of triplet spin magnetic moment with the external magnetic field B_0 via the Zeeman effect can transform a substantial part of the molecular-frame spin polarization. In this mechanism, the triplet reacts rapidly to yield a pair of polarized radicals on a time scale faster than spin-relaxation can quench the Zeeman polarization. The polarization depends on The ratio of the zero field splitting (ZFS) the Zeeman interaction, ω_{ZFS}/ω_0, and the rotational correlation time τ_R in solution are main factors affected on the polarization.

4.2.2   Spin-Correlated Coupled Radical-Pair Mechanism

The spin-correlated coupled radical-pair mechanism (CCRP) is operative when radical pair partners are squeezed by the surrounding matrix and forced into well-defined positions relative to each other. Probing the structure and dynamics of transient intermediates by EPR spectroscopy in photosynthetic reaction center protein complexes or in biomimetic donor–acceptor complexes on the basis of photo-induced electron spin polarization phenomena is one of the promising directions of quantitative photochemistry [5, 6, 8, 12–16] The vector model of S–T conversion is shown in Fig. 4.2 [8].

Fig. 4.2 Vector model of singlet–triplet conversion in radical pairs. S = singlet state; T_0, T_-, T_+ = triplet states; T_1 and T_2 are electron spin relaxation times; S_1 and S_2 correspond to spins of radical 1 and radical 2. The relative dephasing induced by $\Delta\omega$ of the spin precession is sufficient for transitions between the S and T_0 states. Electron spin relaxation T_1 induced transitions between the T_0 and T_-, T_- states [8]

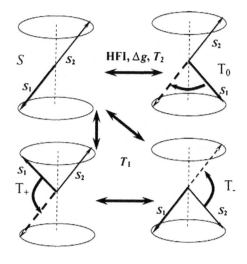

In the general case the Hamiltonian of a radical pair (RP) in a magnetic field B_0 is [8]:

$$H = g_1 \mu_B B_0 \hat{S}_{1z} + g_2 \mu_B B_0 \hat{S}_{2z} + \int_i^a A_i \hat{S}_1 \hat{I}_i + \int_k^b A_i \hat{S}_2 \hat{I}_k - J\left(\frac{1}{2} + 2\hat{S}_1 \hat{S}_2\right) \quad (4.1)$$

where J is the exchange interaction of spins S_1 and S_2, μ_B is the Bohr magneton, g_1 and g_2 are the electron g-factors of radicals 1 and 2, and A_i, Ak are the hyperfine (HFI) constants of electrons coupled to nearby nuclei.

In work [15] a situation where at time $t = 0$ a spin correlated radical pair is formed in the singlet spin state which interacts with a third spin was considered. The theory was applied to the system $^I(P^{\cdot+}Q^{\cdot-})L^{\cdot}$, here $P^{\cdot+}$ is the primary electron donor cation radical and $Q^{\cdot-}$ is the electron acceptor anion radical in the radical pair, and $L^{\cdot}$ denotes the observer spin label in its doublet state. The spin dynamics of this system is governed by the interaction with a constant external magnetic field **Bo**, with a microwave field of frequency ω_o and amplitude B_I, and the exchange and dipole-dipole interactions between paramagnetic particles as well as hyperfine interaction with nuclear spins. The following spin Hamiltonian in the rotating frame was used:

$$\begin{aligned}
\hat{\mathcal{H}}_0 =&(\omega_P - \omega_D)\hat{S}_{P_z} + (\omega_Q - \omega_0)\hat{S}_{Q_z} + (\omega_L - \omega_0)\hat{S}_{L_z} \\
&+ \Gamma_{PQ}\hat{S}_{P_z}\hat{S}_{Q_z} + \Lambda_{PQ}(\hat{S}_{Px}\hat{S}_{Qx} + \hat{S}_{Py}\hat{S}_{Qy}) \\
&+ \Gamma_{QL}\hat{S}_{Q_z}\hat{S}_{L_z} + \Lambda_{QL}(\hat{S}_{Qx}\hat{S}_{Lx} + \hat{S}_{Qy}\hat{S}_{Ly}) \\
&+ \Gamma_{PL}\hat{S}_{Pz}\hat{S}_{Lz} + \Lambda_{PL}(\hat{S}_{Px}\hat{S}_{Lx} + \hat{S}_{Py}\hat{S}_{Ly}) \\
&- \sum_k A_{Pk}\hat{S}_{Pz}\hat{I}_{kz} + \sum_k A_{Qk}\hat{S}_{Qz}\hat{I}_{kz} + \sum_k A_{Lk}\hat{S}_{Lz}\hat{I}_{kz},
\end{aligned} \quad (4.2)$$

where ω_i is the Zeeman frequency of the spins, Sj and i, are the spin operators for electron i and nucleus $\hat{I}_k$, respectively. Γ_{ij} and Λ_{ij}, denote the secular and pseudo-secular couplings between spins i and j, respectively, which are given by,

$$\Gamma_{ij} = -2(J_{ij} - d_{ij}) \text{ and } \Lambda_{ij} = -2(J_{ij} + d_{ij}/2) \quad (4.3)$$

where J_{ij}, is the isotropic exchange coupling between spin i and j and d_{ij} is dipolar coupling between the spins, A_{ik} is the hyperfine coupling constant between spin i and the kth nucleus.

Theory for the mechanisms of chemically induced electron spin polarization (CIDEP) observed in free radical reactions in solution was developed [5, 8, 15, 16]. A heuristic model of chemically induced electron spin polarization that breaks the polarization mechanism into its component steps, with each step governed by an appropriate solution of the diffusion equation, was also developed for a two-dimensional system [16]. The model yields the polarization and its time development for weak to strong singlet-triplet mixing in the radical pairs.

According to the theory, the spin polarization of a radical pair formed in a corre-lated spin singlet (S) or $M = 0$ triplet state (T_0) is controlled by the rapidly varying exchange interaction (4.4).

$$d\rho/dt = -[2Qi + 2J(r)\kappa]x\rho \qquad (4.4)$$

here,

$$Q = \langle T_0|H_m|S \rangle \qquad (4.5)$$

is the singlet–triplet (ST) mixing rate due to the magnetic Hamiltonian H_m and

$$J(r) = J_\sigma \exp -[\lambda(r - \sigma)] \qquad (4.6)$$

is the exchange interaction which splits the S and T_0 states, r is the pair separation, and σ is a collision–reaction separation of the order of a molecular diameter. Here J_σ is of the order of the thermal energy of the interacting radicals. The suggested steps in the polarization process are: ST mixing, exchange polarization, spin exchange depolarization, and repeated polarization sequences.

There is a large body of literature on theoretical and experimental, aspects of spin correlated coupled radical-pairs in chemistry particularly in photosynthesis [8–12, 15–25] and references therein.

4.2.3   Radical-Pair Mechanism

In the radical-pair mechanism (RPM) of CIDEP in liquid solution, the polarization arises during the radical lifetime as a result of differences in the magnetic inter-actions (Zeeman and/or hyperfine) and the exchange interaction between radicals when forming spin-correlated radical pairs under reencounter when diffusing in solution [5, 6].

The radical-pair mechanism includes the following steps: (1) The radical pair partners are created via chemical reaction of an excited molecule (geminate pair) or where the radical pairs are formed by a random encounter of separately generated radicals (free or F-pair), (2) The radicals in the singlet state can form a recombi-nation product, whereas the triplet state is repulsive and does not lead to product formation. The electron-nuclear hyperfine interactions and electron Zeeman inter-actions with the external magnetic field are decisive factors affecting the reactivity of the radical pair, and (3) The radicals which are separated sufficiently so that short-range valence or exchange forces are negligible, will be polarized.

The mechanisms of the RPM process in various systems have been described in detail [6, 26, 27] and references therein.

4.2.4   Radical-Triplet Dynamic Electron Polarization Mechanism

A radical-triplet dynamic electron polarization (RTDEP) mechanism for chemically induced electron polarization (CIDEP) of free radicals in solution was suggested by Blättler et al. in 1990 [28]. The presence of triplet-state molecules and alkyl radicals was found to induce emissive ESR polarizations. The phenomenon was attributed to triplet-state quenching by radicals as well as to quartet-doublet mixing and splitting in radical-triplet encounters through the zero-field splitting (ZFS) interaction of the triplet molecule. Various theoretical and experimental aspects of the problems in the triplet-radical systems were widely discussed (see, for example [29–36]).

For the triplet-doublet system the following Hamiltonian was used [33]:

$$H = H^{ex} + H^{hf} + H^{zfs} \tag{4.7}$$

where H^{ex}, H^{hf}, H^{zfs} are related to the exchange interaction, to hyperfine interaction including Zeeman splitting, and to the zero-field splitting. These terms are given by

$$H^{ex} = 2J(S_R S_T) \tag{4.8}$$

$$H^{hf} = \beta(g_T S_T + g_R S_R)H_0 + \sum_i A_{Ti} I_T^k S_T + \sum_i A_{Ri} I_R^l S_R \tag{4.9}$$

$$H^{zfs} = \frac{D\left(S_{Td1}^2 - 1\right)}{S_T^2} + E\left(S_{d2}^2 - S_{d3}^2\right) \tag{4.10}$$

where R and T represent the radical and triplet molecule, respectively; S is the electron spin angular momentum operator, g is the g-tensor, β is the Bohr magneton, and Ho is the external magnetic field; A_T, and A_R, are the hyperfine tensors of the ith and jth nuclei. I_T, and I_R are the nuclear spin angular momentum operators of the ith nuclei. k and l represent nuclear spin states of triplet and doublet molecules, respectively; D and E are the zero-field splitting constants of the triplet molecule and d_i is the zero-field-splitting axis of the triplet molecule The magnitude of the exchange interaction of pairs of paramagnetic molecules in solution is related to the energy splitting of degenerate spin states such as triplet-singlet splitting in radical pairs and quartet (Q)-doublet (D) splitting in radical-triplet (RT) pairs.

In work [30], the spin exchange Hamiltonian H_{ex} was used to describe the energy difference between D and Q states, $-2J = E_Q - E_D$ and was expressed as,

$$H_{ex} = -\frac{1}{3}J\left(1 + 4\hat{S}_T 4\hat{S}_T\right) \tag{4.11}$$

In RT pairs, the overlap integral of the pair is usually negligibly small and J is proportional to the exchange integral, J_{ex}, of Q and D states. Most RT pairs have an antiferromagnetic interaction in which the Q state is higher in energy than the D state ($J < 0$). In radical ion pairs, when the J value due to Jex is small, the inter-molecular charge transfer (CT) interaction becomes important. In such a system the positive J value becomes predicted. A schematic drawing of the potentials of RT and CT pairs is presented in Fig. 4.3.

A model for quantum mechanical processes in a triplet–doublet (TD) system suggested by Ishii et al. [34] is based on a detailed consideration of the intersystem crossing. According to this model, both the triplet-doublet and triplet-quartet states are populated due to spin orbit coupling with an efficiency that is independent of the sign of the magnetic quantum number of the electron spin state. It was suggested that different populations of spin states with magnetic numbers of different sign can occur due to ZFS-induced mixing of the different spin states.

In work by the S. Yamauchi group [29], systems of three-spin-1/2 systems were considered in detail. The system consists of the following components: electron 1 occupying the lowest unoccupied molecular orbital (LUMO), electron 2 occupying highest occupied molecular orbital (HOMO), respectively of a triplet in an excited state and the third electron 3 locating on the singly occupied molecular orbital (SOMO) of the radical unit. The electron spin polarization in the given system was discussed in terms of a radical–triplet pair spin-state mixing model, including the intersystem crossing processes. Two models were discussed: a model describing kinetic competition between the spin–spin exchange and spin orbital coupling (SOC) mechanisms for ISC and second the model based on the assumption that electron spin relaxation and decay can be spin selective.

In another work, he multispin systems consisting of spin correlated radical pairs (SCRPs) and stable nitroxide radicals, localized in micelles of sodium dodecyl sulfate (SDS), were studied by ESR and pulse laser photolysis techniques [12]. Diagram of the energy states of the SCRPs is presented in Fig. 4.4. In all the systems studied, the stable nitroxide radicals exert no effect on the shape of the ESR spectra of the SCRPs and on the decay kinetics of the ESR signal of the SCRPs,

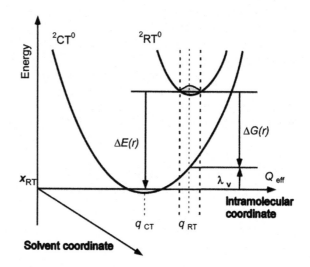

Fig. 4.3 Schematic drawing of the potentials of radical triplet (*RT*) and intermolecular charge transfer (*CT*) pairs (See details in [30])

Fig. 4.4 Diagram of the
energy states of the SCRPs at
different distances r between
the pair radicals **A** and
B (**a**) and the idealized
TR ESR spectrum of the
SCRPs at the spin_adiabatic
occupation of the SCRPs and
the equilibrium triplet state of
the precursor (**b**); ΔS and ΔT
are the spectral shifts of the S
and T components of the
APS, respectively; 2/TS and
2/TT are their widths [12]

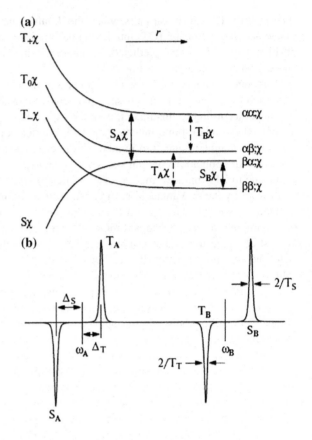

while generate the electron spin polarization in nitroxide radicals. The electron spin polarization transfer from the nonequilibrium electron spin states of the molecular triplets (SCRP precursors) was suggested to be the most efficient mechanism of generation of the radical electron spin polarization.

A theoretical model of a triplet–doublet (TD system) by Kandrashkin and van der Est [37] which consists of a strongly exchange coupled triplet–doublet pair was based on two key assumptions: the triplet-doublet excited electron states of the TD system are initially populated through exchange induced ISC from the singlet-doublet electronically excited states, and the spin orbit coupling (SOC) providing the ISC can be treated in terms of perturbation theory.

4.2.5   Triplet Pair Mechanism of Chemically Induced Dynamic Electron Polarization

A new triplet pair mechanism of chemically induced dynamic electron polarization, TPM CIDEP) in the spin-correlated pairs of triplets which arises due to the spin-selective triplet-triplet annihilation and the subsequent spin dynamics in the spin-correlated pairs of triplets was considered using time-dependent perturbation theory for the case of short-lived pairs of triplets [38]. It was taking in account that Total spin of encounted two triplet states can be 0, 1, or 2 and, due to spin selection rules, the triplet-triplet mutual annihilation can occur only in singlet pairs which possess total spin zero. Since the pairs of triplets are enriched in the states with total spin 1 and 2, the spin-selective annihilation of triplet excitons generates spin correlated pairs of triplets. The suggested spin Hamiltonian for the spin dynamics of the triplet pairs in this system consists of the Zeeman interaction of the triplets with the extremal magnetic field B_0, the zero field splitting (ZFS) terms and the exchange interaction H_{ex} between two triplets [38]:

$$H = \beta g_A B_0 S(A) + S(A) h D_A S(A) + \beta g_B B_0 S(B) + S(B) h D_B S(B) + H_{ex} \quad (4.12)$$

where g_{AB} are the g-factors of the triplets, b is the Bohr magneton, S(A, B) are the vectors of the triplet spins, D is the zfs tensor expressed in frequency units, and he Heisenberg exchange interaction has the form $H_{ex} = -h\, J[1/2 + 2S(A)S(B)]$. The theory predicted that when the exchange integral J < 0, the EPR spectrum should exhibit an emissive form and spin-spin interaction between triplets should occur. It was stressed that the time-resolved EPR spectrum of spin-correlated triplet pairs can be detected only during the lifetime of the triplet pairs and necessary condition for the TPM of the CIDEP formation is the spin-spin interaction between triplets.

Experimental observation of the spin polarization mechanism induced by triplet-triplet annihilation in single crystals of organic charge transfer (CT) complexes was reported in [39]. In two molecular crystals, anthracene-tetracyanobenzene and phenazine-tetracyanoquinodimethane, mutual annihilation of triplet states induces the spin polarization of excited triplet states, which was detected by time-resolved EPR of triplet excitons. The kinetics of the two EPR lines of the triplet exciton spectrum have been measured in the time range up to 200 μs after a laser pulse. Two kind of polarization has been detected: the initial polarization of the lines, due to intersystem crossing, and a long-lasting polarization which is attributed to the triplet-triplet annihilation. The investigation of The dependence of the phenomenon on the intensity of the light pulse and on the orientation of the crystals in the magnetic field was also experimentally observed.

4.3   Dynamic Nuclear Polarization

Dynamic nuclear polarization (DNP) [5, 40–54], results from transferring spin polarization from electrons to nuclei, thereby enhancing the nuclear spin polarization. The alignment of electron and nuclear spins at a given magnetic field and temperature is described by the Boltzmann distribution at thermal equilibrium. Electrons at a given magnetic field and temperature in thermal equilibrium can be aligned to a higher degree by chemical reactions (Chemical-induced DNP, CIDNP), optical pumping and spin injection. Owning to the dynamic nuclear polarization sensitivity of detection of NMR spectra can be fantastically enhanced by 60,000-fold and more [42]. The DNP technique is an effective tool for investigation mechanisms of radical chemical and biological processes and for determining the structure of radicals and transition metal complexes. Several monographs and articles covering the quantum mechanical treatment of electron-nuclear interactions under the influence of irradiation were published [5, 6, 44, 50, 53–55].

The polarization transfer between electrons and nuclei can occur spontaneously, when electron spin polarization deviates from its thermal equilibrium value through electron-nuclear cross relaxation and/or spin-state mixing among electrons and nuclei. At thermal equilibrium, polarization transfer requires continuous microwave irradiation at a frequency close to the corresponding electron paramagnetic resonance (EPR) frequency. The Nuclear Overhauser effect (NOE), the solid-effect (SE), the cross-effect (CE) and thermal-mixing (TM) can be mechanisms for microwave-driven DNP processes.

Dynamic nuclear polarization DNP can be accomplished via different mechanisms [5, 42–50]. The Overhauser effect accounts for the perturbation of nuclear spin level populations observed in metals and free radicals when electron spin transitions are saturated by the microwave irradiation [45]. *The solid effect* occurs when an electron-nucleus mutual spin flip transition in an electron-nucleus two-spin system is excited by microwave irradiation [56, 57]. The *cross effect* requires two unpaired electrons as the source of high polarization [54, 57]. The *thermal mixing effect* is an energy exchange phenomenon between the electron spin ensemble and the nuclear spin, using the electron spin ensemble as a whole to provide hyper nuclear polarization [49, 50].

- In work [58], to describe the DNP effect the general static Hamiltonian was written as

$$
\begin{aligned}
H = H &= H_E + H_N + H_{EN} = \omega_{0E}E_z - \omega_{0N}N_z + H_{EN}^{is} + H_{EN}^{di} \\
&= \omega_{0Z}E_z - \omega_{0N}N_z + K_{SE}(E_Z N_Z + E_X N_X + E_Y N_Y) + K_{PSE}E_X N_z
\end{aligned}
\tag{4.13}
$$

where H_E and H_N are the Hamiltonians for the electron and nucleus respectively; H_{EN} ist he hyperfine coupling, which is separated into the isotropic hyperfine interaction H_{EN}^{is} and the anisotropic dipolar coupling H_{EN}^{di} between the electron and the nucleus; the coefficients K_{SE} and K_{PSE} denote the secular and pseudosecular hyperfine interactions. ω_{0E} and ω_{0N} are the electron and nuclear Larmor frequencies.

Which DNP mechanism occurs, depends on the interplay of two physical parameters, e.g., the populations and sign of the electron energy levels, $\pm\Delta$, and that of the nuclear spin energy levels, $\pm\delta$ [59]. In the Overhauser effect, a relative enhancement of polarization is [45]

$$\eta = -\left(\frac{\Delta}{\delta}\right) = \frac{\gamma_e}{\gamma_n} \qquad (4.14)$$

where γ_e and γ_n are gyroscopic ratios for electron and nucleus, respectively. The solid effect has an inverse field dependence, and requires that the inhomogeneous spread Δ, and homogenous line width δ of the electron spin resonance spectrum should be smaller than the nuclear Larmor frequency: i.e., Δ, $\delta < w_n$. The DNP effect was observed in the solid state in systems where $\Delta > \delta > w_n$ and the irradiation frequency separation between the maximum positive and negative enhancements was less than $2w_n$. The DNP thermal mixing mechanism occurs in the solid state in systems where $\delta > w_n$.

Theoretical descriptions of Overhauser DNP in liquids presented in [43] were based on the Solomon equation [46]:

$$\frac{dI_z}{dt} = -(\rho_I + W^0)(I_z - I_0) - \sigma_{IS}(S_z - S_0), \qquad (4.15)$$

where I and S refer to the nuclear and electron spin, respectively, I_0 and S_0 are their Boltzmann equilibrium values. Energy level diagram for an electron spin S = 1/2 coupled to a nuclear spin I = 1/2 in the context of the Overhauser effect is shown in Fig. 4.5. The relaxation rates $\rho_I = W_0 2 + W_I + W_2$ and $\sigma_{IS} = W_2 - W_0$ are given by the nuclear–electron zero-, one- and two-quantum transitions.

The steady state solution of under continuous microwave irradiation of the allowed electron spin transition leads to the expanded Overhauser enhancement formula [45].

$$\varepsilon = \frac{I_z - I_0}{I_0} = -\xi f s \frac{\gamma_e}{\gamma_n}. \qquad (4.16)$$

Here ε is the enhancement of the nuclear spin polarization, $\xi = \sigma_{IS}/\rho_I$ is the coupling factor describing the efficiency of the cross-relaxation; $f = 1 - T_1/T_1^0$, where T_1 and T_1^0 are the longitudinal nuclear relaxation times in the presence and absence of paramagnetic species, respectively, is the leakage factor, and s the saturation factor.

Fig. 4.5 Energy level diagram for an electron spin S = 1/2 coupled to a nuclear spin I = 1/2. W_I nuclear spin relaxation rate, W_S electron spin relaxation rate, W_2 double quantum relaxation rate, W_0 zero quantum relaxation rate [45]

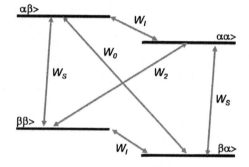

The review [60] focused on recent developments in the field of DNP with a special emphasis on work done at high magnetic fields > 5 T. To understand the polarization transfer mechanisms in a DNP experiment, the authors considered the following Hamiltonian:

$$H = H_S + H_I + H_{SI} = \omega_{0S}S_Z - \omega_{0I}I_Z + H_{SI}^{ISO} + H_{SI}^{dip}$$
$$= \omega_{0S}S_Z - \omega_{0I}I_Z + A(S_ZI_Z + S_YI_Y + S_XI_X) + BS_XI_Z. \quad (4.17)$$

where ω_{0S} and ω_{0I} are the electron and nuclear Larmor frequencies, respectively, S_i and I_i are the electron and nuclear spin operators H_{SI}^{iso} is the isotropic Fermi contact hyperfine interaction between the electron and the nucleus, and H_{SI}^{dip} is the anisotropic dipolar coupling, and the coefficients A and B denote the secular and pseudosecular hyperfine interactions. The energy level diagrams for the Overhauser Effect (OE) and Solid Effect (SE) is also presented....

Theoretical analysis performed in [61] suggested that in a chromophore excited triplet state the nuclear spins are involved in the intersystem crossing process. According to authors, the selective population of the zero-field spin sublevels, obeying to the symmetry selection rules of spin-orbit coupling, causes the high electron spin polarization in an external field. This process was to shown to be a source of oscillatory nuclear spin polarization that gives rise to large signal enhancement in nuclear magnetic resonance. The authors rationalized the formation of the nuclear quantum oscillation as consequence of the following elementary processes: (1) the ISC selectively populates the eigenstates of the extended zero-field Hamiltonian including a hyperfine term; (2) the nuclear spins are aligned along a molecular axis; (3) the spins suddenly experience the external magnetic field which leads to a nonadiabatic change of the spin quantization axis, and to the formation of nuclear quantum oscillations. The time-dependent density matrix $\rho(\Omega, t)$ was given as

$$\rho(\Omega, t) = \exp(-it\mathbf{H}(\Omega)/h) \cdot \rho(\Omega, 0) \cdot \exp(it\mathbf{H}(\Omega)/h) \quad (4.18)$$

Here $\rho(\Omega, 0)$ denotes the initial state at the instant of the light pulse. The spin Hamiltonian $\mathbf{H}(\Omega)$, constrains Zeeman, dipolar, hyperfine, and nuclear Zeeman interactions of the triplet state:

$$H(\Omega) = g\beta\mathbf{B}0 \cdot \mathbf{S} + \mathbf{S} \cdot \mathbf{D}(\Omega) \cdot \mathbf{S} + \mathbf{S} \cdot \mathbf{A}(\Omega) \cdot \mathbf{I} - g_N\beta_N\mathbf{B}0 \cdot \mathbf{I} \quad (4.19)$$

where g, β, $\mathbf{S}$, $\mathbf{I}$, g_N, and β_N are the isotropic g factor of the triplet state, the Bohr magneton, the electron spin operator, the nuclear spin operator, the nuclear g factor, and the nuclear magneton, respectively.

An increase in signal-to-noise ratio of >10,000 times in liquid-state NMR was achieved in *so called "dissolution"* approach, based on phenomenon hyperpolarization of NMR-active nuclei in the solid state under effect of paramagnetic species, Gd^{3+} ions, nitroxide or trityl, for example [41, 42, 53]. At temperature 1.1–1.5 K nuclei of a sample (typically biological object) were hyperpolarized by paramagnetic molecules, followed by fast melting, dissolving in a hot solvent and rapid transferring in water solution. In such a condition the sample preserve the hyperpolarized state. The hyperpolarization approach markedly improves capacity of methods of dynamic polarization for identification of diamagnetic products revealing intermediates of radical processes, identifying the species giving rise to CIDNP, unraveling the connecting pathways, using polarizations as labels to molecules and as labels to individual nuclei.

The kinetics, and mechanism of **1**-Hexene polymerization catalyzed by of zirconium-based metallocenes was studied by ^{13}C NMR using dissolution DNP TEMPOL as the polarizing agent and glycerol as the glass-forming agent. Efficiency of this method in detection of the first steps of in polymerization reactions has been demonstrated [62]. Data on preclinical studies on the dissolution dynamic nuclear polarization with a variety of different substrates (the hyperpolarized [U−^{13}C, U−^{2}H]glucose, and the lactate, for example) have demonstrated the potential of the method to provide insights into tissue metabolism in prostate cancer and other pathology [63]. The principal limitation of the technique is the short half-life of the polarization, which at $\sim$20–30 s in vivo limits studies to relatively fast metabolic reactions.

Recent data on theory and application of DNP are presented in [64–66].

4.3.1   Chemical-Induced Dynamic Nuclear Polarization

Chemical-induced dynamic nuclear polarization (CIDNP) is a non-Boltzmann nuclear spin state distribution produced in thermal or photochemical reactions, usually from colligation and diffusion, or disproportionation of radical pairs, and detected by NMR spectroscopy as enhanced absorption or emission signals [67–73]. The first observations of CIDNP effects were reported in 1967 by Bargon, Fischer, and Johnsen [67], and independently by Ward and Lawler [68]. In the method frame, the signal directions and intensities of CIDNP effects are governed by the magnetic properties of the radicals or radical ions, the initial spin multiplicity of the pair, and by the mechanism of product formation. The CIDNP effect is used for probing the spin multiplicity of intermediates involved in a reaction, the spin density distribution and structure of radical or radical ion intermediates. This method provides insights into biochemical problems, such as protein structure and folding, photosynthetic reaction centers, and the mechanisms of DNA photo-repair. At present it is accepted that the main mechanisms causing spin polarization in free radicals are the triplet mechanism (TM), the radical pair mechanism (RPM) and the radical triplet pair mechanism (RTPM). These mechanisms have been recently reviewed [8].

In 1969 Kaptein and Oosterhoff [69] and independently Closs [70] proposed the radical pair mechanism which relied on the ability of nuclear spin interactions to

Fig. 4.6 Radical pair
mechanism of photo-CIDNP
in liquids. M^* denotes the
spin-polarized molecule of
interest [43]

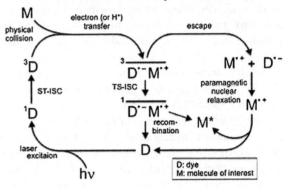

alter the recombination probability in reactions that proceed through radical pairs
(Fig. 4.6). This mechanism is currently accepted as the most common cause of
CIDNP. Two types of CIDNP effect, net and multiplet, were considered.

Figure 4.7 shows a schematic diagram of nuclear spin level populations for net
CIDNP effect for two nuclear spins Ia and Ib.

A theory of spin relaxation effects in photochemically induced dynamic nuclear
polarization spectroscopy of nuclei with strongly anisotropic hyperfine couplings in
strong magnetic fields over 10 T was developed [69]. The effect of nuclear spin
relaxation, cross-relaxation, and cross-correlation induced by the anisotropy of
hyperfine and g-tensors on the geminate photo-CIDNP effect in fluorine-containing
radical pair was explored. Suggesting that all four factors in Kaptein's sign rule for
net nuclear polarization [71] are positive, the authors wrote the following equation:

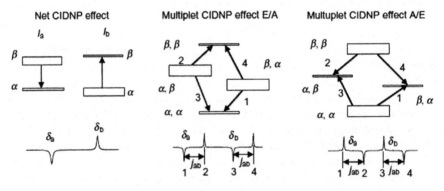

Fig. 4.7 Schematic diagram of nuclear spin level populations for net CIDNP effect for two
nuclear spins Ia and Ib with chemical shifts δa, δb. and multiplet CIDNP effect for two
nonequivalent nuclear spins Ia and Ib, belonging to the same molecule and with spin–spin
coupling jab. The corresponding simulated NMR spectra are shown below (A and are E absorption
and emission processes, respectively [8]

$$\Gamma_{net}(i) = \mu\,\epsilon\,sign(\Delta g)\,sign(A_i) = \left\{ \begin{array}{ll} + & \text{absorptive} \\ - & \text{emissive} \end{array} \right\};$$

$$\mu = \left\{ \begin{array}{ll} + & \text{triplet precursor} \\ - & \text{singlet precursor} \end{array} \right\}; \qquad (4.20)$$

$$\epsilon = \left\{ \begin{array}{ll} + & \text{recombination products} \\ - & \text{escape products} \end{array} \right\}$$

In this equation, Δg is the difference between the g-values of the two radicals and Ai is the hyperfine coupling constant of the nucleus in question. Considering only populations and neglecting coherences and cross-correlations, a classical four-level population dynamics model was analyzed (Fig. 4.8).

The given model includes single-quantum transitions, corresponding to single spin flips (W_n and W_e terms), double-quantum and zero-quantum transitions involving double spin flips (W_0 and W_2 terms). On the basis of second-order time-dependent perturbation theory, it was assumed that the stochastic modulation of the hyperfine coupling in one of the partners of the radical pair causes these transitions. It was concluded that the electron-nuclear cross-relaxation rates W_0 and W_2, across the range of experimentally available correlation times (10 ps–10 ns), influence on the nanosecond to microsecond time scale photo-CIDNP spin dynamics.

The theory of solid-state photo-CIDNP in a very weak field, such as the Earth's magnetic field, was developed [72]. For a simple limiting case, the minimal static spin Hamiltonian in angular frequency units was written as:

$$H_0 = \omega_s \hat{S}_{1z} + \omega_s \hat{S}_{2z} + d\hat{S}_{1z}\hat{S}_{2z} + A\hat{S}_{1z}\hat{I}_z + a_{iso}(\hat{S}_{1x}\hat{I}_x + \hat{S}_{1y}\hat{I}_y) \qquad (4.21)$$

where ω_s is the electron Zeeman frequency, d is the electron-electron dipole-dipole coupling d, a_{iso} is the isotropic hyperfine coupling a_{iso}, $A = a_{iso} + T$ the secular hyperfine coupling and T is the dipole-dipole contribution to the hyperfine

Fig. 4.8 Energy levels and longitudinal relaxation transitions in an electron-nuclear two-spin system. See details in [71]

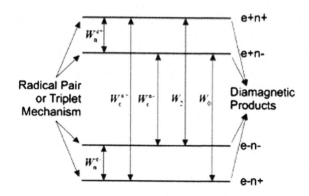

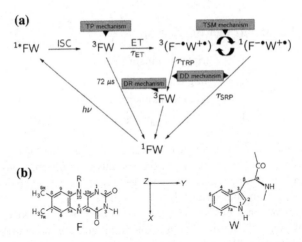

Fig. 4.9 **a** Cyclic reaction scheme for the formation of photo-CIDNP in the LOV2-C450A mutant of phototropin, based on experimental observations by 13C photo-CIDNP NMR and timeresolved optical spectroscopy. **b** *F*: FMN cofactor. *W*: tryptophan-491 in the apoprotein. The following mechanisms are pointed out: triplet (TM), three-spin mixing (TSM), differential decay (DD) and differential relaxation (DR) [73]

coupling. On the basis of theoretical calculations, it was suggested that efficient coherent three-spin mixing in the S-T$^-$ or S-T$^+$ manifold is caused by isotropic hyperfine couplings of about 350 kHz and that dipolar coupling between two electron spins at distances of around 30 Å matches the electron Zeeman interaction in the Earth's magnetic field.

Methods of ^{13}C photo-CIDNP and time resolved optical spectroscopy were exploited for mechanistic study of the formation of photo- CIDNP in the LOV2-C450A mutant of phototropin [73]. Figure 4.9 presents cyclic reaction scheme for the formation of the singlet tryptophan—flavin ion-radical pair (TFRP), which can occur by the following mechanisms: TP triplet polarization (TP), three-spin mixing (TSM), differential decay (DD), and differential relaxation (DR). Experimental data and theoretical consideration have been proved that a triplet state is the major source for photo-CIDNP in a photoactive protein. For TFRP values of exchange integral *Jex* = −38.4 MHz and life time tr = 20 ns were detected.

Solid-state mechanisms including a novel triplet mechanism CIDNP in LOV2-C450A of Avena sativa phototropin with specifically ^{13}C-labeled flavins was also suggested [74]. The intermediates ^{3}(TrpNH• +) and ^{3}[TrpNH• +) − TyrO•] were detected. The rate constants for intramolecular electron transfera in oxidized peptides were measured by time resolved CIDNP techniques the microsecond scale.

The present status of knowledge of the dynamic nuclear polarization mechanism was recently concisely reviewed [43]. Figure 4.10 schematically illustrated the widely used DNP mechanisms.

Fig. 4.10 Overview of
methods that perturb the
longitudinal magnetization of
nuclear spins in solution

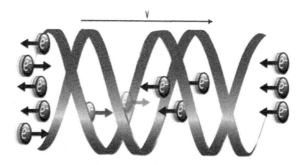

4.4   Chiral-Induced Spin Selectivity. Spin Filter

Spin filter is a device in which spins with the desired spin orientation are allowed to
pass through without decay, while .spins with undesired orientation are not allowed.
The spin polarization or spin selectivity (S) was defined as

$$S = (I^+ - I^-)/(I^+ + I^-) \qquad (4.22)$$

in which I^+ and I^- are the intensities of the signals corresponding to the spin
oriented parallel and antiparallel to the electrons' velocity, respectively. For
instance, chiral-induced spin selectivity (CISS), obtained from low-energy photoelectron transmission (LEPET) spectroscopy in chiral Langmuir-Blodgett films,
arises in ordered films of chiral organic molecules on surfaces and can act as
electron spin filters at room temperature (Fig. 4.10) [75]. In the LEPET experiment
circularly polarized light was used to eject spin-polarized photoelectrons from the
underlying a substrate. It was shown that LEPET signal is correlated with the spin
polarization, indicating spin selectivity in the photoelectrons' transmission through
the chiral monolayers.

     In another example, the original experimental device for measuring the con-
ductance of a nanoparticle—dsDNA—nickel molecular junction was reported [76].
The device includes the permanent magnet, underneath of the nickel substrate splits
the sub-bands of the Ni and the tip of the AFM, which is placed in contact with the
gold nanoparticle. The current was measured between the nickel substrate and the
tip of the AFM, which is placed in contact with the gold nanoparticle. The CISS
effect opens the way for employing of chiral systems in spintronics devices and for
investigation of detail mechanisms of spin-selective processes in biology. Various
aspects of theory, experiments and application spin filters on the base of principles
of the Chiral-induced spin selectivity were discussed in [77–80].

     For information about the CIDNP effect beyond those already cited the reader is
referred to the following references [81–92].

References

1. A.B. Gasper, V.M. Ksenofontov, M. Seredyuk, P. Gullich, Coord. Chem. Rev. **249**, 2661 (2005)
2. R. Jain, K. Kabir, J.B Gilroy, K.A Mitchell, Wong KC, Hicks RG, High-temperature metal-organic magnets. *Nature* 2007, **445**(7125), 291–294
3. C-Y. Kao, Y Lu, J-W. Yoo, V.N. Prigodin, A.J. Epstein, Appl. Phys. Lett. **99**, 153–503 (2011)
4. A.F. Collings, C. Critchley. *Artificial Photosynthesis: From Basic Biology to Industrial Application* (Wyley-VCH, 2005)
5. H. Hayashi, *Introduction to Dynamic Spin Chemistry. Magnetic Field Effects upon Chemical and Biochemical Reactions* (World Scientific, 2004)
6. K. Mobius, W.W. Lubitz , A. Savitsky, App. Magn. Reson. **41**, 113 (2011)
7. A. Schweiger, G. Jeschke, *Principles of Pulse Electron Paramagnetic Resonance*, 1st edn. (Oxford University Press, 2001)
8. L.J. Berliner, E. Bagryanskaya, in Misra, K. *Multifrequency Electron Paramagnetic Resonance*, ed. by S. Miscra (Willey-VCH, 2011), p. 947
9. R.W. Fessenden, R.H. Schuler, Electron spin resonance studies of transient alkyl radicals. J. Chem. Phys. **39**, 2147–2196 (1963)
10. G.R. Eaton, K.M. Salikhov, S.S. Eaton, *Foundations of Modern EPR* (World Scientific, Singapore, 1998)
11. A.I. Shushin, Chem. Phys. **133**, 044505 (2010)
12. V.R. Gorelik, V.F. Tarasov, S.R. Shakirov, E.G. Bagryanskaya, *Russian Chemical Bulletin, International Edition*, vol. 57, pp. 14–16 (2008)
13. K.M. Salikhov, Y.N. Molin, R.Z Sagdeev, A.L. Buchachenko, *Spin polarization and magnetic field effects in radical reaction* (Elsevier, 1984)
14. M.R. Brustolon, *Principles and Applications of Electron Paramagnetic Resonance Spectroscopy* (Blackwell Publishers, 2008)
15. K.M. Salikhov, G. Zech, D. Stehlic, Mol. Phys. **100**, 1–1321 (2002)
16. F.J. Adrian, Chem. Phys. **376**, 23 (2010)
17. A. Maliakal, W. Weber, N.J. Turro, M.M. Green, S.Y. Yang, S. Pearsall, M.-J. Lee, Macromolecules **3**, 9151 (2002)
18. Y.E. Kandrashkin, K.M. Salikhov, D. Stehlik, Appl. Magn. Reson. **12**, 141 (1997)
19. P.J. Hore, D.A. Hunter, Chem. Phys. Lett. **137**, 495 (1987)
20. A.J. Hoff, J. Deisenhofer, Phys. Rep. **1997**(287), 2–247 (1997)
21. K.L Ivanov, A. Yurkovskaya, H.-M. Vieth, J. Phys. Chem. **128**, 154701 (2008)
22. K. Salikhov (ed.), *Electron Paramagnetic Resonance Spectroscopy: From Fundamental Research to Pioneering Applications* (AXAS Publishing Ltd, Wellington, New Zealand, 2009)
23. A. Savitsky, K. Mobius, Helv. Chim. Acta V. 2544 (2006)
24. V. Rozenstein, A. Berg, E. Stavitski, H. Levanon, L. Franco, C. Corvaja, Phys. Chem. A **109**, 11144 (2005)
25. S. Santabarbara, I. Kuprov I, W.V. Fairclough, P.J. Hore, S. Purton, P. Heathcote, M.C.W. Evans, *Biochemistry* **44**, 2119 (2005)
26. K.A. McLauchlan CIDEP theory, technique and its application to chemical reactions, in A. J. Hoff (ed.), *Advanced EPR. Applications in Biology and Biochemistry*. Elsevier, Amsterdam, Oxford, New York, Tokyo (1989)
27. E. Bagryanskaya, M. Fedin, D.E. Forbes, J. Phys. Chem. A **109**, 5064 (2005)
28. C.F. Blättler, F. Jent, H. Paul, Chem. Phys. Lett. **166**, 375 (1990)
29. V.F. Tarasov, S.S.M. Islam, Y. Ohb, D.E. Malcolm, M.D.E. Forbes, S. Yamauchi,. Appl. Magn. Reson. **41**, 175 (2011)
30. Akio Kawai, Kazuhiko Shibuya, J. Phys. Chem. A **111**, 4890 (2007)
31. G.I. Likhtenstein., K. Ishii, S. Nakatsuji, Photochem. Photobiol. **83**, 871–881 (2007)
32. M. Porel, S Jockusch, M.F. Ottavani, N.J. Turro, V Ramamurthy, Langmuir **27**,10548 (2011)
33. A. Kawai, T. Okutsu, K. Obi, J. Phys. Chem. **95**, 130 (1991)

34. K. Ishii, J.-I. Fujisawa, A. Adachi, S. Yamauchi, N. Kobayashi, J. Am. Chem. Soc. **120**, 3152 (1998)
35. M. Porel, S Jockusch, M.F Ottaviani, N.J. Turro, V. Ramamurthy,. Langmuir **27**, 10548 (2011)
36. G.I. Likhtenstein, K. Ishii, S. Nakatsuji, Photochem. Photobiol. **83**, 871 (2007)
37. Y. Kandrashkin, A. van der Est, Appl. Magn. Reson. **40**, 189 (2011)
38. K.M. Salikhov, Appl. Magn. Reson. **26**, 135 (2004)
39. C. Corvaja, l. Franco, K.M. Salikhov, V.K. Voronkova, Appl. Magn. Res. **28**,181 (2005)
40. A. Abragam, M. Goldman, Rep. Prog. Phys. **41**, 395 (1976)
41. J.H. Ardenkjar-Larsen, B. Fridlund, A. Gram, G. Hansson, L. Hansson, M.H. Lerche, R. Servin, M. Thaning, K. Golman, Proc. Natl. Acad. Sci. U.S. A. **100**, 10158 (2003)
42. L. Lumata, A.K. Jindal, M.E. Merritt, C.R. Malloy, A.D. Sherry, Z. Kovacs, J. Am. Chem. Soc. **133**, 8673 (2011)
43. J.H. Lee, Y. Okuno, S. Cavagnero, ELS **241**, 18 (2014)
44. V. Atsarkin, Dynamic nulcear polarization: yesterday, today and tomorrow. J. Phys. **324**, 1 (2011)
45. A.W. Overhauser, Phys. Rev. 92, 411 (1953)
46. I. Solomon, Phys. Rev. 99, 559 (1955)
47. A. De Luca, A. Rosso, Phys. Rev. Lett. **115**, 080401 (2015)
48. D.A, Bordignon, E Joseph, B. Tschaggelar, R. Tschaggelar, G. Jeschke, J. Magnet. Res. **222**, 34 (2012)
49. S. Jannin, A. Comment, J.J. Klink, Appl. Magn. Reson. **43**, 59 (2012)
50. K.H. Sze, Q. Wu, H.S. Tse, G. Zhu, Dynamic nuclear polarization: new methodology and applications. Top. Curr. Chem. **326**, 215–242 (2012)
51. A. Karabanov, D. Wiśniewski, I. Lesanovsky, W. KöckenbergerPhys, Rev. Lett. **115**, 020404 (2015)
52. S.-J. Lee, J.H. Shim, K. Kim, K.K. Yu, S.-M. Hwang, J. Magn. Reson. **255**, 114 (2015)
53. U.L. Gunther, Top. Curr. Chem. **335**, 23 (2013)
54. J.H. Lee, Y. Okuno, S. Cavagnero, J. Magn. Reson. **241**, 18 (2014)
55. G.R. Eaton, S.S. Eaton, D.P. Barr, R.T. Weber, *Quantitative EPR* (Springer, 2010)
56. C.D. Jeffries, Phys. Rev. **106**, 164 (1957)
57. D. Shimon, Y. Hovav, A. Feintuch, D. Goldfarb, S. Vega, Phys. Chem. Chem. Phys. **14**, 5729 (2012)
58. K.H. Sze, Q. Wu, H.S. Tse, G. Zhu, *Topics in Current Chemistry*, vol. 326, p. 215 (2012)
59. N. Chandrakumar, J. Indian Inst. Sci. **90**,133 (2010) www.journal.library.iisc.ernet.in
60. N. Maly, G.T. Debelouchina, V.S. Bajaj, K.-N. Joo, C.G. Hu, M.L. Mak-Jurkauskas, J.R Sirigiri, J.R. van der Wel, J. Herzfeld, R.J. Temkin, R.G Griffin, J. Chem. Phys. **128**, 052211 (2008)
61. G. Kothe, T. Yago, J.-U. Weidne, G. Link, M. Lukaschek, T.-S. Lin, J. Phys. Chem. B **114**, 14755 (2010)
62. C.-H. Chen, W.-C. Shih, C. Hilty, J. Am. Chem. Soc **137**, 6965 (2015)
63. K.M. Brindle, J. Am. Chem. Soc. **137**, 6418 (2015)
64. Y. Hovav, A. Feintuch, S. Vega, D. Goldfarb, J. Magn. Reson. **238**, 94 (2014)
65. D. Le, F. Ziarelli, T.N.T. Phan, G. Mollica, P. Thureau, F. Aussenac, O. Ouari, D. Gigmes, P. Tordo, S. Viel,. Macromolecular Rapid Communications (2015). Ahead of Print
66. Q. Chappuis, J. Milani, B. Vuichoud, A. Bornet, A.D. Gossert, J. Bodenhausen, S. Jannin, J. Phys. Chem. Lett. **6**, 1674 (2015)
67. J. Bargon, H. Fischer, U.Z. Johnsen, Naturforsch A Phys. Sci. **22**, 1551 (1967)
68. H. Ward, R.J. Lawler, Am. Chem. Soc. **89**, 5518 (1967)
69. R. Kaptein, L.J. Oosterhoff, Chem. Phys. Lett. **4**, 214 (1969)
70. G.L. Closs, J. Am. Chem. Soc. **91**, 4552 (1969)
71. R. Kaptein, J. Chem. Soc. D, 732 (1971)
72. G. Jeschke, B.C. Anger, B.E. Bode, J. Matysik,| J. Phys. Chem. A **115**, 9919 (2011)

73. G. Kothe, M. Lukaschek, G. Link, S. Kacprzak, B. Illarionov, M. Fischer, W. Eisenreich, A. Bacher, S. Weber, J. Phys. Chem. B **118**, 11622 (2014)
74. O.B. Morozova, A.V. Yurkovskaya, J. Phys. Chem. B **119**, 140 (2015)
75. K. Ray, Ananthavel, S.P.D.H. Waldeck, R. Naaman, Science **283**, 814 (1999)
76. R. Naaman, D.H. Waldeck, J. Phys. Chem. Lett. **3**, 2178 (2012)
77. W.J.M Naber, S. Faez, W.G. van derWiel, J. Phys. D: Appl. Phys. **40**, R205 (2007)
78. I. Bergenti, V. Dediu, M. Prezioso, A. Riminucci, Organic spintronics. Philos. Trans. R. Soc. Lond. Ser. A **369**, 3054 (2011)
79. M. Urdampilleta, S. Klyatskaya, J.-P. Cleuziou, M. Ruben, W. Wernsdorfer, Nat. Mater. **10**, 502 (2011)
80. S. Sanvito, Nat. Mater. **10**, 484–485 (2011)
81. J.A. Weil, J.R. Bolton, *Electron Paramagnetic Resonance: Elementary Theory and Practical Applications*, 2nd edn. (Wiley, 2007)
82. L.T. Kuhn, Top. Curr. Chem. **276**, 125 (2007)
83. M.V. Fedin, E.G. Bagryanskaya, H. Matsuoka, S. Yamauchi, L. Veber, K. Yu Maryunina, E. V. Tretyakov, V.I. Ovcharenko, R.Z. Sagdeev, J. Am. Chem. Soc. **134**, 16319 (2012)
84. O.B Morozova, R. Kaptein, Yurkovskaya, J. Phys. Chem. B **116**, 8058 (2012)
85. G.J. Janssen, E. Roy, J. Matysik, A, Alia, Appln. Magn. Reson. **42**, 57 (2012)
86. K.L. Ivanov, R.Z. Sagdeev, Dokl. Phys. Chem. **409**, 221 (2006)
87. A. Buchachenko, *Magneto-Biology and Medicine* (Nova Biomedical, New York, 2015)
88. W.R. Hagen, *Biomolecular EPR Spectroscopy* (CRC, 2008)
89. B.C. Gilbert, M.J. Davies, K.A. McLauchlan (eds.), *Electron Paramagnetic Resonance*, a *Specialist Periodical Report*, vol. 17 (Royal Society of Chemistry, 2010)
90. G.M. Smith, K. Keeble, O Schiemann (eds.), *Introduction to Modern Epr Spectroscopy* (CRC, 2010)
91. A.L. Buchachenko, J. Phys. Chem. B **117**, 2231 (2013)
92. R. Kaptein, Photo-CIDNP studies of proteins. Biol. Magn. Res. **4**, 145–191 (1982)

Chapter 5
Experimental Methods of Investigation of Electron Spin Interactions Based on ESR Phenomena: Continuous Wave EPR Measurements

Abstract The method of electron spin resonance (ESR, Sect. 3.1) has proven to be the leading approach in the investigation of electron spin interactions in chemistry, biology and physics. Structural and dynamic properties of molecular objects of interest can be measured and characterized by suitable parameters through an arsenal of experimental ESR methods including conventional Continuous wave (CW ESR), electrically (ED ESR) and optically (OD ESR) detected ESR and advanced pulse techniques. The CW technique is used for collecting information of interest through the analysis of ESR spectra or saturation curves, obtained from electron-electron double resonance (ELDOR) or double electron-electron resonance (DEER), electron-nuclear double resonance (ENDOR) or multifrequency ESR (MF ESR) techniques. OD ESR makes it possible to acquire good ESR, ELDOR and ENDOR spectra using the simple and very sensitive luminescence technique. Application of such methods as two-dimensional ESR (2D ESR), (fried) two-dimensional electron-electron double resonance (2D-ELDOR), ENDOR with circularly polarized radiofrequency fields (CP-ENDOR), electron-nuclear-nuclear resonance (double ENDOR), proton-electron double resonance imaging (PEDRI), and electron-nuclear-nuclear triple resonance (TRIPLE) can be more informative than simple CW ESR experiments for appropriate systems. Methods of reaction yield detection magnetic resonance (RYDMR) and magnetically affected reaction yield (MARY) can also be used in the context of CW and pulse techniques. This chapter briefly describes theoretical grounds and typical application of the various methods of modern CW ESR spectroscopy, which continue to evolve synergetic and at a rather rapid pace.

5.1 Introduction

The main ESR features are: (1) Integral intensity and amplitude of the ESR spectrum; (2) The position of the spectral features depending on the value of the g-factor according to the resonance condition (3.5), which is a quantitative characteristic of the magnetic moment of the paramagnetic species under investigation; (3) The ESR

© Springer International Publishing Switzerland 2016
G. Likhtenshtein, *Electron Spin Interactions in Chemistry and Biology*,
Biological and Medical Physics, Biomedical Engineering,
DOI 10.1007/978-3-319-33927-6_5

spectral line shape which can be either homogenous (single spin packet) or heterogeneous (overlap of packets); (4) Spin lattice (T_1) and spin-spin (T_2) relaxation times (Sect. 3.2); (5) The ESR spectra hyperfine splitting (hfs) attributed to the contact (3.35) and dipolar (3.36) interactions of electron spin with nuclear spin I; (6) The ESR spectral fine splitting (fs) caused by electron spin electron spin exchange (Sect. 1.2) and dipolar (Sect. 3.3) interactions; (7) The response of the ESR signal to progressive saturation; (8) The degree of electron spin polarization. Measurement of these features in combination with other chemical, physico-chemical and biochemical methods provides a powerful tool for the detailed investigation of the structure and dynamics of molecular objects including biological systems of high complexity. An extensive bibliography covering the fundamentals and various theoretical and experimental aspects of ESR technique including recent advances is available [1–14].

Continuous Wave ESR (CW ESR) is an experimental technique during which a sample is continuously illuminated with microwave radiation at a fixed frequency and the strength of the applied magnetic field is swept, while observing changes in the microwave absorption. Usually, for technical reasons, the collected spectrum is plotted in an X-Y format where the X-axis displays the magnetic field strength in Gauss and the Y-axis displays the first derivative of the absorption in arbitrary units (Fig. 5.1).

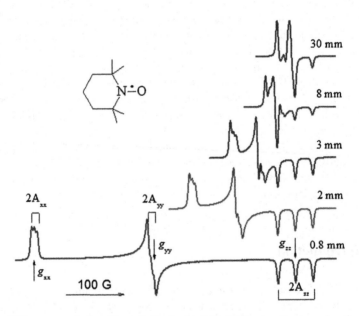

Fig. 5.1 EPR spectra of nitroxide radical calculated for different spin precession frequencies (microwave linewidths) [14]. http://hf-epr.awardspace.us/index.htm

5.2 Spin Electron Relaxation

5.2.1 Determination of Spin-Spin Relaxation Rates from the Form of the ESR Spectrum

Spin-spin exchange and dipolar interactions directly affect the spin lattice (longitudinal) $1/T_{1e}$ and spin-spin (transfer) $1/T_{2e}$ relaxation rates (Sects. 3.2 and 3.3.2) [15–21].

For the simplest homogeneous ESR spectrum (spin packet)

$$\frac{1}{T_{2e}} = \frac{4\pi\Delta H_{1_2}}{\sqrt{3}} \tag{5.1}$$

where $\Delta H_{1/2}$ is the spectral line width at half the maximum amplitude. Commonly, ESR spectra are heterogeneous, e.g. are composed of many overlapped spin packets. The CW ESR spectrum can be thought of as a sum of Lorentzian lines centered at a distribution of resonant field values. Deconvolution of such heterogeneous spectra allows one to evaluate the spin packet parameters and, therefore, calculate $1/T_{2e}$.

New methods of deconvolving ESR spectra have been developed that are based upon on a generic algorithm at the first step and the Marquard-Levenberg algorithm at the second step [15]. Other methods include global analysis using spectra obtained at multiple frequencies subject to constraints [16, 17].

5.2.2 Determination of Spin-Lattice Relaxation Rates from ESR Spectral Saturated Curves

The ability of ESR signals to become saturated in an applied microwave MW field is determined by the value of the MW amplitude H_1 and the spin relaxation parameters (T_{1e} and T_{2e}). The dependence of the ESR absorption signal (A) on H_1 is described by the formula for a saturated line shape:

$$A = \frac{a_2 T_{2e} H_1}{1 + b_2 T_{2e}^2 + c_2 H_1^2 T_{2e} T_{1e}} \tag{5.2}$$

where a_2, b_2 and c_2 the included magnetic parameters of paramagnetic center If (5.2) is applied to a spin packet line with no nuclear hyperfine splitting, then values of T_{1e} and T_{2e} could be obtained.

Portis [18] and Castner [19] in their pioneering works showed that measurement of the saturation factor of paramagnetic centers can give spin-lattice and spin-spin relaxation times. Special algorithms have been developed by the Lebedev and Muromtsev group [20] to analyze the saturation curves of non-homogeneous ESR spectra. The method can be effective for the determination of very short relaxation times $T_{1,2} < 10^{-9}$ s [21–23] Based on the theories of Portis and of Castner, a non-linear least-squares procedure for analyzing the microwave-power dependence of inhomogeneously broadened lines using all data points on a saturation curve has been developed [24]. The extreme sensitivity of the saturation curves to week dipolar and exchange interactions is due to their dependence on the spin relaxation rate $(1/T_{1e})$ which is commonly significantly slower than the spin-spin relaxation rate $(1/T_{2e})$ Therefore, as was first demonstrated by the Likhtenshtein and Kulikov groups [25, 26] measurement of $(1/T_{1e})$ and the saturation curves parameters permits to determine long distance between paramagnetic centers and slow spin exchange rate.

- The first attempts to employ CW ESR techniques for the investigation of location of radicals in macromolecules on the base of measurement of spin relaxation rate $(1/T_{1e})$ were made in the early 1970s [21, 22, 26]. The determination of the distance of closest approach of a radical and a paramagnetic species (usually a transition metal ion complex) was performed by examining the effect of the dipole-dipole interaction between the radical and transition metal ion distributed uniformly in a vitrified matrix on the ESR saturation curves of the radicals [23] and references therein A similar approach based on examination of the kinetics of spin-lattice relaxation for paramagnetics in solids was developed by Bowman and Norris [27].
- The observed relaxation rate enhancement for a radical, as a consequence of dipole-dipole interaction with a paramagnetic complex that is distributed uniformly in the vitrified matrix, consists of a combination of multiple individual curves as indicated by (5.3) [23].

$$W = \Delta(1/T_{1s}) = 1/T_{1s} - 1/T_{1s}^{0} = \sum w_{rp} \qquad (5.3)$$

where w_{rp} is the relaxation rate induced by dipole-dipole interaction for an individual radical-paramagnetic ion pair, $1/T_{1s}^{0}$ and $1/T_{1s}$ are spin-lattice relaxation rates of the radical. Using this model, algorithms for estimating the depth of radical immersion have been developed for low and high concentrations of the transition metal ion complex. This method was applied to the determination of the immersion depth of nitroxide segments of spin probes in biological membranes and enzymes [23] and references therein.

Pulse methods of detection of the relaxation rate are briefly described in details [3, 9].

5.3 Exchange Interaction

Exchange interactions (Sect. 1.1, (1.6–1.12)) can be studied via CW-ESR spectra [3, 9, 14, 27–29]. In the simplest case of interaction of two identical paramagnetic centers in the ferromagnetic (parallel spins, triplet) state with S = 1/2 and exchange integral $J \ll k_B T$, the position and total line intensity of the ESR spectrum does not depends on J (Fig. 5.2). When J and $k_B T$ becomes comparable the line intensity increases because of partial transformation of the ferromagnetic state to the anti-ferromagnetic (antiparallel, singlet) state. The T → S transformation is complete when $J \gg k_B T$. For two paramagnetic centers with different spins S_1 and S_2, and different resonance energy (Δv) the position and intensity of the ESR spectrum depend on the relation of J and Δυ. When Δv > J, each level is split by an amount approximately equal to J. Increasing J at fixed Δυ to Δv ≈ J, leads to a diminishing intensity of the extreme external line component and its separation, and an increasing intensity and closer approach of the internal components. At Δv ≪ J, the external pair component disappears, and internal components merge transforming the ESR spectrum to a singlet. Similar trends occur for radical spectra with hyperfine structure. For example, in the triplet signal of nitroxides in solution, the lines broaden at an exchange frequency of about $v = 10^7$ s^{-1}, and at $v = 10^8$ s^{-1} the lines merges to form a broad singlet. Further increase of the exchange frequency leads to narrow singlet.

Fig. 5.2 Position and intensities of the ESR lines of two spins, S1 and S2, upon their approach as functions of the magnitude of the exchange integral (J), energy splitting between the lines (ΔE_v) and thermal energy $k_B T$: **a** $\Delta E_v \neq 0$. **b** $\Delta E_v = 0$ [14]

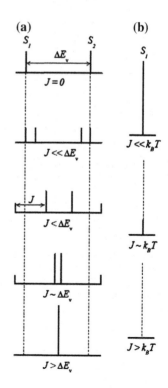

Values of the exchange integral can be derived from dynamic exchange interactions in solution [14, 29–33] (Sect. 1.2.3, (1.27–1.30)) and on static and dynamic exchange in biradicals [34]. Analysis of experimental spectra of various paramagnetic species allows one to evaluate the exchange integral value [3, 9, 14, 35–37].

As an example, Fig. 5.3 shows results of quantitative analysis of spin exchange interactions between spin labels attached to amyloid fibrils formed by the Ure2 prion domain [36].

Values of exchange integrals can be also derived from data on dynamic spin exchange in solution (Sect. 1.2.3).

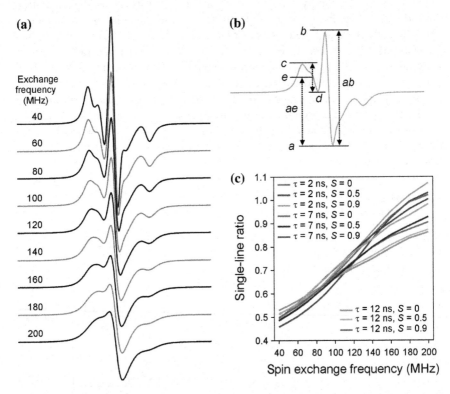

Fig. 5.3 Quantitative analysis of spin exchange interactions with single-line ratio. **a** Simulated EPR spectra with various spin exchange frequencies. Anisotropic motion with correlation time (τ) of 7 ns and order parameter (S) of 0.5 was used for these spectra. **b** Calculation of single-line ratio from the EPR spectrum. Point e is the mid-point between c and d. Single-line ratio = ae/ab. **c** Plot of single-line ratio versus spin exchange frequency for a range of motional parameters as indicated [36]

5.4 Multiquantum ESR Spectroscopy

In multiquantum ESR (MQ-ESR) spectroscopy, two or more microwave frequencies derived from a common oscillator are incident on the sample [38, 39]. The separation of the frequencies should be much less than the homogeneous linewidth. Successive absorption and emission of photons from both fields by the spin system leads to oscillation of the spin population at k (ω1 − ω2) and to the appearance of intermodulation sidebands at $\omega_0 \pm (k + 1/2) (\omega_1 - \omega_2)$, where ω_0 is the average frequency and k is a positive integer. These new microwave frequencies produced by the spin system are detected. Benefits of MQ-ESRare: (1) detection of pure absorption lines, rather than derivative-like shapes, (2) spectral intensities proportional to T_1, and (3) reduced linewidths. It was shown that the signal intensity of 3-quantum spectra is proportional to the spin lattice relaxation time T_1, while its linewidth in a frequency difference sweep is $1/T_1$. An example of multiquantum EPR spectra is shown in Fig. 5.4

Three- and five-quantum absorption and dispersion multiquantum ESR spectra of a spin-labeled arrestin have been obtained at Q-band (35 GHz). Enhanced sensitivity to T_1 and T_2 and to lineshape variation relative to absorption was also evident in the spectra. The authors concluded that the feasibility and the practicality of the use of the MQ Q-band technique have been demonstrated on a biologically relevant protein at realistic volumes and concentrations and that MQ EPR will be a valuable technique to study protein structure and dynamics.

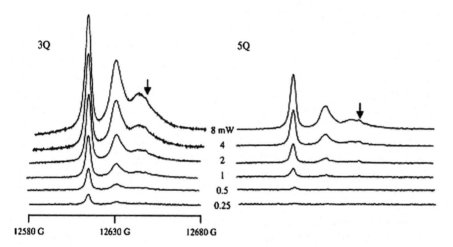

Fig. 5.4 Multiquantum EPR spectra of arrestin K267C-MTSL obtained at Q-band. 3Q and 5Q spectra were recorded of spin-labeled arrestin at varying microwave powers. The protein concentration of these 30-nL single scan samples was 300 μM, with a 100-G scan width, 120-s scan time, 0.2-s time constant, and $\Delta f = 10$ kHz. Powers indicated are per arm and the spectra were recorded in air. The *arrows* indicate the second motional component, which is more evident in the 5Q spectra [39]

5.5   Electron Spin Electron Spin Electron Dipolar Interaction

The dipole-dipole interaction of two spins with fixed distance (r) between them, assuming that both have isotropic g-factors (or gyromagnetic ratios), will depend on the distance and on the angle θ between the direction of the magnetic field and the line connecting the given pair of spins (Sect. 3.3.2.1, (3.11–3.14)). The interaction manifests as a dipole splitting of the ESR spectrum line (the Pake doublet, (3.13.14), Fig. 3.2). When there is a distribution of orientations of the spins, the ESR spectrum will be inhomogeneous. Section 3.3.2.2 describes theories of spin-spin anisotropic interactions. Evaluation of the parameters characterizing dipolar interaction in the Pake doublet such as spin-spin distances, angle θ, the angles defining the orientation of paramagnetic species 1, and three angles relating the orientation of paramagnetic 2 to paramagnetic 1 has been published in the literature [40–44]. The elements of the various tensors for two nitroxide radicals that are specified by Euler angles is shown in Fig. 3.3.

5.6   Hyperfine Interaction in ESR Spectra

5.6.1   Hyperfine Interaction Parameters in CW-ESR Spectra

The hyperfine interaction between the electron and the nuclear spins consists of the isotropic Fermi contact interaction and the anisotropic dipole-dipole interaction (Sect. 3.41, (3.34–3.38)). The hyperfine interaction is manifested by characteristic splittings of ESR spectra. For example, the ESR spectra of the rapidly tumbling radical with ^{14}N nuclei (spin $I_n = 1$) are characterized by three components with corresponding quantum members of +1, 0, and −1 and the isotropic hyperfine splitting constant a_n ((3.38), Figs. 5.1, 5.7). Determination of a_n allows one to estimate the spin density on the given atom.

For asymmetric molecules, the dipolar interaction depends on the electron spin electron nuclei distances and on their orientations, which links interacting spins referred to the direction of the applied magnetic field (3.36). This interaction can be described by a matrix of parameters A_{xx}, A_{yy} and A_{zz}, corresponding to the principal axes of the radical. The appearance of the spectrum will change depending on the orientation of the molecule relative to the magnetic field (Fig. 5.5).

The ESR spectrum of the immobilized radical is composed of a great number of microscopic signals corresponding to various orientation (Figs. 5.1 and 5.5). The values of the magnetic parameters, the g-factors and the hyperfine splitting constants can be evaluated by analyzing the position and shape of the ESR signal lines. Such an analysis is particularly effective by using multifrequency ESR spectroscopy (Fig. 5.1).

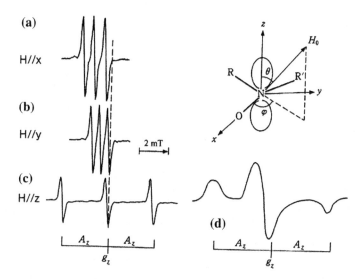

Fig. 5.5 ESR spectra of di-t-butyl nitroxide in a diamagnetic matrix depending on **a–c** the magnetic field direction and **d** it powder pattern. The insert shows the principle axis system in the N–O bond [4]

According to the electron magnetic resonance condition (3.4), the appearance of an ESR spectrum is strongly dependent on the resonance frequency and magnetic field. At conventional X-band ESR spectroscopy (9.5 GHz, 3 cm), small g-factor differences of different paramagnetic species and small g-factor anisotropies of anisotropic paramagnetics lead to strongly overlapping ESR lines. Based on the pioneering works of professor Y.S. Lebedev and his colleagues [45, 46] who developed high-field high frequency, high resolution 148 GHz—2 mm ESR technique, spectral resolution has been remarkably improved (Fig. 5.1) [47–49]. In high-frequency ESR spectroscopy strong magnetic fields (up to 9 T), based on cryogenically cooled superconducting solenoids and standard microwave technology at high frequencies are used. Significant progress in HFHF ESR has been achieved with the use of millimeter-wave quasi-optics techniques, permitting, e.g., the construction of a 9-T, 250-GHz (1.2 mm) spectrometer (Fig. 5.1) [50, 51].

Figures 5.1 and 5.6 demonstrate advantages of the 2 mm ESR spectrometer in the separation of spin-probe ESR spectra over the traditional X-band 30 mm *ESR* spectrometer. Temperature dependences of g- and A-tensors of the 2-mm band ESR spectrum of spin-labeled lysozyme are presented in Fig. 5.6 [14]. As is seen in the figure, the 2 mm ESR spectroscopy not only determined the spectral parameters but also followed their temperature dependencies in a wide range of temperatures.

The high-frequency technique significantly enhances the resolution of CW-ESR spectroscopy, and HFHF spectra clearly show anisotropic g-factors and hfs position and values (Fig. 5.1), particularly for those systems where g-strain is not an issue.

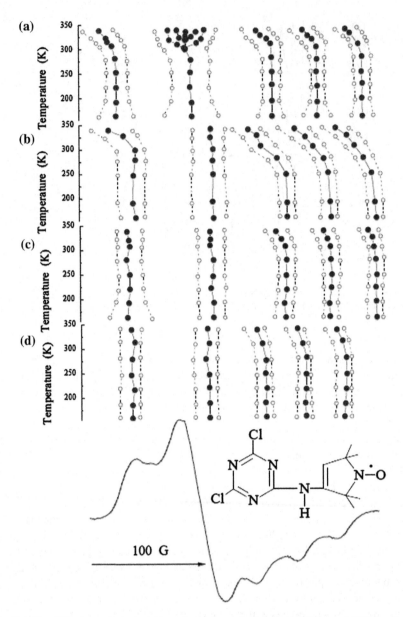

Fig. 5.6 The EPR spectra of nitroxide radical in egg lysozyme at 0.04 **a** and 0.96 **b** relative humidity recorded at the 2-mm wave band and 130 K and the plot of the position (*solid lines*) and half-width (*dashed lines*) of its EPR spectral components versus temperature [14]

Since the 1970s highfrequency ESR spectroscopy has been successfully applied to the investigation of a number of chemical systems [8, 13, 48–51] and references therein.

5.6.2   Electron Spin Nuclear Spin Interaction and Dynamical Phenomena

Radical motion (rotation, tumbling, libration, wobbling) can partially or completely average the dipolar interaction depending on frequency, anisotropy and mechanism of the motion. Combination of multifrequency radiospectroscopy with other MW techniques allows one to characterize the molecular motion of a nitroxide over a wide range of correlation times ($\tau_c = 10^6$–10^{-10} s). The stochastic Liouville equation is commonly used for describing the effect of motion on the line shape of ESR spectra [52, 53].

For X-band (30 mm) spectra, one can estimate rotational correlation times in the region of the fast motion ($\tau_c = 5 \times 10^{-9}$–$10^{-10}$ s) via the following equation [52]

$$\frac{1}{\tau_c} = \frac{3.6 \times 10^9}{\left(\sqrt{\frac{h_0}{h_{-1}}} - 1\right)\Delta H_0} \tag{5.4}$$

where h_0, h_{-1} are the heights of the ESR spectrum hyperfine components, respectively, and ΔH_0 is the line width of the middle hyperfine component. The detailed theory of nitroxide ESR spectra for nitroxide motion in the slow motion region ($\tau_c = 10^{-7}$–10^{-8} s) was developed by Freed [53] using the stochastic Liouville equation. In the slow motion region the following equation can be used at X-band (30 mm):

$$\tau_c = ax\left(1 - \frac{A_{zz}}{A_{zz}^0}\right)^b \tag{5.5}$$

where A_{zz}^0 and A_{zz} are the z-components of A-tensor for immobilized and mobile nitroxide, respectively. The coefficient a was found to be 5.4×10^{-10} and 2.6×10^{-10} s for systems modeling isotropic and anisotropic Brownian diffusion, respectively, and b was found to be -1.36 and -1.39 for the respective models. Recently, progress has been made for simulating ESR spectra directly from molecular dynamics simulations of a spin-labeled protein [54–57]. In order to perform such computations it is necessary to compute a large number (hundreds or thousands) of relatively long (hundreds of nanoseconds and subnanosecond) trajectories. For example, from analysis of these trajectories, accurate stochastic models of the spin label dynamics were constructed [54] and references therein.

The advantages of the multifrequency ESR technique for investigations of the mobility of stable radicals were demonstrated by Freed group [55–57]. It was shown that rapid subnanosecond dynamic modes of internal fluctuations were best resolved at higher frequencies while nanosecond or longer global tumbling, large bending motions, and nanosecond conformational distortions would appear to be frozen at frequencies as high as 236.6 GHz but could affect the low-frequency (9.5 GHz) spectra. The increased sensitivity to the g tensor allows high-frequency EPR to distinguish rapid rotations about the gx, gy, or gz tensor directions so that at high frequencies one can resolve from the spectrum the nature of anisotropic motions (Fig. 5.7).

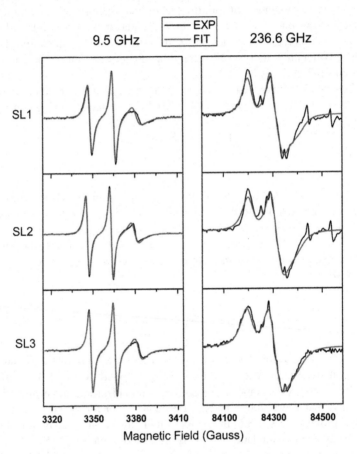

Fig. 5.7 Combined simulations of 9.5 and 236.6 GHz spectra to provide evidence of the anisotropic diffusion tensor of the probe provided through the parameters for different spin labels *SL1*, *SL2*, and *SL3*. See details in [57]

In work [57], EPR at 236.6 and 9.5 GHz was used for probing the tumbling of nitroxide spin probes in the lower stem, in the upper loop, and near the bulge of mini c TAR DNA (Fig. 5.7). High-frequency 236.6 GHz EPR, was found to be sensitive to fast, anisotropic, hindered local rotational motion of the spin probe, occurring approximately about the NO nitroxide axis.

The effect of saturation transfer between the component of the ESR spectra of nitroxide radicals provide the basis for an approach to the study of very slow motion, with correlation time $\tau_c = 10^{-2}-10^{-6}$ s [58–60]. Saturation transfer affects the dependence of the second harmonic dispersion signal (V_2) on the field strength. Saturation transfer electron paramagnetic resonance (ST-EPR) spectroscopy has been employed to characterize the very slow microsecond to millisecond rotational dynamics of a wide range of nitroxide spin-labeled proteins and other macromolecules in the past three decades [61, 62] and references therein. Work in this field can be carried out on spectrometers that operate at X-band (~ 9 GHz), Q-band (~ 34 GHz), K-band (22.0 GHz) and 94–250 GHz microwave frequencies. Techniques for studying aqueous samples, overcoming the significant absorption losses that are common in such samples, are now available at the higher frequencies as well. In the absence of electron spin-spin interactions the features of saturation transfer electron paramagnetic resonance spectra depend on the electron spin-lattice relaxation time (T_{1e}) and nuclear spin-lattice relaxation time (T_{1n}). The relaxation processes compete with saturation effects, and rotational motion transfers saturation effects. Thus characteristic times of saturation transfer should be comparable to (T_{1e}), (T_{1n}) and the correlation time of rotational motion. For example, for nitroxides the value of T_{1e} at room temperature and in nonpolar solvents ranges from 3 to 5 ms.

Saturation transfer is typically studied by recording the spectrum of the second harmonic dispersion (V_2') [58–62]. The influence of frequency on V_2' ST-EPR spectra calculated using the uniaxial rotational diffusion (URD) model is shown in Fig. 5.8 [61, 62] The spectra show the range of lineshape changes that are predicted to occur as a function of the correlation time for URD for $\tau_\parallel$ from 1 μs to 100 ms at microwave frequencies of 9.8, 94, 140, and 250 GHz. According to the calculation, the magnitude of the overall change in lineshape increases monotonically as a function of increasing microwave frequency and there is increasing sensitivity to longer correlation times as the microwave frequency is increased.

Lazarev and Lebedev [63] have developed a procedure to measure superslow ($\tau_c = 10^{-2}-10^6$ s) reorientation of radicals. In their procedure, the intense polarized radiation destroys radicals whose electron transition moment coincides with the polarized vector of the incident light. The light thus causes a dip in the ESR signal intensity. The time of recovery of the dip corresponds approximately to the rotational correlation time.

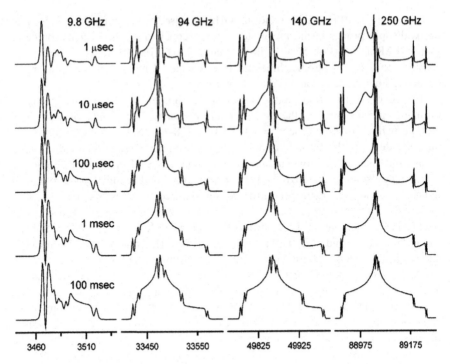

Fig. 5.8 Calculated V_2' signals at 9.8, 94, 140, and 250 GHz (see details in [61])

5.7 CW Double Resonance Techniques

5.7.1 *Electron Nuclear Double Resonance*

Electron nuclear double resonance (ENDOR) was introduced by George Feher [64]. In a steady state ENDOR experiment, an EPR transition is partly saturated by microwave radiation of amplitude B_1 (Fig. 5.9) [65–70]. A driving radio frequency (rf) field of amplitude B_2 induces nuclear transitions. While the magnetic field is swept through the EPR spectrum, the rf frequency follows the Zeeman frequency of the nucleus. Transitions occur at frequencies v_1 and v_2 and obey the NMR selection rules $\Delta M_I = \pm 1$ and $\Delta M_S = 0$. These NMR transitions are detected by ENDOR via the intensity changes to the simultaneously irradiated EPR transition. The ENDOR spectrum can reveal both the hyperfine coupling constant (a) and the nuclear Larmor frequencies (v_n)

$$v_1 = |v_n - a/2|$$
$$v_2 = |v_n + a/2| \tag{5.6}$$

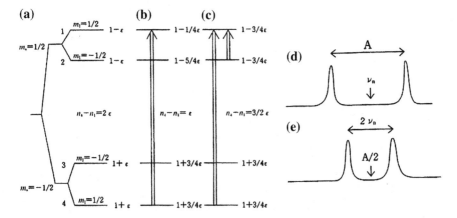

Fig. 5.9 ENDOR sublevels for S = 1/2 and I = 1/2. **a** At thermal equilibrium. **b** After ESR transition 4 → 1. **c** Added by NMR transition. The population difference for $n_4 - n_1 = 2\varepsilon$ (**a**) and $n_4 - n_1 = \varepsilon$ (**b**) [4]

The Hamiltonian for the electron spin nuclear spin system can be written as

$$\mathcal{H}_0 = \mathcal{H}_{EZ} + \mathcal{H}_{NZ} + \mathcal{H}_{HFS} + \mathcal{H}_Q \qquad (5.7)$$

The four terms in this equation describe the electron Zeeman interaction, the nuclear Zeeman interaction, the hyperfine interaction (hfs), and the nuclear quadrupole interaction (Q) respectively. The nuclear quadrupole interaction is present only in nuclei with I > 1/2. Thus, ENDOR spectra contain information on the type of nuclei in the vicinity of the unpaired electron, on the distances between nuclei, on the spin density distribution, and on the electric field gradient at the nuclei (Q).

An example of effective application of Q-band ^{1}H and ^{14}N Mims ENDOR, in combination with the transient EPR, magnetophotoselection experiments, ^{14}N ESEEM, ^{14}N HYSCORE, and density functional theory (DFT) calculations was presented in excellent work [68]. Q-band ENDOR spectra of a linear butadiyne-linked porphyrin dimer are shown in Fig. 5.10. These data prove the delocalization of the triplet state over both porphyrin units. The orientations of the zero-field splitting tensors with respect to the molecular frame in both porphyrin monomer and dimer have been also established.

5.7.2   Continuous Wave Electron–Electron Double Resonance

The phenomenon of continuous wave electron–electron double resonance (CW ELDOR) was independently discovered in 1968 by Hyde and Freed [71], and

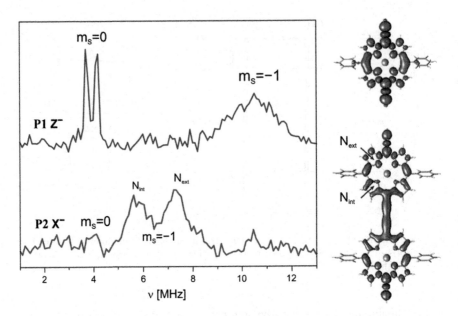

Fig. 5.10 Q-band Davies ENDOR spectrum for P1 and Mims ENDOR spectrum for P2 recorded at the Z − (1233.2 mT) and X − (1235.0 mT) field positions, respectively [68]

Bendersky and Blumenfeld [72] groups. In the CW ELDOR technique, one part of the EPR spectrum of a paramagnetic sample is irradiated with an intense saturating microwave pump field and the effect of this intense field on other parts of the spectrum is assessed by a second weak microwave probe field (Fig. 5.11). When the two frequencies are separated either by an integral number of hyperfine energy differences or by a g-factor difference of two or more paramagnetic species, the EPR signals detected by the weak microwave probe field is reduced in intensity due to saturation transfer effects. In some cases this reduction is as much as 40 %.

The following interactions can be identified by CW ELDOR techniques: electron–nuclear dipolar and contact interaction, electron spin electron spin dipolar and exchange interactions. From the 1970s through the 1990s these mechanisms were intensively investigated in liquid and frozen solution [72], in biological membranes [73], and in various systems [74]. Some technical aspects of the CW ELDOR spectroscopy are discussed in the literature [75].

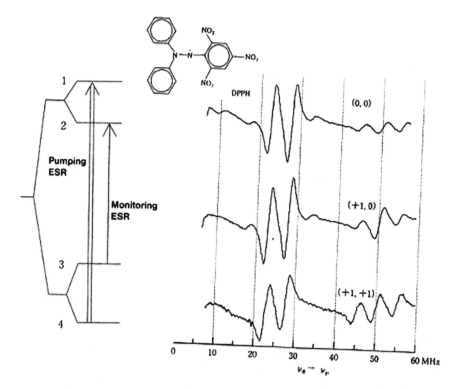

Fig. 5.11 ELDOR sublevels of pumping and monitoring ESR (S = 1/2 and I = 1/2) [4]

5.8 Methods Based on the Effect of an Applied External Magnetic Field

5.8.1 General

There are processes, for example, the recombination of a singlet radical or ion-radical pairs, which are affected by the probability of singlet-triplet conversion in an applied external magnetic field (Fig. 5.12). In the absence of a magnetic field ($H_0 = 0$), three degenerate triplet levels are generated. When a magnetic field is applied, the level T splits into three levels: T_{+1}, T_0 and T_{-1} (Fig. 9.3. Sect. 9.3). The singlet-triplet characteristic conversion time depends on the difference in g-factors of the radicals (Δg), the a magnetic field strength and energy of electron nuclear contact interaction (hyperfine constant a) [76]

$$1/\tau_{ST} = [\Delta g(H_0 \pm a)] \qquad (5.8)$$

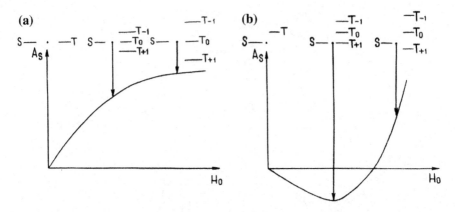

Fig. 5.12 The effect of the field strength (H_0) on the population (A_S) at the singlet level, which depends on the rate of the S → T transition (k_{ST}) for a radical pair: **a** without exchange interaction (J = 0) and **b** with exchange interaction (J > 0). The *vertical arrows* indicate qualitatively the correspondence between the triplet splitting and the associated behavior of the static magnetic field (H_0) [76]

The magnetic isotopic effect manifests itself through the quantity a.

Figure 5.12 illustrates the effect of the external magnetic field strength (H_0) on a process, the rate of which is proportional to the population of the singlet level of a radical pair. According to the principle of spin conservation, the transition $S_0 \rightarrow T_0$ is favored compared to transitions to other triplet levels. As the field strength increases, the T_+ and T_- levels are progressively displaced from the T_0 level and, therefore, their influence on the overall rate becomes progressively less, while the relative population of the T_0 level and, therefore, the S levels increase and the rate of recombination becomes faster. In the high field limit, the transition $S \rightarrow T_+$, T_- are insignificant and the recombination rate becomes insensitive to any further increase in H_0.

When strong exchange is operative, in the absence of an applied magnetic field, the energy level of the system splits into a singlet and triplet levels with a difference equal to twice the value of the exchange integral (J). When a magnetic field is applied, the T level is split. As the strength of the magnetic field is increased, the T_+ level becomes closer to the S level and the probability of an $S \rightarrow T_+$ transition increases, hence, the rate of recombination from the level S decreases. Further increase of the magnetic field leads to coincidence of the S and T_+ levels, which provides a mechanism for a recombination bottleneck. Further increase of the magnetic field leads to increased splitting of the S and T_+ levels, which enhances the recombination rate. Spin orbital and superfine interaction can also affect on the magnetic field effect.

5.8.2   Optically Detected ESR

Microwave-optical double resonance (MODR) was pioneered by Field et al. [77]. Converting the resonance absorption of microwave photons into a change of optical signal observed either by laser-induced fluorescence or by laser-polarization has enhanced the sensitivity by several orders of magnitude. The optically detected (OD) EPR technique which involves the detection of spectra through the changes in luminescence intensity of the products of recombination of radical ion pairs was invented and implemented by the Molin and Salikhov group [78–80] and the Frankevich group [81] The principal advantage of OD EPR is the high sensitivity of OD EPR, exceeding that of conventional EPR by several orders of magnitude. This technique allows selective detection of the spectra of short-lived radical ions, characterized by lifetimes as short as tens of nanoseconds, in liquid solution at ambient temperatures.

The efficiency of OD EPR was clearly demonstrated on an example of investigation of short living spin-correlated radical ion pairs $D^{\cdot+}/A^{\cdot-}$ which are produced in the following reactions⁻

$$S \rightarrow S^{\cdot+} + e^-,$$
$$A + e^- \rightarrow A^{\cdot-},$$
$$D + S^{\cdot+} \rightarrow D^{\cdot+} + S, \qquad (5.9)$$

where S is a solvent molecule. The $D^{\cdot+}/A^{\cdot-}$ pair is formed in the singlet spin state, and hence its recombination gives singlet-excited molecules which can emit light:

$$D^{\cdot+} + A^{\cdot-} \rightarrow D^* + A(\text{or} D + A^*) \rightarrow D + A + h\nu. \qquad (5.10)$$

In a magnetic field of an OD EPR spectrometer, under resonance microwave (mw) pumping, the EPR transitions, defined by the hyperfine coupling and the difference in the g values of the radical ions, occur between the T_0 and T_+, T_- states of the pair. This leads to a decrease in the population of the singlet state of the pair and, as a consequence, to a drop in the fluorescence intensity, thus giving rise to the OD EPR signal. Under conditions far from saturation, the OD EPR spectrum is a superposition of EPR spectra of both recombining radical ions with equal integral intensity.

An example of OD ESR spectra observed under the photoionization of a polystyrene film is presented in Fig. 5.13 [82].

The theory of optically detected electron nuclear double resonance was developed in [80]. Information about instrumentation and application of OD ESR can be found in recent publication [82–84].

Fig. 5.13 **a** OD ESR signal
from VUV-irradiated film of
4-pentyl-40-cyanobiphenyl
(5CB) applied on a quartz
substrate at room temperature.
b OD ESR signal from an
X-irradiated solution of
8×10^{-3} M 5CB in squalene
with 2 M benzene added [82]

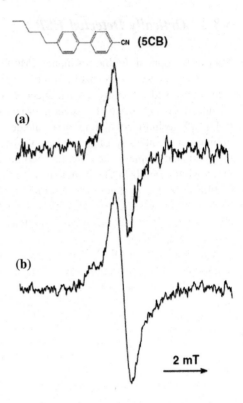

5.8.3   Reaction Yield Detected Magnetic Resonance

The idea of reaction yield-tested magnetic resonance (RYDMR) to register electron paramagnetic resonance spectra of paramagnetic species involved in spincorrelated chemical processes has been launched by Frankevich and coauthors [85]. The reaction yield detected magnetic resonance RYDMR method detects reacting radical pairs by monitoring the electron spin resonance of its component radicals [85–89]. The inversion of one of the electron spins in a radical pair while the other is left unperturbed induces a change in the total spin state of the radical pair and the resonance effect induces a change in product yields.. Several different variants of RYDMR methods including fluorescence-detected, transient absorption-detected, ESR-detected, and photoconductivity-detected methods have been reported.

The reaction yield detected magnetic resonance (RYDMR) experiments were performed on solutions of chrysene (Chr), pyrene (Py) or their perdeuterated analogues (1 mM) with 1,2-, 1,3- or 1,4-dicyanobenzene (DCB) (20 mM) in 1: 9 acetonitrile: cyclohexanol solvent mixture at 201 C [88]. For the fixed radiofrequency of 36 MHz with a static magnetic field of up to 4 mT RYDMR spectra were envisaged where the static field strength B_0 is varied (Fig. 5.14). The following consequence of processes were suggested after photoexcitation of Chr or Py:

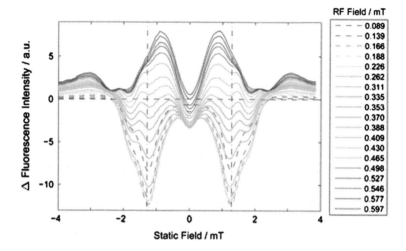

Fig. 5.14 Experimental RYDMR spectra of Py-d10/1,3-DCB for a 36 MHz oscillating field applied orthogonal to the static field. RYDMR-B_0 [88]

generation a singlet excited state which forms an exciplex with DCB, electron transfer to DCB which generates a radical ion pair in the singlet state, and back electron transfer and exciplex fluorescence for radicals that re-encounter in the singlet state.

5.8.4 Magnetically Field Effects on Reaction Yield

The dependence of reaction yield on the external magnetic field strength is called a magnetically affected reaction yield (MARY) spectrum [90–93]. Through this effect, the evolution of the overall spin in a spin correlated radical pair, Zeeman and hyperfine interactions may control the reaction yields in different reaction channels. MARY spectra of a radical pair reaction show a monotonic field dependence which saturates at high fields because the population of the radical pair singlet state practically does not depend on the magnetic field strength in this limit. MARY spectroscopy has proven to be an efficient method for studying radical ion reactions in liquid solutions in the presence of ionizing irradiation (see also Sect. 9.4, Fig. 9.7).

One object of MARY spectroscopy studies are resonance-like features occurring at the magnetic field effect curve at fields where the energy levels of the radical ion pair cross (so-called MARY lines) [90–93] (see also Section X) The position and shape of the lines bear information on both the hyperfine structure of the radical ions present in the system and on the rates of processes in which they are involved. J-resonance line splitting in the MARY spectrum for the radical ion/biradical ion three-spin system has been predicted in [91]. The method used in this work is based on affecting the radical stage of the reaction with an external magnetic field, which is further detected using the magnetic field dependence of the intensity of recombination fluorescence. For the latter case, it is sufficient to have magnetic nuclei only in the radical ion. The radical cation $(D^{+\cdot})$/ biradical anion $(A^{-\cdot} - R^{\cdot})$ three-spin system [91]

$$D^{+\cdot}...A^{-\cdot} - R^{\cdot}, \tag{5.11}$$

was studied. The presence of the exchange interaction gives rise to an additional line in the MARY spectrum in a field equal to the strength of the interaction J (J resonance). The origin of this line is the mixing of the singlet spin state S of the pair of spins and with one of their triplet states T in the magnetic field $B \approx J$. A theoretical description of spin evolution taking into account the exchange interaction for a three-spin system is based on the Hamiltonian:

$$\hat{H} = g\beta^{-1}B(S_{Dz} + S_{Az} + S_{Rz}) + JS_AS_R + \sum_{i=1}^{N_A} a_i S_A I_i^A$$
$$+ \sum_{k=1}^{N_D} a_k S_D I_k^D + \sum_{l=1}^{N_R} a_l S_R I_l^R. \tag{5.12}$$

Here, the external magnetic field B is aligned along the z axis, the g factors of all three electrons are taken to be equal, J is the strength of the exchange interaction, SD,A, R are spin operators, and $a_{i,k,l}$—are hyperfine coupling (hpc) constants for the three electron spins. MARY spectra with various choices for the parameter values in (5.12) were calculated. The shape of the J-resonance line in the spectra bears detailed information on hyperfine couplings in both the radical ion and the biradical ion partners of the pair, namely, the values of the HFC constants, their signs, and the lifetimes of the radical ion species. At magnetic fields where MARY spectra are saturated, it is possible to record the magnetic resonance spectra of the radical pairs by monitoring the chemical reaction yield as a function of microwave frequency at constant field or as a function of the magnetic field at constant microwave frequency.

Figure 5.15 illustrates the effects of degenerate electron exchange (DEE) in coherent spin phenomena present in short-lived radical ion pairs reactions on MARY spectra [94]. It was shown that the DEE process leads to relaxation-type changes in the spin dynamics of the ion radial pairs and to broadening of the MARY lines at zero field as a result of relaxation-type changes in the spin dynamics of the ion radial pairs.

Fig. 5.15 MARY spectra of radical ion pairs undergoing degenerate electron exchange DEE at different times τ_{dee} (see details in [94]

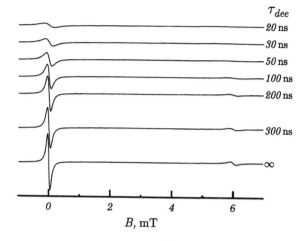

5.8.5   Time-Resolved Magnetic Field Effect

The method of time-resolved magnetic field effect (TR MFE) is based on the effect of dynamic transitions between the singlet and triplet states of a spin-correlated radical ion pair and on the yield of the singlet-excited (fluorescing) products of the pair recombination [95–98]. The TR MFE is defined as the ratio of recombination fluorescence kinetics measured with ($I_B(t)$) and without ($I_0(t)$) an applied magnetic field. This ratio is expressed as follows [94]:

$$\frac{I_B(t)}{I_0(t)} = \frac{\theta \rho_{ss}^B(t) + \frac{1}{4}(1 - \theta)}{\theta \rho_{ss}^0(t) + \frac{1}{4}(1 - \theta)} \tag{5.13}$$

where θ is the fraction of pairs formed in the spin-correlated singlet state, $\rho_{ss}(t)$ is a function describing the evolution of the population of the singlet state of such pairs, and indices "B" and "0" denote the values in a high and zero magnetic field, respectively. The evolution of the spin pair population was interpreted in terms of the ESR spectral features arising from the interplay of hfc constants, g-factor, and paramagnetic relaxation times. These features can be interpreted either from simulation of the TR MFE or analysis of experimental data.

Radical cations of heptane and octane isomers, as well as several longer branched alkanes, were detected in irradiated n-hexane solutions at room temperature by TR MFE [96]. (DFT) approach was used for analytical solution in the case of hfc with two groups of equivalent magnetic nuclei based on the density functional theory was used for identification and describing radical cations. In the case of hfc with two groups of equivalent magnetic nuclei, the contributions of the 2,2,3-trimethylbutane and 2,2-dimethylpentane radical cations to the spin dynamics of the radical ion pair was described by analytical solution. The TR MFE curves

Fig. 5.16 Experimental and calculated ratio of fluorescence intensities in high ($I_B(t)$) and zero ($I_0(t)$) magnetic fields recorded at room temperature for irradiated n-hexane solutions of 0.1 M 2,2,3-trimethylbutane in magnetic fields of 0.1 T (*curve 1*) and 1.1 T (*curve 2*) [96]

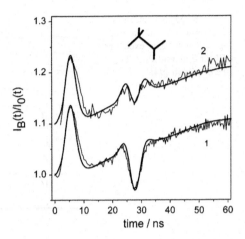

assigned to the dynamic singlet–triplet transitions were observed on the time scale of the phase relaxation time $T_2 < 60$ ns and $T_1 > 500$ ns. The experimental and calculated ratio of fluorescence intensities in high ($I_B(t)$) and zero ($I_0(t)$) magnetic fields is presented in Fig. 5.16.

The paper [97] demonstrated the utility of TR MFE using the model system hexamethylethane radical cation whose EPR spectrum in solution is determined by an isotropic hfc with 18 equivalent protons. The hfc constant was determined from the position of peaks of the time-resolved magnetic field effect of the hexamethylethane and perdeuterated p-terphenyl solution in n-hexane at ambient temperature. The difference in the g-values of (hexamethylethane)$^{·+}$ and (p-terphenyl-d_{14})$^{·-}$ radical ions was found by analyzing the shape of the magnetic field effect curve as the magnetic field strength was varied. The method of time resolved magnetic field effect was also used for determination of the times of longitudinal and transverse paramagnetic relaxation.

5.8.6 Singlet–Triplet Oscillations

- The dynamic singlet–triplet transitions between the spin states of recombining radical pairs occur on a nanosecond time scale and are harmonic in nature. The external magnetic field and the internal fields created by the magnetic nuclei of radicals determine the transitions frequencies. Under certain conditions this transition is accompanied by spin oscillations (spin beats) [99, 100]. The problem of quantum beats in radical recombination was discussed by Klein and Woltzm [100], who detected nonmonotonic time variations of magnetic field effects in radical-ion recombination. The first successful observations of singlet–triplet oscillations in an ensemble of radical ion pairs in a spin-correlated singlet state in solution were performed by Anisimov et al. [101] using a time-resolved investigation of the recombination of the geminate singlet radical-ion pairs (tetramethyl-ethiene)$^+$/(paraterphenyl d_{14})$^-$ generated by radiolysis. The decay of recombination fluorescence from the sample induced by a short pulse of ionizing radiation was detected. According to the authors, oscillations in a high external magnetic field arise if spins precess at different Larmor frequencies due to the difference ΔH_{loc} of local magnetic fields induced by hyperfine couplings and g-value differences. It was shown that the pairs synchronously oscillate between their singlet and triplet states when the distance between the interacting members of the pairs was typically about 10 nm. The values of the hyperfine constant a for spin electron—spin proton interaction was calculated by the equation T = $2\pi/\gamma$a.

- A model of the quantum beats of radical pairs induced by a microwave field resonant for one of the component radicals was considered [102]. According to this model, the following elementary events lead to the quantum beats when a radical pair (RP) in the T$_+$ ($\alpha\alpha$) spin state is irradiated by a MW pulse: (1) the α-spin on the radical a, which is resonant to the MW, starts to rotate and turns into a β-spin, (2) the RP has $\alpha\beta$ spins (ST$_0$ mixed state) and can recombine, (3) the β-spin rotates and returns to an α-spin by continued irradiation of the MW field, (4) the RP recovers to an $\alpha\alpha$ spin state (T$_{+1}$ state) again. Thus, the MW pulse drives the periodic changes in the spin-state populations of the RP, the quantum beat. The oscillation frequency depends on the resonant MW field strength, B$_1$ = ω_1/γ_e (ω_1 is the frequency of oscillation, and γ_e is the gyromagnetic ratio), and the difference of the resonant frequencies of the radicals a and b, $|\omega_a - \omega_b|$. The frequency of oscillation between T$_{\pm1}$ and T$_0$ states is predicted to be $2\omega_1$. The probability of finding the ST$_0$ mixed state given an initial T$_{+1}$ state is expressed by (5.14)

$$P_{ST_0}(t) = |\langle S|\exp(-i\mathcal{H}t)|T_{+1}\rangle|^2 + |\langle T_0|\exp(-i\mathcal{H}t)|T_{+1}\rangle|^2 \qquad (5.14)$$

In (5.14) the corresponding Hamiltonian in the rotating frame with the angular frequency of the MW field was used.

The quantum beats were experimentally observed as an alternation of the yields of the component radicals by a nanosecond time-resolved optical absorption in the presence of an X-band (9.15 GHz) resonant microwave pulse [102] This technique was applied to the photochemical reaction of benzophenone (BP) derivatives benzophenone, benzophenone-d10, and benzophenone-carbonyl-^{13}C in a sodium dodecylsulfate micellar solution with a step-by-step increase of the resonant microwave pulse width. Figure 5.17 demonstrates the calculated quantum beats in the of the populations of the ST$_0$ mixed state for the initial T$_{+1}$ state in this system. See details in the figure capture.

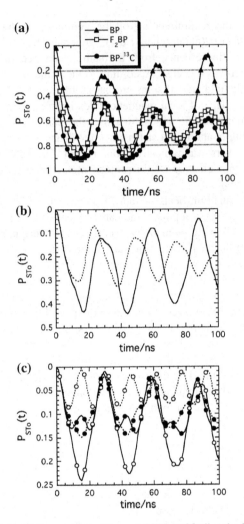

Fig. 5.17 **a** The calculated time profiles of the probabilities of finding the ST_0 mixed state for the initial T_{+1} state under continuous resonant MW irradiation with $B_1 = 1.2$ mT. **b** The calculated time profiles of the probabilities of finding the ST_0 mixed state for center and side hyperfine lines of the F2BP ketyl radical. *Solid* and *dotted lines* show the cases of the center hyperfine line and the sum of two side hyperfine lines due to the HFC of fluorine atoms, respectively. **c** The calculated time profiles of the probabilities of finding the ST_0 mixed state for four combinations of hyperfine lines of SDS alkyl and BP-^{13}C ketyl radicals (benzophenone-carbonyl-13C) in sodium dodecylsulfate (SDS) in closed and *open circles* show the cases in which the hyperfine lines of the SDS alkyl radical exist in the low- and high-field side, respectively. Solid and dotted lines show the cases of M_I +1/2 and −1/2 due to the ^{13}C atom [102]

References

1. A.J. Hoff (ed.), *Advanced EPR. Applications in Biology and Biochemistry* (Elsevier, Amsterdam, 1989)
2. C.P. Poole, *Electron Spin Resonance: A Comprehensive Treatise on Experimental Techniques* (Dover Publcation, 1997)
3. G.R. Eaton, K.M. Salikhov, S.S. Eaton, *Foundations of Modern EPR* (World Scientific, Singapore, 1998)
4. J. Yamauchi, in Likhtenshtein GI, Yamauchi J, Nakatsuji S, Smirnov A., Tamura R. *Nitroxides: Application in Chemistry, Biomedicine, and Materials Science* (WILEY-VCH, Weinhem, 2008), p. 71
5. A.I. Smirnov, in Likhtenshtein GI, Yamauchi J, Nakatsuji S, Smirnov A., Tamura R. *Nitroxides: Application in Chemistry, Biomedicine, and Materials Science* (WILEY-VCH, Weinhem, 2008), p. 121
6. M.R. Brustolon, *Principles and Applications of Electron Paramagnetic Resonance Spectroscopy* (Blackwell Publishers, 2008)
7. W.R Hagen, *Biomolecular EPR Spectroscopy* (CRC, 2008)
8. Savitsky A, Mobius K High-Field EPR *Spectroscopy on Proteins and Their Model Systems* (Wiley-VCH Verlag GmbH, 2009)
9. G.R. Eaton, S.S. Eaton, D.P. Barr, R.T. Weber, *Quantitative EPR* (Springer, 2010)
10. B.C. Gilbert, M.J. Davies, K.A. McLauchlan (eds.), *Electron Paramagnetic Resonance, a Specialist Periodical Report*, vol. 17 (Royal Society of Chemistry, 2010)
11. G.M. Smith, J. Keeble, O. Schiemann (eds.), *Introduction to Modern EPR Spectroscopy* (CRC, 2010)
12. A. Belay, *Fluorescence and Electron Paramagnetic Resonance (EPR) Spectroscopy* (LAP LAMBERT Academic Publishing, 2011)
13. S.K. Misra (ed.), *Multifrequency Electron Paramagnetic Resonance* (Wiley-VCH, New York, 2011)
14. G.I. Likhtenshtein, *Biophysical Labeling Methods in Molecular Biology* (Cambridge University Press. Cambridge, NY, 1993)
15. A. Brunetti, O. Baffa, Radiat. Meas. **32**,361 (2000)
16. E. Husted, A. Beth, in *Magnetic Resonance in Biology*, vol. 19, ed. by L. Berliner, S. Eaton, G. Eaton (Kluwer Academic Publishers, Dordrecht 2000), p. 155
17. W. Xiao, Y.-K. Shin, in *Magnetic Resonance in Biology*, vol. 19, ed. by L. Berliner, S. Eaton, G. Eaton (Kluwer Academic Publishers, Dordrecht 2000), p. 249
18. M. Portis, Phys. Rev. **91**, 1071 (1953)
19. T.G. Jr, Castner. Phys. Rev. **115**, 1506 (1959)
20. Ya. S, Lebedev, and V.I. Muromtsev, *ESR and Relaxation of Stable Radicals* (Khimiya, Moscow, 1972)
21. A.V. Kulikov, G.I. Likhtenstein, Biofizika **19**, 420 (1974)
22. A.V. Kulikov, G.I. Likhtenstein, Adv. Molecul. Relax. Proc. **10**, 47 (1977)
23. G.I. Likhtenstein, in *Magnetic Resonance in Biology*, vol. 19, ed. by L. Berliner, S. Eaton, G. Eaton (Kluwer Academic Publishers, Dordrecht 2000), p. 309
24. A. Lund, E. Sagstuen, A. Sanderud, J. Maruani, Radiat. Res. **172**, 753 (2009)
25. G.I. Likhtenshtein, A.V. Kulikov, A.I. Kotelnikov, V.R. Bogatyrenko, Photobiochem. Photobiol. **3**, 178 (1982)
26. A.V. Kulikov, Mol. Biol. (Moscow) **10**, 109 (1976)
27. K.M. Salikhov, J. Magn. Reson. **63**, 241 (1985)
28. Shojiro Kimura, Yuya Sawada, Yasuo Narumi, Kazuo Watanabe, Masayuki Hagiwara, Koichi Kindo, Hiroaki Ueda, Phys. Rev. B **92**, 144410 (2015)
29. K.I. Zamaraev, Yu.N. Molin, K.M. Salikhov, *Spin Exchange. Theory and Physicochemical Application* (Springer-Verlag. Heidelberg, 1981)

30. G.I. Likhtenstein, Y.B. Grebentchikov, P.K. Bobodzhanov, Y.V. Kokhanov, Molec. Biol. (Moscow) **4**, 782 (1970)

31. J.S. Hyde, H.M Swartz, W.E. Antholine, The spin probe-spin label methods, in *Spin Labeling. Theory and Application*, vol. 2, ed. by L. Berliner (Academic Press, New York, 1976), p. 72

32. C. Altenbach, W. Froncisz, J.S. Hyde, W.L. Hubbell, Biophys. J. **56**, 61183 (1989)

33. K.M. Salikhov, A.B. Doctorov, YuN Molin, K.I. Zamaraev, Spin relaxation of radicals and complexes upon encounters in solution. J. Magnet. Reson. **5**, 189 (1971)

34. V.N. Parmon, A.I. Kokorin, G.M. Zhidomirov, *Stable Biradicals* (Nauka, Moscow, 1980)

35. G.R. Eaton S. S. Eaton, Biol. Magn. Reson. **8**, 339 (1989)

36. S. Ngo, V.V. Chiang, Z. Guo, J. Struct. Biol. **180**, 374 (2012)

37. V. Paredes-García, R.C. Santana, R. Madrid, B. Baldo, A. Vega, E. Spodine, J. Inorg. Biochem. **114**, 75 (2012)

38. H.S. Mchaourab, S. Pfenninger, W.E. Antholine, W.E. Antholine, C.C. Felix, J.S. Hyde, P. M. Kroneck, Biophys. J. **64**, 1576 (1993)

39. C.S. Klug, T.G. Camenisch, W.L. Hubbell, J.S. Hyde, Biophys. J. **88**, 3641 (2005)

40. E.J. Hustedt, A.H. Beth, Ann. Rev. Biophys. Biomol. Struc. **28**, 129 (1999)

41. C. Riplinger, J.P.Y. Kao, G.M. Rosen, V. Kathirvelu, G.R. Eaton, S.S. Eaton, A. Kutateladze, F. Neese, J. Am. Chem. Soc. **131**, 10092 (2009)

42. S.K Misra, J.H. Freed, in *Multifrequency Electron Paramagnetic Resonance*, ed. by S.K Misra (Wiley-VCH, New York, 2011), p. 545

43. A.W. Kittell, E.J. Hustedt, J.S. Hyde, J. Magn. Res. **221**, 51 (2012)

44. A. Berliner., S. Eaton, G. Eaton (eds.) *Magnetic Resonance in Biology. Distance Measurement in Biological Systems by ESR*, vol. 19 (Kluwer Academic Publishers, Dordrecht, 2000)

45. O.Y. Grinberg, A.A. Dubinskii, V.F. Shuvalov, L.G. Oranskii, V.I. Kurochkin, YaS Lebedev, Dokl. Akad. Nauk SSSR **230**, 884 (1976)

46. Ya. S. Lebedev, Appl. Magn. Res. **7**, 339 (1994)

47. V.I. Krinichnyi, *2-mm Wave Band EPR Spectroscopy of Condensed Systems* (Boca Raton, CRC Press, 1995)

48. O.Y. Grinberg, L.J. Berliner (eds.) *Very High Frequency (VHF) ESR/EPR*, vol. 22. Series Biological Magnetic Resonance (Kluwer Academic Plenum Publishers, 2004)

49. V.I. Krinichnyi, in *Spectroscopy of Polymer Nanocomposites*, ed. by D. Ponnamma, D. Rouxel, S. Thomas (Elsevier, 2015), p. 202

50. D.E. Budil, K.A. Earle, W.B. Lynch, J.H. Freed, in *Advanced ESR: Application in Biology and Biochemistry*, ed. by A.J. Hoff (Elsevier, Amsterdam, 1989)

51. J.H. Freed, Ann. Rev. Phys. Chem. **51**, 655 (2000)

52. D. Kivelson, J. Chem. Phys. **33**, 1094 (1960)

53. J.H. Freed, in *Spin Labeling. Theory and Applications*, vol. 1, ed. by L. Berliner (Academic Press, New York, 1976)

54. A. Polimeno, V. Barone, J.H. Freed, in *Computational Spectrocopy*, Chapter 12, ed. by V. Barone (Wiley, New York, NY, 2012)

55. S.K. Misra, J.H. Freed, Molecular motions, in *Multifrequency Electron Paramagnetic Resonance*, ed. by S.K Misra (Wiley-VCH, New York, 2011), p. 497

56. Z. Zhang, M.R. Fleissner, Z. Liang, D.S. Tipikin, K. Moscicki, K.A. Earle, W.L. Hubbell, J. H. Freed, J. Phys. Chem. B **114**, 5503 (2010)

57. Y. Sun, Z. Zhang, V.M. Grigoryants, W.K. Myers, F. Liu, K.A. Earle, J.H. Freed, C. P. Scholes, Biochemistry **51**, 8530 (2012)

58. M.D. Smigel, L.R. Dalton, J.S. Hyde, J.S. Dalton, Proc. Nat. Acad. Sci. USA **71**, 1925 (1974)

59. A. Kusumi, J.S. Hyde, Biochemistry **21**, 5978 (1982)

60. P. Fajer, D.D. Thomas, J.B. Feix, J.S. Hyde, Biophys. J. **50**, 1195 (1986)

61. E.J. Hustedt, A. Beth, Biophys. J. **86**, 3940 (2004)

62. A.H. Beth, Biophys. J. **103**, 1109 (2012)

63. G.G. Lazarev, Ya.S. Lebedev, Khimicheskaya Fizika **7**, 1159 (1988)
64. G. Feher. Phys. Rev. 834 (1956)
65. J.S. Hyde, J. Chem. Phys. **43**, 1806 (1965)
66. H. Kurreck, B. Kirste, W. Lubitz, *Electron Nuclear Double Resonance Spectroscopy of Radicals in Solution* (VCH Publishers, New York, 1988)
67. D.M. Murphy, R.D. Farley, Chem. Soc. Rev. **35**, 249 (2006)
68. C.E Tait,. P. Neuhaus, H.L. Anderson, C. R. Timmel, J. Am. Chem. Soc. **137**, 6670 (2015)
69. B. Katterle, R.I. Gvozdev, N. Abudu, T. Ljones, K.K. Andersson, Biochem. J. **363**, 677 (2002)
70. L. Kulik, W. Lubitz, Photosynth. Res. **102**, 391 (2009)
71. J.S. Hyde, J.C.W. Chien, J.H. Freed, J. Chem. Phys. **48**, 4211 (1968)
72. V.A. Benderskii, L.A. Blyumenfel'd, P.A. Stunzhas, E.A. Sokolov, Nature **220**, 365 (1968)
73. J.J. Yin, J.B. Feix, J.S. Hyde, Biophys. J. **52**, 1031 (1987)
74. L.D. Kispert, Biol. Magn. Res. (Biomedical EPR, Part B) **24**, 165 (2005)
75. E. Reijerse, F. Lendzian, R. Isaacson, W. Lubitz, J. Magn. Res. **214**, 237 (2012)
76. A.L. Buchachenko, *Chemical Nuclear and Electron Polarization* (Nauka, Moscow, 1974). (in Russian)
77. R.W. Field, R.S. Bradford, D.O. Harris, H.P. Broida, J. Chem. Phys. **56**, 4712–4714 (1972)
78. O.A. Anisimov, V.M. Grigoryants, V.K. Molchanov, YuN Molin, Chem. Phys. Lett. **66**, 265 (1979)
79. K.M. Salikhov, Y. Sakaguchi, H. Hayashi, Chem. Phys. **220**, 355 (1997)
80. YuN Molin, O.A. Anisimov, V.M. Grigoryants, V.K. Molchanov, K.M. Salikhov, J. Phys. Chem. **84**, 1853 (1980)
81. E.L. Frankevich, S.I. Kubarev, *Triplet State Optical Detection Magnetic Resonance Spectroscopy*, ed. by R.H. Clarke (John Wiley & Sons, Inc., New York, 1982)
82. O.A. Anisimov, V.N. Verkhovlyuk, S.B. Zikirin, A.G. Matveeva, S.I. Trashkeev, Yu.N. Molin, Appl. Magn. Reson. **45**, 881 (2014)
83. M. Vyushkova, P. Potashov, V. Borovkov, V. Bagryansky, Y.N. Molin, in *Selectivity, Control, and Fine Tuning in High-Energy Chemistry*, ed. by D.V. Stass, V.I. Feldman (2011), p. 191
84. A. Dreau, M. Lesik, L. Rondin, P. Spinicelli, O. Arcizet, J.-F. Roch, V. Jacques, Phys. Rev. B: Condens. Matter Mater. Phys. **84**, 195204/1 (2011)
85. E.L. Frankevich, A.I. Pristupa, V.I. Lesin, Chem. Phys. Lett. **54**, 99 (1978)
86. Masaharu Okazaki, Takesui Shiga, Product yield of magnetic-field-dependent photochemical reaction modulated by electron spin resonance. Nature **323**, 240–243 (1986)
87. T. Miura, A. Kageyama, S. Torii, H. Murai, J. Phys. Chem. B **114**, 14550 (2010)
88. C.J. Wedge, J.C.S. Lau, K.-A. Ferguson, S.A. Norman, P.J. Hore, C.R. Timmel, Phys. Chem. Chem. Phys. **15**, 16043 (2013)
89. S.N. Batchelor, K.A. McLauchlan, I.A. Shkrob, Time-resolved reaction yield detected magnetic resonance (RYDMR). Chem. Phys. Lett. **181**, 327 (1991)
90. C.A. Hamilton, J.P. Hewitt, K.A. McLauchlan, U.E Steiner, Mol. Phys. **65**,423 (1988)
91. V.N. Verkhovlyuk, N.N. Lukzen, J.B. Pedersen, D.V. Stass, Y.N. Molin, Dokl. Phys. Chem. **417**, 311 (2007)
92. K.B. Henbest, E. Athanassiades, K. Maeda, I. Kuprov, P.J. Hore, C.R. Timmel, Molec. Phys. **104**, 1789 (2006)
93. V.N. Verkhovlyuk, N.N. Lukzen, J.B. Pedersen, D.V. Stass, Y.N. Molin, Dokl. Phys. Chem. **417**, 311 (2007) (Pleiades Publishing, Ltd)
94. K.L. Ivanov, D.V. Stass, E.V. Kalneus, R. Kaptein, N.N. Lukzen, Appl. Magn. Reson. **44**, 217 (2013)
95. N.L. Lavrik, YuN Molin, Int. J. Quan. Chem. **40**, 387 (1991)
96. P.A. Potashov, L.N. Shchegoleva, N.A. Gritsan, V.A. Bagryansky, Y.N. Molin, J. Phys. Chem. A **116**, 3110 (2012)
97. V.A. Bagryansky, V.I. Borovkov, Y.N. Molin, Phys. Chem. Chem. Phys. **6**, 924 (2004)
98. M.E. Michel-Beyerle, H.W. Krüger, R. Haberkorn, H. Seidlitz, Chem. Phys. **42**, 441 (1979)

99. Y.N. Molin, Mendeleev Commun. **3**, 85 (2004)
100. J. Klein, R. Voltz, Can. J. Chem. **55**, 2103 (1977)
101. O.A. Anisimov, V.L. Bizyaev, N.N. Lukzen, V.M. Grigoryants, YuN Molin, Chem. Phys. Lett. **101**, 131 (1983)
102. Y. Kitahama, Y. Sakaguchi, J. Phys. Chem. A **112**, 176 (2008)

Chapter 6
Experimental Methods for the Investigation of Electron Spin Interactions Based on ESR Phenomena. Pulse EPR Measurements

Abstract This chapter has described the remarkable progress in the development of modern pulse ESR techniques in quantitative investigations of electron spin dipolar and exchange interactions. Measurements of distances between spins up to at least 80 Å as well as their spatial distributions, detection of the scale and frequencies of different motions of spin-bearing molecules and their segmental motions over a wide range correlation time from milli- to nanoseconds are available by these techniques. Pulse ESR in combination with CW ESR, advanced NMR techniques, synthetic chemistry and biochemistry opens new horizons for detailed structural and molecular dynamics studies of various objects including such complicated systems as biological molecules and their complexes.

6.1 Introduction

The basis of all magnetic resonance experiments involves the perturbation of an equilibrium magnetization, typically established in a static magnetic field. In a pulse experiment the magnetization is typically perturbed by applying a brief, intense, sinusoidal magnetic field. Such a pulse is called a microwave pulse, but the relevant frequency of the pulse can be anywhere in the MHz to GHz region of the electromagnetic spectrum, depending on the static applied field and the nature of the spin couplings in the system. Once the pulse is removed, one can then measure the emitted microwave signal which is created by the sample magnetization as it recovers to its initial value. Over the years, a vast variety of pulse sequences have been developed. Using them, it is possible to gain extensive knowledge of the structural and dynamical properties of paramagnetic compounds.

The development of these pulse sequences, in turn, has been spurred by the development of powerful, coherent amplifiers with large instantaneous bandwidths. Some of the relevant applications are discussed in the [1–29]. Some of the more important developments include: Fourier transform ESR (FT ESR), two-dimensional Fourier transform ESR (2D FT ESR), electron spin echo (ESE), electron spin echo envelope modulation (ESEEM), hyperfine sublevel correlation spectroscopy

© Springer International Publishing Switzerland 2016   123
G. Likhtenshtein, *Electron Spin Interactions in Chemistry and Biology*,
Biological and Medical Physics, Biomedical Engineering,
DOI 10.1007/978-3-319-33927-6_6

(HYSCORE), double nuclear coherence transfer hyperfine sublevel correlation electron spin echo DONUT-HYSCORE), electron spin echo detected magnetization transfer (ESE MT), two-dimensional electron spin echo correlation spectroscopy (2D SECSY), pulse electron-nuclear double resonance (pulse ENDOR), pulse electron-electron double resonance (PELDOR), electron spin transient nutation (ESTN), two-dimensional electron spin transient nutation (2D ESTN), phase inverted echo-amplitude detected nutation (PEANUT), saturation recovery ESR (SR ESR), pulse multifrequency ESR (PMF ESR), magnetic isotope effect (MEF), quantum beats effect (QBE), double quantum coherence pulsed ESR (DQC ESR), ESR Spectra Hole Burning.

Many qualitative insights into the behavior of EPR-active spins can be gleaned from elementary considerations, such as precession phenomena. Larmor precession describes the precession of the magnetic moments of electrons, atomic nuclei, and atoms about an external magnetic field, which is often taken to be static, although it need not be (Fig. 3.1) (https://en.wikipedia.org/wiki/Larmor_precession). The magnetic field exerts a torque $\vec{\Gamma}$ on the magnetic moment,

$$\vec{\Gamma} = \vec{\mu} \times \vec{B} = \gamma \vec{J} \times \vec{B} \qquad (6.1)$$

where $\vec{\mu}$ is the magnetic dipole moment, $\vec{J}$ is the angular momentum vector, $\vec{B}$ is the external magnetic field and γ is the gyromagnetic ratio. Given that the torque is the time rate of change of the angular momentum vector $\vec{J}$, (6.1) gives an equation of motion for the sample magnetization. The angular momentum vector $\vec{J}$ precesses about the external field axis with an angular Larmor frequency ω_L.

The effects of the magnetic field on the spin magnetization vector can be considered in the laboratory frame as well as in the rotating frame In the laboratory frame the static magnetic field B is assumed to be parallel to the z-axis and the microwave field B_1 parallel to the x-axis (Fig. 6.1).

Because electron spins of 1/2 are characterized by two quantum mechanical states, one parallel and one antiparallel to applied static magnetic field B_0, with relative populations determined by the Boltzmann distribution, more electron spins can be found in the parallel state of lower energy. This results in a net magnetization (M_0), which is the vector sum of all magnetic moments in the sample, parallel to the z-axis and the magnetic field. EPR experiments usually use a linearly polarized microwave field B_1, perpendicular to the much stronger applied magnetic field. The linearly polarized B_1 may be resolved into two counter-rotating, circularly polarized components. In the rotating frame, the B_1 component rotating with frequency w_0, say, will be on resonance with the precessing magnetization vector M_0 if

$$\omega_L = -\omega_0 \qquad (6.2)$$

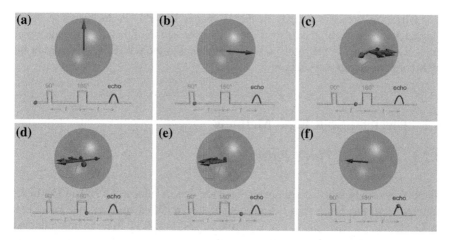

Fig. 6.1 Hahn spin echo sequence in the rotating frame. https://upload.wikimedia.org/wikipedia/commons/9/99/SpinEcho_GWM_stills.jpg

The time dependence of the M_0 vector is then determined by the stationary field B_1 in the rotating frame, as described by (6.1), leading to precession of M_0 around B_1 at the frequency ω_1

$$\omega_1 = -\gamma B_1 \qquad (6.3)$$

The magnetization vector will rotate around the x-axis in the zy-plane for the duration of the applied B_1 field. The tip angle α by which M_0 is rotated is given by:

$$\alpha = -\gamma |B_1| t_p \qquad (6.4)$$

Here t_p is the duration for which B_1 is applied, also called the pulse length. It is important to note that the axis along which the B_1 field is applied can be changed by altering the phase relationship in the laboratory frame. The end position of the magnetization vector M_0 at the end of a B_1 pulse thus depends on the length, the magnitude and the choice of axis in the lab frame of the microwave pulse B_1.

In reality EPR spectra have a distribution of different resonant frequencies and not all of them can be exactly on resonance simultaneously. For example, a $\pi/2$ (90°) pulse leaves magnetization in the xy-plane, but since the microwave B_1 field does not have the same frequency as the precessing magnetization vector for all components of the spectrum, the magnetization vector rotates in the xy-plane, either faster or slower than the microwave magnetic field B_1 when viewed in the rotating frame. The electron spins interact with their surroundings, the magnetization in the xy-plane will decay by , or T_2-type spin relaxation mechanisms, and eventually the magnetization M_0 will recover to its equilibrium orientation along the z-axis due to spin lattice, or T_1-type processes. In the laboratory frame, this corporate precession of the magnetization generates a microwave signal, the   is maximized if the magnetization vector is exactly in the xy-plane. The FID is the response of the magnetization in the time domain. Fourier transformation of this time domain signal gives rise to the FT-ESR signal, which is identical to the CW-ESR spectrum in the high temperature approximation.

6.2   Electron Spin Echo Methods

6.2.1   Two and Three Pulse Methods

The first two pulse nuclear spin-echo detection was developed by Hahn [31]. In the simplest case the Hahn spin echo sequence (as modified by Carr and Purcell) is $\pi/2 - \tau - \pi - \tau$−echo using non-selective pulses (Fig. 6.1). In a rotating reference frame at the resonance frequency, a non-selective $\pi/2$ pulse is applied that turns the magnetization vector into the x-y plane. Due to local magnetic field inhomogeneities arising from the applied magnetic field, or from dipole-dipole interactions, as the magnetization precesses, some spins precess at a lower Larmor frequency due to a smaller net local field strength, while some components of the magnetization precess at a higher Larmor frequency due to a larger net local field strength. As a result, different components of the magnetization precess at different rates for a time τ and become dephased, or defocused. τ. A $180°$ (π) pulse is applied so that in the x-y plane the more slowly precessing magnetization components are leap-frogged ahead of the more quickly precessing components (cf. Fig. 6.1). At this point the quickly precessing magnetization components catch up to the more slowly precessing components and the various components start to rephase, just as they were immediately following the non-selective $\pi/2$ pulse. After a time τ, maximal refocussing has occurred and the refocused spins are observable as a spin echo.

Monitoring the decay of the spin echo intensity

$$I(2\tau) = \exp(-2\tau/T_M) \tag{6.5}$$

allows one to determine T_M, the phase memory time. There are several mechanisms that can contribute to this process and the spin-spin relaxation time T_2 is one of them. One mechanism for a T_2 process is stochastic modulation of intermolecular dipole-dipole interactions. We also note that the recovery of the magnetization towards its initial state would be an instance of a spin-lattice, or T_1 relaxation process.

The Hahn three-pulse sequence $\pi/2 - \tau - \pi/2 - T - \pi/2\square\tau-$ echo comprises two $90°$ pulses with time τ after the first pulse, $90°$ pulse after time T and detection of the so-called stimulated spin echo after time τ [31]. After the first $90°$ pulse, the magnetization vector spreads out forming a "disk" in the x-y plane. The spreading continues for a time τ, and then a second $90°$ pulse is applied such that the "disk" is now in the x-z plane. After a further time T a third pulse is applied and a stimulated echo is observed after waiting a time τ after the last pulse. The decay of the echo intensity is fitted to an exponential function of T as

$$I(2\tau + T) = I_0 \exp(-T/T_1) \tag{6.6}$$

The stimulated echo is useful for those systems where the T_1 relaxation time is much longer than T_2, as the magnetization is 'stored' along the z axis for a time T until it is read out by the final $\pi/2$ pulse.

6.2.2 Spin Echo Detected ESR

Echo-detected (ED) spectra from the two-pulse, primary spin-echo (pulse sequence: $\pi/2 - \tau - \pi - \tau - $ echo) experiment are used to detect rapid angular motions, on the time scale of the phase memory time (T_{2M}), that is, in the nanosecond regime [26, 32, 33]. Echo-detected spectra from the three-pulse, stimulated spin-echo (pulse sequence: $\pi/2 - \tau - \pi/2 - T - \pi/2 - \tau - $ echo) experiment are used to detect slower angular motions, due to its dependence on the longer T_1 process.

For the two-pulse primary echo experiment and the three-pulse stimulated echo detected experiments, the EPR line shape is obtained when the magnetic field B is scanned across the resonance line. In the limit of small-amplitude motion, the ED-EPR line shape of a nitroxide radical is given by (6.7) [26]

$$ED'(2\tau + T, B)$$
$$= \sum_m \frac{1}{4\pi} \iint \sin\theta \, d\theta \, d\varphi f(B - \omega_m(\theta, \varphi)/\gamma) E(2\tau + T, \theta, \varphi) \tag{8}$$

$$\tag{6.7}$$

where

$$E(2\tau + T, \theta) = \exp\left(-\frac{\tau}{\tau_c}\right)\left[1 + \frac{1}{2\tau_c R}\sinh 2R\tau\right.$$
$$\left. + ;\frac{1}{2\tau_c^2 R^2}\sinh^2 R\tau - \frac{\Delta\omega^2(\theta, \varphi)}{R^2}\sinh^2 R\tau\left(1 - \exp\left(-\frac{T}{\tau_c}\right)\right)\right] \tag{5}$$

$$\tag{6.8}$$

and $\Delta\omega(t)$ is the stochastic fluctuation of the resonance frequency, the angular brackets indicate a time average, τ_c is the correlation time for motion. Equation X is valid for the two-pulse primary echo experiment (pulse sequence: $\pi/2 - \tau - \pi - \tau - $ echo) and the three-pulse stimulated echo (pulse sequence: $\pi/2 - \tau - \pi/2 - T - \pi/2 - \tau - $ echo), a when $R^2 = (2\tau_c)^{-2} - \Delta\omega^2(\theta, \varphi)$ and $\Delta\omega^2\tau_c^2 \ll 1$.

An additional mechanism, which induces distortions in the line shape of the ED-EPR spectra, is instantaneous diffusion that arises from dipolar spin–spin

Fig. 6.2 *Top Three-pulse,*
stimulated-echo detected
spectra of 14-PCSL in DPPC
bilayer membranes at 200 K.
Spectra are recorded with
$\tau = 216$ ns and different
values of the *T*-delay
(increasing from *top* to
bottom) and are corrected for
spin–lattice relaxation.
Bottom Spectra simulated
with the two-site model
according to (6.5) and (6.8).
Parameters used in the
simulation are $\tau_c^* = 0.8$ μs
and $\alpha_0^* = 1.7 \times 10^{-2}$ rad
[26]

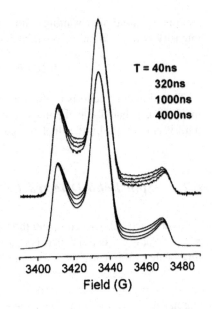

T = 40ns
320ns
1000ns
4000ns

3400 3420 3440 3460 3480
Field (G)

interactions between different spins. As an example, the dynamics of spin-labeled lipid chains in the low-temperature phases of dipalmitoyl phosphatidylcholine (DPPC) membranes, with and without equimolar cholesterol, have been investigated measuring echo-detected spectra from the two-pulse (detection of rapid angular motions) and the three-pulse, spin-echo (detection of slow angular motions) on the nanosecond time scale of the phase memory time (T_{2M}) was measured [26]. The echo-detected EPR spectra simulated with an isotropic librational model are shown in Fig. 6.2.

Continuous-wave electron paramagnetic resonance (CW EPR), echo-detected (ED) EPR, and field-step electron-electron double resonance (FS ELDOR) were simultaneously applied to study molecular motions of nitroxide spin probes of two different types in glassy *o*-terphenyl [32]. The motion was attributed to the β-relaxation process that is temperature-independent between 160 and 265 K.

6.3 Electron Spin Resonance Methods Based on Pulse ELDOR

6.3.1 Three-Pulse ELDOR

Echo-detected ELDOR (pulse ELDOR, PELDOR, double electron-electron resonance DEER) was invented by Milov, Salikhov and Tsvetkov [34] and was further widely successfully applied by the Zvetkov and other groups [1, 18, 27, 35–49]. In further work, the effect of "instantaneous diffusion" treated by Klauder and

Anderson [37] in a system of two spins (observed and pumped), which is caused by the effects of diffusion coupled with the action of the applied pulse on the precession frequency of the observed spin, was accounted for consistently.

In three pulse PELDOR, the ESR spectrum is excited by two pulses and an additional pumping pulse. For a system of two spins A and B, the experiment consists of the following consequence: (1) two $\pi/2$ pulse electron spin echo (sub-sequence with a mixed interpulse delay τ at the observer frequency w_a for the spin A and (2) a pump π pulse at frequency w_b for the spin B, after the first ESE pulse. The pumping pulse duration should satisfy the following relation for all three pulses

$$t_{pi}|\omega_a - \omega_b| \gg 1, \tag{6.9}$$

where t_{pi} is the duration of the ith mw pulse. The local magnetic fields at the sites of spins A. is change by the pulse on spins B, caused a change in the z-projections of spins B. The magnitude of this change, determined by the dipole–dipole (d–d) interaction of spins A and B, which effects on the magnitude of the ESE signal compared to what is seen in the absence of the pumping pulse.

Thus information on dipole-dipole couplings in the spin system can be derived from an analysis of the decay function V(T) of the ESE signal (6.10), when the time interval T between the first ESE pulse and pumping pulse is varied. The kinetics of V(T) decay strongly depends upon distance, mutual orientation within the interacting spin pairs and the spatial distribution of radicals throughout the sample. According to [38], for spins A and B a fixed distance between, coupled by the d-d interaction at a distance R and given the angle between the direction of the external magnetic field and the vector which connects the paramagnetic centers (Θ), the ratio of the primary echo signal amplitude $V_p(2\tau)$ with the pumping pulse to the ESE signal amplitude $V_0(2\tau)$ without the pumping pulse is

$$V(T) = \frac{V_p(2\tau)}{V_0(2\tau)} = 1 - p_B[1 - \cos(DT)]. \tag{6.10}$$

where p_B is the probability of a spin B flip under the action of the pumping pulse; D is the splitting (in rad/s) of the resonance spin A line due to the interaction with spin B, and has the analytical form

$$D = \frac{\gamma^2 \hbar}{R^3}(1 - 3\cos^2 \Theta) + J, \tag{6.11}$$

which includes the dipolar and J exchange interaction. This method allows one to detect rather weak d-d interactions

$$|D| \approx \frac{1}{\tau} \ll |\omega_a - \omega_b|. \tag{6.12}$$

The lower limit on the determination of the D value is set by the phase relaxation rate $1/T_2$. The use of the appropriate orientation of the external magnetic field with respect to the vector connecting the two spin labels allow one to determine from the Pake doublet [39] (For systems in which the distance between paramagnetic centers is not fixed and may be described by a distribution function n(r) the PELDOR signal may be calculated using the equation:

$$V(T) = 1 - p \int 2\pi F(T)n(r)r^2 \, dr,$$

$$p = \langle p_B \rangle_{\omega_{rb}}.$$

(6.13)

Applications of PELDOR can be illustrated by the following examples. The use of the pulse electron double resonance method to determine the effective saturation factor of nitroxide radicals for dynamic nuclear polarization experiments in liquids was proposed [40]. The facility of the method to measure electron spin relaxation times was demonstrated on the nitroxide radical TEMPONE-D,[15]N. Data on the structure of Photosystem II studied by PELDOR were presented in [41]. X-Band pulsed electron–electron double resonance (PELDOR) spectroscopy was used to investigate the magnetic dipole–dipole interaction between spin labels for frozen glassy methanol solutions at 77 K of double spin-labeled, medium-length peptaibiotics, namely, tylopeptin B and heptaibin [42]. A comparative study of series of model DNA duplexes, 5′-spin-labeled with various Triarylmethyl (trityl, TAM) rfdicals and nitroxides using DEER/PELDOR) was repoted [43]. The experiments showed that TAMs allow obtaining narrower spin-spin distance distributions by a factor of 3 and potentially more precise distances between labeling sites because the conformational disorder is noticeably smaller for TAM compared to nitroxide due to specific interaction of TAM [43]. Figure 6.3 shows the obtained DEER time traces, and distance distributions for duplexes I–IV with piperazine linkers.

Since the pioneering work of Kulik et al. the relaxation-induced dipolar modulation enhancement (RIDME) experiment has been significantly advanced [45]. A comparative study of MTSSL-labeled cytochrome P450cam mutant C58R1 employing two independent pulse methods (PELDOR) and relaxation-induced dipolar modulation enhancement (RIDME) were performed (Fig. 6.4) [46]. In addition, several methods for the suppression of ESEEM artifacts in the RIDME time trace were compared to each other.

The PELDOR technique was used for detail investigation of peptide conformation and aggregation of r mono and double TOAC substituted analogs of [Glu(OMe)] alamethicin F50/5 (Alm′) peptaibiotic on the surface of the Oasis HLB sorbent [47] Two effects: (1) overlapping of the ESR (EPR) spectra of paramagnetic spin 1/2 particles (spin labels) in pairs and (2) overlapping of the excitation bands by the pump and echo-forming pulses were taking in consideration in new theory of the three and four-pulse electron double resonance (PELDOR) [48, 49].

A "211" electron spin echo method developed by Raitsimring and Tsvetkov is complementary to the pulse ELDOR [50, 51]. The difference between the PELDOR

Fig. 6.3 a ED EPR spectra of duplexes **I–IV**. Spectral positions of pump and observe pulses are indicated for each duplex. **b** DEER time traces for each duplex after removal of relaxation background. **c** Distance distributions obtained for each duplex. Regularization parameter 100 (see details in [43])

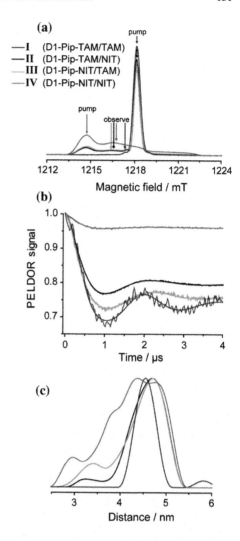

(a)

—**I** (D1-Pip-TAM/TAM)
—**II** (D1-Pip-TAM/NIT)
.....**III** (D1-Pip-NIT/TAM)
—**IV** (D1-Pip-NIT/NIT)

pump

pump observe

1212 1215 1218 1221 1224

Magnetic field / mT

(b)

PELDOR signal

1.0
0.9
0.8
0.7

0 1 2 3 4

Time / μs

(c)

3 4 5 6

Distance / nm

and "2 + 1" methods consists in the choice of the carrier frequency of the second microwave pulse. The pulse sequence of the "2 + 1" method is as follows. The primary ESE signal is formed by the first and third m.w. pulses with the carrier frequency ω_{01}, separated by the time interval t. The amplitude of the ESE signal is observed as a function of the position t' of the second m.w. pulse with the carrier frequency ω_{02}. The "2 + 1" pulse train benefits are an ability to measure dipole-dipole interactions between paramagnetic centers which are substantially weaker than those that can be measured by the ordinary two-pulse train and to determine the dipole interactions between spins with overlapping EPR spectra. The method was successfully used for the study of the structure of biological objects, photosynthetic systems in particular.

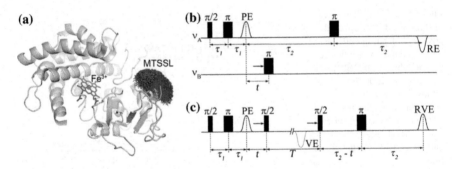

Fig. 6.4 a Model of MTSSL-labeled cytochrome P450cam mutant C58R1 using the crystal structure of the native form of the protein (PDB 3L61). The Fe^{3+}–MTSSL distances were measured using **b** the PELDOR and **c** the five-pulse RIDME experiments [46]

More elaborate variants of PELDOR, which address some of these shortcomings, are discussed below.

6.3.2 Four-Pulse PELDOR

In the advanced four-pulse PELDOR experiment invented and applied by the Jeschke group [15, 52], the following detection sequence $\pi/2 - \tau_1 - \pi - \tau_1 - echo_1 - \tau_2 - \pi$ was applied at a microwave frequency ν_A (Fig. 6.5). Then pulse at frequency ν_B during the time interval T_2 flips spins (spins B), which are in resonance with this second frequency is introduced. Incrementing the pump pulse from $t = 0$ to $t = T_2$ leads to an oscillation of the amplitude V of the refocued eco., which is dependent on the time position t of the inversion pulse, if spins A and B are coupled, and on the dipolar coupling ν_{dd} between both spins. The four pulse technique allows one to avoid a number of artifacts (see details in [15]. Data on the method's application to nitroxide biradical are also presented in [15].

Then pulse at frequency ν_B during the time interval T_2 flips spins (spins B), which are in resonance with this second frequency is introduced. Incrementing the pump pulse from $t = 0$ to $t = T_2$ leads to an oscillation of the amplitude V of the refocued eco., which is dependent on the time position t of the inversion pulse, if spins A and B are coupled, and on the dipolar coupling ν_{dd} between both spins. The four pulse technique allows one to avoid a number of artifacts (See details in [15]).

In work [16] application of the four pulse DEER experiment to long range distance measurements between two nitroxide spin labels, a nitroxide spin label and a Gd^{3+} ion, and two Gd^{3+} ions up to 6 nm was demonstrated (Fig. 6.6).

Thus, application of the four pulse PELDOR makes it possible to get improved structural information from various system under investigation including complex biological objects.

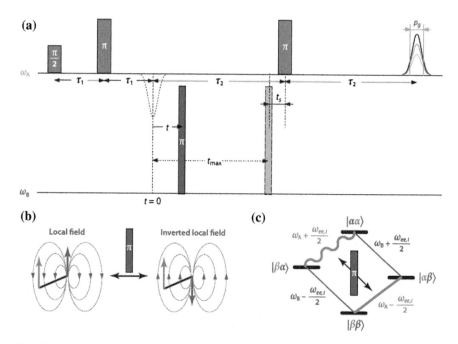

Fig. 6.5 Four-pulse DEER experiment. **a** Pulse sequence. Time t is varied from $t < 0$ to t_{max}, and variation of the integral echo intensity in the window of length pg is recorded. **b** Local field picture. The π pump pulse at frequency ωB inverts the state of spin B (*gray*), thus inverting the local field imposed by spin B at the site of spin A (*black*). **c** Energy level diagram. Inversion of the local field at spin A exchanges coherence between the two transitions of spin A that differ in frequency by $\omega_{ee,i}$ [15]

6.3.3 5 Pulse ELDOR (DEER 5)

The novel five-pulse DEER method is intended to develop into a widely useful technique to study nanoscale systems, where long or more accurate distances need to be measured [28-30]. This technique could also help to considerably shorten spectral acquisition times by offering greater sensitivity than could be achieved using standard DEER methods. Five-pulse DEER has four dipolar pathways and second pump pulse, 5, after the pulse 4 shifts dipolar modulation toward pulse 4, (Fig. 6.7c).

On the basis of an analysis of possible dipolar pathways in DEER-5, the intramolecular contribution of the dipolar signal in five-pulse DEER has the following form [28]

$$V_{\text{intra}}(t) \propto \langle q_3 q_5 + q_3 p_5 \cos[A(\tau - \delta T)] + p_3 q_5 \\ \cos[A(t - \tau)] + p_3 p_5 \cos[A(t - \delta T)]\rangle_{\text{orientations}} \tag{6.14}$$

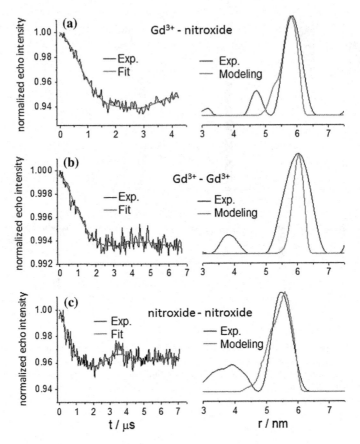

Fig. 6.6 2 DEER results of mixed labeled ERP29 chaperon dimers at 10 K. *Left panel* DEER traces after background removal with fits obtained with the distance distributions shown on the right. *Right panel* Distance distributions (see details in [16])

The first and the second terms in (6.14) are constant in t, the third term is the residual of the DEER-4 dipolar signal that exists in the absence of the fifth pulse. The fourth term gives the new DEER-5 specific dipolar signal. The angular brackets denote averaging over all orientations and Euler angles for all magnetic tensors.

The application of DEER 5 to investigation of spin-labeled T4 lysozyme is illustrated in Fig. 6.8. It was shown that for spin-labeled T4 lysozyme at a concentration less than 50 µM, as an example, the useful evolution time increases by a factor of 1.8 in protonated solution and 1.4 in deuterated solution to 8 and 12 µs, respectively. This enables a significant increase in the measurable distances, improved distance resolution, or both as the details of the V function may be traced with improved accuracy. The DEER 5 application allowed to measure distances between spin labels with good agreement with the enzyme X-ray model.

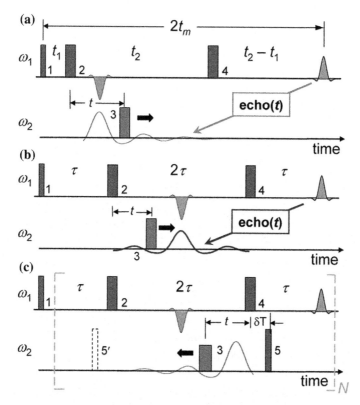

Fig. 6.7 a Standard four-pulse DEER sequence with the respective dipolar modulation pattern plotted in *green*. **b** The four-pulse sequence modified for $t_2 - 2t_1 - 2\tau$ to minimize nuclear spin diffusion **c** Placing the second pump pulse, 5, after the pulse 4 shifts dipolar modulation toward pulse 4, thereby recovering the full time span, 2τ (see details in [28])

The different versions of pulse ELDOR spectroscopy are widely employed in structural and molecular dynamical investigation of various systems including biradicals [53, 54], polymers [55], peptides [56], proteins [57], enzymes [58], Photosystem II [59], metal-containing systems [60], membranes [61]' oligonu-cleotides [62], DNA [63], RNA [64], and various biological systems [65, 66]. A number of reviews on theory, instrumentation and applications of various versions of the pulse ELDOR experiment were recently published [3, 5, 6, 8, 10, 11, 13, 18].

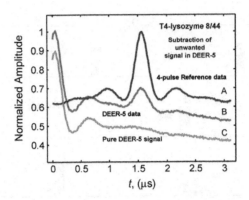

Fig. 6.8 Isolation of pure DEER-5 signal. The reference, A, was recorded in the absence of the fifth pulse and used in the removal of the incompletely suppressed dipolar signal of A-type from the raw signal, B, recorded in the presence of the fifth pulse. C indicates the pure five-pulse dipolar signal produced by subtracting scaled down A from B. A–C are normalized to unit amplitude at their maxima, with C shifted by −0.1 for clarity [28]

6.4   Two-Dimentional Fourier Transfer ESR

6.4.1   COSY (Correlation Spectroscopy), SECSY (Spin–Echo Correlation Spectroscopy)

Two pulse COSY and SECSY experiments are similar, although different in execution and analysis from Hahn's original two $\pi/2$ echo experiment were used (Fig. 6.9). In the standard COSY experiment, the first $\pi/2$ pulse creates transverse magnetization (± 1 coherences) [67]. The signal is measured after a second $\pi/2$ pulse. The COSY signal is given by the coherence pathway: $0 \rightarrow +1 \rightarrow -1$ for each orientation of the dipolar vector. The SECSY signal is related to the COSY signal by the transformation $t_2 \rightarrow t_2 + t_1$ and inhomogeneities are refocused in t_1. Hence the SECSY experiment allows one to determine the homogeneous single quantum linewidths along.

The SECSY experiment (Fig. 6.9) with a hard pulse has an advantage over the conventional Hahn echo method with a soft pulse in that the homogeneous line with across the spectrum can be obtained in a single experiment [68]. From The linear dependence T_2^{-1} versus τ_R^a, which is is the value of rotational correlation time in the slow motion region ($10^{-7} - 10^{-8}$ s), can be determined 2D FT ESR thus allows one to distinguish between the limiting models of very slow motion with little or no ordering versus that of very high ordering but substantial rotational diffusion rate. The determination of the hyperfine tensor element for protons also available for the SECSY technique.

The determination of the hyperfine tensor element for protons located at about electron spin also available for the SECSY technique.

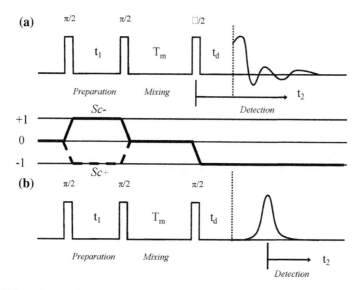

Fig. 6.9 The pulse sequences for **a** the standard 2D-ELDOR experiment and **b** SECSY format 2D-ELDOR experiments [68]

A detail theory for interpreting two-dimensional Fourier transform (2D-FT) electron spin resonance (ESR) experiments which encompasses the full range of motional rates from fast through very slow motions, and provides for microscopic as well as macroscopic molecular ordering was developed by Freed with coworkers [67]. Using the stochastic Liouville equation, the general properties of the pulse propagator superoperator, which describes the microwave pulses in Liouville space, were analyzed in terms of the coherence transfer pathways of 2D-FT ESR. In the framework of this approach, COSY, SECSY and 2D-ELDOR (electron–electron double resonance) sequences can be also considered a in terms of either the free-induction decay (FID)-like or echo decay-like spin response. The theory applied to simulate ESR spectra of n&oxide spin labels in membrane vesicles,

6.5 Two-Dimensional Electron-Electron Double Resonance (2D-ELDOR)

Two-dimensional electron-electron double resonance (2D-ELDOR) technique was invented, developed and applied by the Freed group [68]. The two-pulse primary echo experiment (pulse sequence: $\pi/2 - \tau-\pi - \tau -$ echo) and the three-pulse stimulated echo (pulse sequence: $\pi/2 - \tau-\pi/2 - T-\pi/2 - \tau -$ echo) were described.

The following time periods in a three pulse 2D-ELDOR experiment were established: (1) the preparation period beginning with the first $\pi/2$ pulse to generate

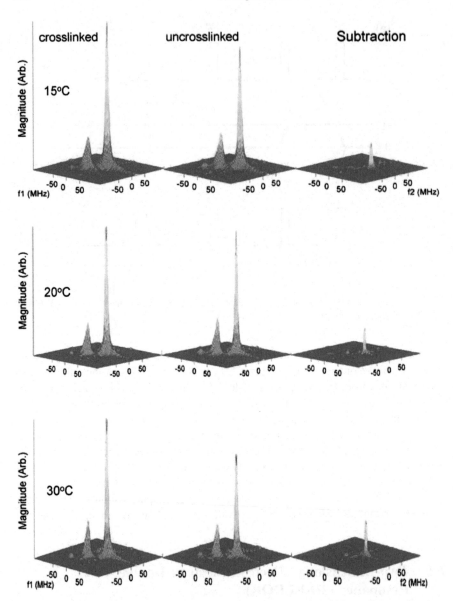

Fig. 6.10 Experimental 2D-ELDOR spectra for spin label 1-palmitoyl-2-(16 doxyl stearoyl) phosphatidylcholine (16PC) in cross-linked and un-cross-linked in plasma membrane vesicles (PMV) and their respective difference spectra at temperatures of 15, 20, and 30 °C, shown in the standard magnitude mode for convenience of display. (Here, $T_m = 50$ n [68])

the initial transverse magnetization, after which during the evolution period of length t_1; the "spin packets" evolve with their characteristic frequency, (2) The mixing period with time Tm follows the second $\pi/2$ pulse, and during this time, in the presence of pseudo secular hyperfine interactions, the longitudinal magnetization components of each hyperfine line may be exchanged, (3) The final $\pi/2$ pulse rotates the magnetization again into the xy plane for detection, and (4) the free induction decay (FID) signal is observed as a function of the detection time t_2, and the FID, obtained during the evolution period, t_1 is collected as a function of t_2 (Fig. 6.14). The experiments are repeated for a series of mixing time, Tm, at several temperatures. For the two-pulse primary echo and the three-pulse stimulated echo experiments, equations for the decay of the echo amplitude are given in [68].

In work [68], the simple case of a spin probe species characterized by inhomogeneous magnetic interactions was considered. It was suggested that in the absence of slow-motional effects each spin packet has a distinct resonance frequency with a well-defined transverse relaxation time T_2. As an example, experimental 2D-ELDOR spectra for 16PC in cross-linked and un-cross-linked plasma membrane vesicles are presented in Fig. 6.10.

The stochastic Liouville equation for which the director field is treated as a multidimensional Gaussian process was used for describing a nitroxide spin motion which occurs on two different time scales, the faster molecular motion relative to the local director, and the slower collective fluctuations of the director field [69]. Examples of the bending elastic moduli and transverse molecular relaxation rates extracted from 2D-ELDOR experiments using a slow-motional model on 1,2-dipalmitoyl-sn-glycero-phosphatidylcholine (DPPC) vesicles were also determined in [69].

Thus 2D-ELDOR has proven to be a technique that is sensitive to the fast and slow dynamical processes affecting spin in complex fluid environments [69] and references therein,

6.6 Double-Quantum Coherence ESR

An ESR version of multiple quantum coherence (MQC) is analogous to the NMR phenomenon [70]. The ESR MQC was predicted by Tang and Noris [71] and observed in experiments on spin-correlated pairs formed by laser-induced dissociation by Dzuba et al. [72]. A detailed theory for the analysis of a wealth of experimental data based on a novel pulsed ESR technique for the detection of double quantum coherence (DQC), which yields high quality dipolar spectra for distance measurements, was developed by Freed with coworkers [73, 74]. The DQC experiment extends the range of measurable distances compared to other techniques and also provides the high quality of the obtainable dipolar spectra to extract distance distributions,

An energy level diagram for a system of two coupled spins 1/2 showing single (SQT), double (DQT) and zero (ZQT) quantum transitions as well as the six-pulse DQC sequence are presented in Figs. 6.11 and 6.12, respectively.

Fig. 6.11 Energy level diagram for the system of two coupled spins of 1/2 showing single (SQT), double (DQT) and zero (ZQT) quantum transitions [75]

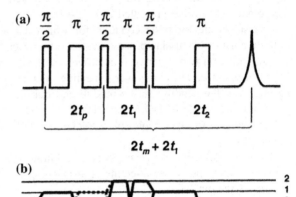

Fig. 6.12 The six-pulse DQC sequence **a** is shown with the corresponding coherence pathways (**b**). In this sequence the sum, $t_m \equiv t_p + t_2$, is kept constant (t_1 is also set constant). The echo amplitude is recorded versus $t_\xi \equiv t_p - t_2$ (**c**), and it is symmetrical with respect to $t_\xi = 0$. Cosine Fourier transformation of (**c**) produces the dipolar spectrum (**d**) [75]

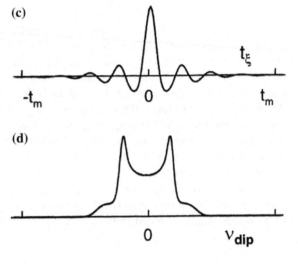

The double (DQT) and zero (ZQT) quantum transitions correspond to transitions with simultaneous flips or flip-flops of the spins, respectively. In the six-pulse sequence (Fig. 6.12a), the evolution of electron spin coherences due to the dipolar interaction is selectively detected. The dipolar spectrum of a biradical has the shape of a Pake doublet with a splitting of $4D/3$ between the turning point peaks indicative of a weak coupling, which corresponds to distances sufficiently exceeding 20 Å. The echo envelope is collected e as a function of $t_\xi = t_p - t_2$. In the limit of full excitation, the DQC dipolar signal takes on a simple form:

$$-\sin a t_p \sin a \left(t_m - t_p\right) = 1/2[\cos a t_m - \cos a t_\xi], \qquad (6.15)$$

where

$$t_\xi \equiv t_p - t_2 (c) \qquad (6.16)$$

and

$$a = 2D(1 - 3 \cos^2 \theta)/3 \quad (3) \qquad (6.17)$$

with the dipolar spin–spin interaction constant $D = 3\gamma_e^2 \hbar/2r^3$.

The distances between dipolar coupled electron spins are determined via the constant of dipolar interaction, D. For comparison purposes, it is possible to measure distances of up to about 20 Å by CW-ESR and to about 60–80 Å by pulsed ESR techniques

The main virtues of DQC ESR are the following [19]: (1) detection of weak dipolar interactions between paramagnetic molecules in the ESR signal and reliable measurement of distances and distance distributions between them up to 80 Å routinely and for longer distance in favorable cases; (2) determination of the angular geometry of the biradical; (3) determination of the fully asymmetric g-factor and hyperfine tensors; (4) characterization of many-body spin systems and (5) investigation of conformers important for elucidating molecular dynamic processes. Detailed comparison of DQC ESR with other pulse ESR technique was also presented.

The six-pulsed DQC technique was used for distance measurements between nitroxides in eight T4 lysozyme (T4L) mutants, doubly labeled with methanethiosulfonate spin-label methanethiosulfonate spin-label (MTSSL) at 9 and 17 GHz [75]. It was shown that the distances between labels span the range from 20 Å for the 65/76 mutant to 47 Å for the 61/135 mutant (Fig. 6.13). The high quality of the dipolar spectra also allows the extraction of distance distributions, the width of which can be used to set upper and lower bounds for geometry refinement. The distances and distributions found in this study was rationalized in terms of the known crystal structure, the characteristic conformers of the nitroxide side chains, and molecular modeling. This study sets the stage for the use of DQC-ESR for

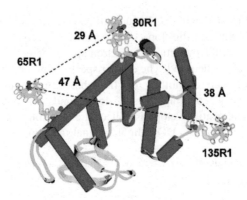

Fig. 6.13 Distances in spin labeled T4 lysozyme [75]

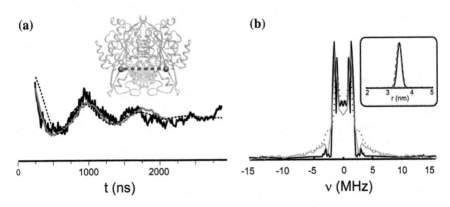

Fig. 6.14 The comparison of the Cu2$^+$DEER data (*gray solid line*) with the experimental (*black solid line*) Cu2$^+$DQC data (**a**) The time traces, (**b**) The DEER and DQC spectra, the distance distribution functions are provided in the inset Echo detected (ED) EPR spectra, Two-Pulse ESEEM, 2H-Mims ENDOR Four-Pulse DEER [76]

establishing the tertiary structure of large proteins with just a small number of long-distance constraints.

On an example of measurements of the Cu^{2+}–Cu^{2+} distance in the complex of restriction endonuclease EcoRI with its cognate DNA a simple way to minimize the low frequency nuclear peaks in the Double quantum coherence spectrum and resolve the dipolar interaction between the two with high sensitivity was reported [76]. The DQC data were found to be in good agreement with the Cu2$^+$-DEER data (Fig. 6.14).

Further examples on the use of the DQC technique for distance measurements in immobilized spin labeled proteins and membranes in liquid solution can be found in [19, 74, 75] and references therein.

6.7 Electron Spin Echo Envelope Modulation

Electron spin echo envelope modulation (ESEEM), pioneered by Mim [77] arises from a coupling between the EPR-active electron being probed and nuclei within the system when there is state mixing of the hyperfine levels. In ESEEM experiments, the modulation of the echo intensity is monitored as a function of the delay between the microwave pulses. ESEEM gives information complementary to ENDOR. ESEEM is suited for measuring weak hyperfine couplings, e.g. of the order of the free nuclear Larmor frequency, while continuous wave ENDOR is better suited for strong hyperfine couplings [13, 21, 78]. The resulting ENDOR spectrum is an ESR-enhanced NMR spectrum of the nuclei that are coupled to the electron being probed. In ESEEM experiments, time-resolved data acquisition is achieved on ESR timescales, that is, up to 1000 times faster than NMR.

The two most common forms of ESEEM experiments use either a two pulse ($\pi/2 - \tau - \pi - \tau -$ echo) or three pulse ($\pi/2 - \tau - \pi/2 - T - \pi/2 - \tau +T -$ echo) sequences. In 2-pulse ESEEM experiments a typical Hahn Echo sequence is used and the π pulse is added to longer periods of τ [13]. The amplitude or integrated area of the echo is measured for each τ, giving rise to a time domain ESEEM signal. The modulation envelope arises from nuclear transitions that are partially allowed during an electron spin flip. Such nuclear spin flips can occur when the hf interaction is anisotropic, or if the nuclear quadrupole interaction is non-negligible.

The rotating-frame spin Hamiltonian, in angular frequency units, of an electron spin (S = 1/2) coupled to N spin I nuclei ($I_1 = \cdots = I_{Ni} = 1/2$) using the high-field approximation is given by [78].

$$H_0 = \Omega_S S_z + \sum_{q=1}^{N_1} (\omega_{I,q} I_{z,q} + A_q S_z I_{z,q} + B_q S_z I_{x,q}),$$ (6.18)

where $\Omega_S = \omega_S - \omega_{mw}$, where ω_S is the electron resonance frequency ω and ω_{mw} is the frequency of the applied microwave (mw) pulse, $\omega_{I,q}$ is the Larmor frequency of nucleus q, and Aq and Bq are the secular and pseudo secular hyperfine interactions, respectively. This equation is valid when the time scale of the ESEEM experiment is short compared to the transverse electron spin relaxation and the spin Hamiltonian describes a closed multi-spin system. The oscillation in the system was described by Hamiltonian

$$H_0^{\text{diag}} = U H_0 U^\dagger \quad U = \prod_{q=1}^{N_1} e^{-i(\eta_{\alpha,q} S_\alpha + \eta_{\beta,q} S_\beta) I y R}.$$ (6.19)

where h_{aq} and $\eta_{\alpha q}$ are the nuclear quantization axes for a($m_s = +\frac{1}{2}$) and b($m_s = -\frac{1}{2}$) correspondingly. The resonance frequencies of nucleus q also associates with the a and b.

In the frame of the ESEEM theory, transverse relaxation was also taken into account [78]. The transverse relaxation, can be induced by fluctuating hyperfine fields of matrix protons, with the fluctuations arising from nuclear spin diffusion. Experiment and theory show that in this situation the echo intensity has a quadratic time dependence, $E(\tau) = \exp[-2(\tau/T_m)^2]$, where the phase memory time T_m is proportional to the reciprocal number density of protons in the matrix.

Two-, three-, four-pulse ESEEM, and HYSCORE techniques were considered in details in a comprehensive review [79]. In the three-pulse ESEEM experiment two pulses, separated by time τ, are applied followed by a third pulse after time T, and the stimulated echo is observed at time t after the third pulse [79]. The echo envelope, obtained as T is incremented, is modulated by the nuclear transition frequencies of nuclei coupled with the electron spin; Advantages of the three-pulse over the two-pulse procedure include: modulations of the stimulated-echo envelope persist for longer times, leading to narrower lines in the frequency-domain spectrum; the spectral lines in three-pulse ESEEM spectrum correspond to the individual nuclear frequencies, and this usually simplifies the spectrum. The four-pulse ESEEM was proved to be effective to measure 'combination frequencies', i.e. the sum ($\varpi a + \varpi b$), or difference ($\varpi a - \varpi b$) of basic nuclear transition frequencies in different MS manifolds of the electron spin.

A new type of ESEEM was observed in a three-pulse stimulated echo experiment on a nitroxide in a molecular glass [21]. With fixed interval T between the second and the third pulses and variable interval τ between the first and the second ones, τa response appears in the frequency domain manifesting as a broad peak around ~ 5 MHz. Time domain stimulated ESEEMs for R1 in toluene at 77 K with $T = 3\,\mu s$ and $T = 50\,\mu s$ was observed. The rate of growth of this peak with increasing T reports on the methyl group reorientation rate, which was found to be strongly temperature dependent. In an extension of the method, four-pulse ESEEM was used to measure 'combination frequencies [80]. The four pulse method, which exploits additional filtering of the relevant signal is advantageous for measuring combination lines compared to two-pulse ESEEM. Another variant of ESEEM that aims to improve the ESEEM modulation depth is five-pulse ESEEM, with the sequence $(\pi/2 - \tau_1 - \pi - \tau_1 \Box \pi/2 - T - \pi/2 - \tau_2 - \pi - \tau_2 -$ echo has also been proposed.

Efficiency of advance ESEEM techniques can be illustrated by recent examples. To provide solvent accessibility of Gd^{3+-}ADO3A-labeled melittin a combination of modern pulse ESR techniques has been employed [81]. The following sequences were used: Echo detected EPR ($\pi/2 - \tau - \pi - \tau$–echo), the saturation ($t_{sat} - t - /2 - \tau - \pi - \tau -$ echo), ESEEM, ($\pi/2 - \tau - \pi - \tau -$ echo), ^{2}H-Mims ENDOR ($\pi/2 - \tau - \pi/2 - T - \pi/2 - \tau -$ echo), and Four-Pulse DEER, $\pi/2(\nu_{obs}) - \tau - \pi(\nu_{obs}) - \tau + t - \pi(\nu_{pump}) - (T - \tau - t) - \pi(\nu_{obs}) - (T - \tau) -$ echo). The Gd^{3+} ESEEM and ENDOR results were found to be in reasonable agreement with each other, 82. To get insight into the structure and the chemical environment of the [4Fe–4S] cluster of HydF from the hyperthermophilic organism *Thermotoga neapolitana*, a combination of CW-EPR, three-pulse ESEEM, and HYSCORE spectroscopies has been exploited [83]. In order to characterize the proton

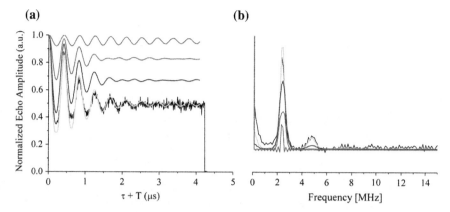

Fig. 6.15 X-band 3p-ESEEM quotient traces of HydFT.n. H/D exchange recorded at a magnetic field corresponding to g = 1.90 (*black*), before **a** and after **b** Fourier transform; T = 10 K. Simulation as derived from the contributions of matrix protons (*dark yellow*), protons I (*blue*), and protons II (*red*) is reported as an *orange bold line* [83] (color online)

environment of the [4Fe–4S] cluster of HydFT.n and possibly assign the fourth ligand to a protonated species, proton/deuteron (H/D) exchange experiments were performed The H/D exchange experiments allowed to authors measuring the weak magnetic interactions between the FeS cluster and the nearby deuterium/proton nuclei. The closest to the cluster a Fe–H distance was estimated as 2.1–3.0 Å. X-band 3p-ESEEM spectra of HydFT.n are presented in Fig. 6.15.

5-pulse ESEEM and 6-pulse HYSCORE examined in [20] were characterized with two benefits: they can recover signals from weakly modulating nuclei and yield spectra that are less affected by cross suppression. Figure 6.16 shows 2, 3, 4, 5, 6 pulse ESEEM and HYSCORE.

Hyperfine sublevel correlation spectroscopy (HYSCORE), introduced by Hofer et al. [82]. is a two-dimensional ESEEM technique based on a four-pulse sequence (Fig. 6.16). The echo intensity is recorded as a function of the interpulse time between the second and third pulses, t_1, and the time between the third and fourth pulses, t. The nuclear coherences exchanged by the π-pulse evolve during the time intervals t_1 and t_2. Correlations between nuclear transitions from the different M_S manifolds are provided by the frequency-domain HYSCORE, obtained after Fourier transforming the time-domain spectrum in two dimensions, e.g. t_1 and t_2, contains cross-peaks Accordingly, in the case of one I = 1 nucleus, a total of 18 cross-peaks are expected. Due to the high spectral resolution, HYSCORE experiments can be performed on samples with a low concentration of paramagnetic species.

DONUT-HYSCORE (double nuclear coherence transfer hyperfine sublevel correlation) is a two-dimensional experiment, which was designed to obtain correlations between nuclear frequencies belonging to the same electron spin manifold [84]. The DONUT-HYSCORE experiment is designed to improve the resolution of

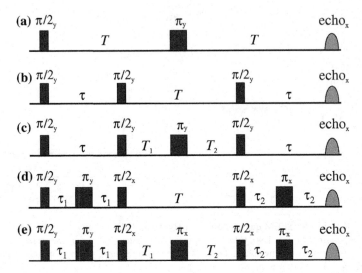

Fig. 6.16 ESEEM pulse sequences (variable T, T1, T2 and fixed s, s1, s2: **a** 2-pulse ESEEM, **b** 3-pulse ESEEM, **c** standard (4-pulse) HYSCORE, **d** 5-pulse ESEEM, **e** 6-pulse HYSCORE [21]

the ESEEM frequencies and to resolve and simplify the assignment of ESEEM frequencies for the case of an electron spin, $S = {}^1/_2$, interacting with a number of nuclear spins with $I \geq 1$ with nuclear quadrupole interactions This method is complementary to the standard HYSCORE experiment which generates correlations between nuclear frequencies belonging to different M_S manifolds. The sequence employed is $\pi/2 - \tau_1 - \pi/2 - t_1 - \pi - \tau_2 - \pi - t_2 - \pi/2 - \tau_1 -$ echo, and the echo is measured as a function of t_1 and t_2 whereas τ_1 and τ_2 are held constant. The DONUT is manifested as cross-peaks in a 2D spectrum obtained by Fourier transform with respect to t_1 and t_2. As an illustration, the DONUT-HYSCORE experiment was successfully applied to detect nuclear frequencies spectra in systems with ^{14}N nuclei in both a single crystal and frozen solution.

The combination of high-field electron paramagnetic resonance and ESEEM with specific site-directed spin labeling (SDSL) techniques was used to reveal subtle changes of the polarity and proticity profiles in proteins embedded in membranes [85]. This information was obtained by high-field EPR resolving principal components of the nitroxide Zeeman (g) and hyperfine (A) tensors of the spin labels. The W-band (95 GHz) high-field ESEEM method was subsequently used for determining with high accuracy the ^{14}N quadrupole tensor principal components of a nitroxide spin label in disordered frozen solution. Figure 6.17 shows the W-band microwave pulse sequence for the stimulated high-field ESEEM experiment on the nitroxide radical, examples of a nuclear modulation echo decay trace at the indicated magnetic field position, and the Fourier transformed spectrum of the ESEEM decay. The quadrupole components of a five-ring pyrroline-type

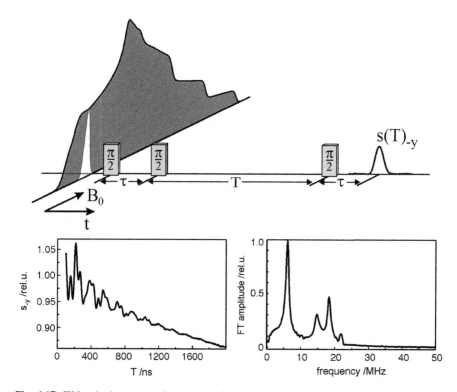

Fig. 6.17 W-band microwave pulse sequence for the stimulated high-field ESEEM experiment on the five-ring pyrroline-type nitroxide radical. *Top* The echo detected EPR spectrum of the R1-14N radical as well as the microwave excitation bandwidth for typical microwave pulse-length settings are shown for clarity. *Bottom left* A representative example of a nuclear modulation echo decay trace at the indicated magnetic filed position is shown. *Bottom right* The Fourier transformed spectrum of the ESEEM decay example is given, the cutoff frequency of the high-pass filter was set to 1.5 MHz [85]

nitroxide radical in glassy *ortho*-terphenoyl and glycerol solutions have been determined by W-band ESEEM.

Recent application of ESEEM in combination with other pulse ESR techniques are reported in [81–83] and references therein.

6.8   Pulse ENDOR

In CW ESR a strong hyperfine interaction is usually observed when the nucleus is in close proximity of the unpaired electron. The interactions with more distant nuclei can be investigated by means of ESEEM, continuous wave or pulse ENDOR [13, 21–23]. Pulse ENDOR and ESEEM spectra contain information on the type of nuclei, on the distances between nuclei and on the spin density distribution (hyperfine interaction) and on the electric field gradient at the nuclei caused by nuclear quadrupole interaction. In addition, in the frame of pulse-ENDOR technique, the microwave pulse sequence can generate an electron spin echo which can be modulated by manipulation of the nuclear spins due to the radio-frequency pulses. Therefore, the ENDOR spectrum yields an ESR-enhanced representation of the nuclear transition frequency spectrum, which might otherwise be too weak to observe directly. The basic principles of continuous wave and pulse ENDOR were reviewed in [23].

In Davies ENDOR of an $S = 1/2$, $I = 1/2$ system, a microwave (mw) inversion-recovery pulse sequence $(\pi - T - \pi/2 - \tau - \pi - \tau - echo)$ is used (Fig. 6.17) [13, 85]. First, one EPR transition is inverted by the preparation pulse. Subsequently the mw π-pulse inverts the positive initial thermal equilibrium polarization, Therefore, Davies ENDOR is useful for systems with large HFIs. During the T interval, the rf pulse changes the population of the nuclear sublevels, and thereby the polarization of the EPR transition is partially restored. This effect is detected by the echo intensity, i.e., by the final part of the pulse sequence $\pi/2 - \tau - \pi - \tau - echo$.

In Mims ENDOR [86], bEPR transitions are excited by the applied stimulated echo mw pulse sequence $(\pi/2 - \tau - \pi/2 - T - \pi/2\pi - \tau - echo)$. The precession frequency during the first and the second evolution times differ by the value of a was initiated by the rf π-pulse flips the nuclear spin m_I, which in turn changes the frequency of the electron spin Larmor precession No ENDOR effect is observed when $a\tau = 2\pi n$, where n is an integer number. applications of this method are limited to relatively small HFI constants $(B_1 > a)$.

Pulse Mims and Davies ENDOR experiments (Fig. 6.18) were performed for detail investigation of the photoexcited triplet state of the carotenoid peridinin in the

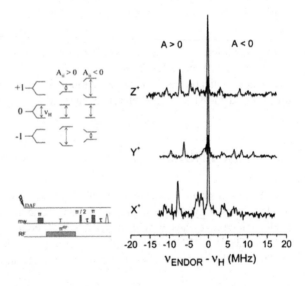

Fig. 6.18 *Top left* Spin energy levels of a triplet state, showing positive and negative hyperfine interaction with one nucleus with spin I = 1/2 for the field parallel to one of the canonical axes of the ZFS tensor *Bottom left* Pulse scheme for Davies-ENDOR experiment for photoexcited triplet states. *Right* Davies ENDOR spectra of HPCP from H. pygmaea for the X, Y and Z field position at T = 20 K. The frequency scale gives the deviation of the resonance line from mH. For experimental conditions see [22]

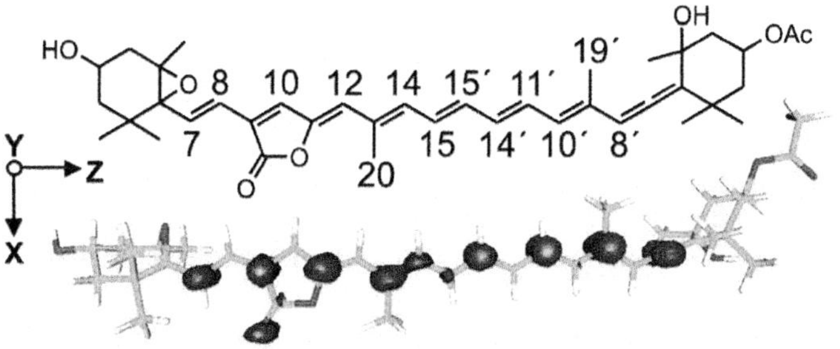

Fig. 6.19 Pulse ENDOR spectra of the short-lived photoinduced spin-polarized triplet state of the carotenoid peridinin in the PCP (peridinin–chlorophyll–protein) antenna of *A. carterae*. Molecular structure of the carotenoid peridinin and spin density plot of peridinin 614 in its excited triplet state is also shown [88]

peridinin–chlorophyll a–protein (PCP) of the dinoflagellate *Heterocapsa pygmaea* [22].

Mehring et al. [87] introduced a time-domain pulsed method in which polarization transfer ENDOR schemes can be performed with selective as well as nonselective mw pulses. In these experiments, the mixing period consists of two rf $\pi/2$ pulses separated by a variable time interval T'. Hyperfine-correlated electron nuclear double resonance spectroscopy (HYEND), where the nuclear transition frequencies are correlated with the corresponding hyperfine frequencies, leads to a considerable simplification of the interpretation of the spectra.

Orientation-selective pulse Q-band ENDOR spectroscopy (34 GHz) and density functional theory (DFT) calculations were used for investigation of the triplet state of the carotenoid peridinin in the refolded N-domain peridinin-chlorophyll-protein (PCP) antenna complex from *Amphidinium carterae* created by triplet–triplet transfer from ^{3}Chl *a*, generated by illumination at 630 nm (Fig. 6.19) [88]. Thirteen proton hyperfine coupling (hfc) tensors are deduced for the peridinin triplet state.

ENDOR spectroscopy can provide detailed information on multiple, magnetically active nuclei ($^{1,\ 2}$H, ^{13}C, ^{57}Fe, ^{95}Mo) from paramagnetic centers in a bioinorganic system [89]. Thus pulsed ENDOR allows identification of the constituent metal ions and their formal oxidation states, and identifies the nature of the substrate analog or inhibitor binding to the clusters including the iron–molybdenum cofactor in nitrogenase.

6.9 Pulse Saturation and Inversion Recovery

A detailed description of the continuous wave microwave saturation recovery technique was presented in publications of the Hyde group [90]. The theory of saturation recovery was developed by Freed and Wang [91]. A disadvantage of this technique is that the sensitivity of CW detection is less than the 2-pulse echo repetition method. An advantage of the CW detection scheme is that it can be used in the case of very short T_2 and deep echo envelope modulation. The theory of relaxation, applied to saturation and double resonance in ESR spectra was developed in [91]. The saturation recovery literature prior to 2000 has been reviewed comprehensively by the Eatons [92]. Data on relaxation times, T_1 and T_2 for this period were also collected in this review. Experimental, and theoretical aspects, as well as applications of the saturation recovery technique were discussed in [93, 94, 96] and references therein.

As an example of a recent application of saturation phenomena, electron spin relaxation times of perdeuterated tempone (PDT) 1 and of a nitronyl nitroxide (2-(4-carboxy-phenyl)-4,4,5,5-tetramethylimidazoline-3-oxide-1-oxyl) 2 in aqueous solution at room temperature were measured by 2-pulse electron spin echo (T_2) or 3-pulse inversion recovery (T_1) in the frequency range of 250 MHz to 34 GHz [93]. At 9 GHz values of T1 measured by long-pulse saturation recovery were in good agreement with values determined by inversion recovery. The data obtained on T_1 and T_2 relaxation were used for detailed analysis of nitroxide motion in the fast tumbling regime. It was shown that a thermally activated process makes a significant contribution in the fast tumbling regime.

The inversion recovery three pulse sequence ($\pi - T - \pi/2 - \tau - \pi - \tau -$ echo) includes two Free Induction Decay and a three pulse echo [92]. The sequence steps are: inversion of spins followed by relaxation during time T, generation of the spin echo by the second and third pulses and pulse phase cycling to eliminate unwanted echoes. This sequence allows one to observe the effect of spectral diffusion because the inverting pulse is short. In the so called "picket fence" sequence a series of $\pi/2$ pulses with spacing greater than a few spin echo dephasing times, Tm, is used to saturate the spin system followed by sampling its recovery with a 2-pulse echo [92]. The echo repetition rate pulse method makes it possible to determine T_1 by measuring the amplitude of a 2-pulse echo as a function of the repeating pulse sequence ($\pi/2 - \tau - \pi -$ echo $- \tau > T_1$) and increasing the pulse repetition rate. The T_1 value can be found from the dependence of the echo amplitude on the pulse repetition rate [92]. In this method, a 2-pulse echo is formed before the z magnetization completely recovered and the echo forms with a progressively smaller amplitude.

Information on recent applications of the inversion recovery technique is presented below.–lattice relaxation times, T_1, for EPR signals created by irradiation of tooth enamel or carbonate-doped hydroxyapatite were studied by three-pulse inversion recovery and long-pulse saturation recovery [97]. The dominant component in the EPR signal for tooth samples is assigned to CO_2^-. The dose independence of T_1 and T_2 at ambient temperature provides the basis for using dose-independent microwave powers to record dosimetric tooth signals at microwave powers above the linear response regime. Inversion recovery (T_1) and microwave power saturation studies have also been performed, between 4 and 25 K, on the EPR signal from the stable tyrosyl radical, $Y_D^{\cdot}$, in photosystem II core complex preparations from higher plants [95]. Measurements were performed for the dark stable S1 and first turnover S2 states of the photosystem catalytic Mn cluster and in two cryoprotectant regimes; sucrose and ethylene glycol/glycerol [97].

The inversion recovery kinetics in the dark stable S1 and first turnover S2 states of the photosystem catalytic Mn cluster show a dominant, non exponential decay component which is well described by a through space dipolar relaxation model. The background rate showed an S state temperature dependence, consistent with an interaction between $Y_D{}^{\bullet}$ and the Mn cluster in the multiline S_2 state, over a distance of ~ 30 Å. The temperature dependence of the electron spin-lattice relaxation time T_1 was measured for the S0 state of the oxygen-evolving complex in photosystem II and for two dinuclear manganese model complexes by pulse EPR in temperature range 4.3–6.5 K using the inversion-recovery method [97]. It was shown that for [Mn(III)Mn(IV)(μ-O)2bipy4]ClO$_4$, the Raman relaxation process dominates at temperatures below 50 K, while Orbach type relaxation was found for [Mn(II)Mn (III)(μ-OH)(μ-piv)2(Me$_3$tacn)$_2$](ClO$_4$)^2 between 4.3 and 9 K.

6.10   ESR Spectra Hole Burning

For inhomogeneously broadened electron paramagnetic resonance spectra of solids and viscous liquids, spectral diffusion can be studied by selective hole burning (Dzuba et al.) [99]. The width of a selective hole can be comparable to the homogeneous linewidth. The hole shape is detected using nonselective pulses to form free-induction decays and spin echoes. In the hole burning method, a so called "soft pulse" technique has been developed by Schweiger et al. [99]. The soft pulse 'burns' a hole in the inhomogeneously broadened EPR spectrum that is detected using a free-induction decay (FID). When the hole has a width comparable to that of the ESR homogeneous linewidth (0.5–0.05 G) the hole is called a selective hole. The pulse sequence for the hole-burning experiment is $\pi - \tau - \pi/2 - \tau - \pi/2$. Applications of the method are partitioned into two broad classes with respect to the origin of perturbations that influence the hole shape, after the action of the hole-burning pulse. A so-called internal perturbation is caused by spin-lattice relaxation, by spectral diffusion or by driving of forbidden transitions, which change the ESR spectral line shape, while external perturbations include rapid stepping of the magnetic field, change of the orientation of the external magnetic field or sample rotation.

Sources of spectral diffusion after the selective hole burning are [100]: (1) chemical exchange which can be induced by chemical reactions, by transitions between different molecular conformations or isomers, and by rotation about some particular molecular bond; (2) molecular tumbling in a viscous liquid; (3) spin-lattice relaxation which can induce a stochastic shift of the resonance lines for individual spins, which can then change the orientation along the magnetic field; (4) spin diffusion, a process of mutual flip-flops in the spin system leading to a spatial diffusion of the Zeeman energy. Spectral diffusion is characterized by two important parameters, the scale of spectral diffusion $\Delta\omega$ and the correlation time τc [98]. For a slow tumbling molecule, $\Delta\omega$ is of the same order as the full ESR spectra linewidth $\Delta\Omega$. In the case of fast spectral diffusion, the spectral diffusion lineshape collapses into a single narrow line. For a slow tumbling molecule, $\Delta\omega$ is of the same order as the full ESR spectra linewidth $D\Omega$.

The possibility of studying the spatial distribution of paramagnetic species in solids was discussed in detail by Dzuba and Kawamori [100]. This approach is based on the spectral diffusion effect, which arises from fluctuations of dipole-dipole interactions induced by spin-lattice relaxation. The frequency shift of the A spin caused by the dipole-dipole interaction with the **B** spin a pair was described by the function

$$\in (r, \theta) = \gamma_A g_B \beta (1 - 3\cos^2\theta)/r^3 \tag{6.20}$$

where r is the distance from the A to the B spin and Θ is the angle between the z axis and the radius vector r. When the B spin flips, the resonance frequency for the A spin is shifted because the position of the resonance line is determined by the product $e(r,\Theta)mB$; where mB ($+1/2$, $-1/2$) is projection of the B spin onto the z axis.

For selective hole-burning relaxation-induced dipolar modulation enhancement (RIDME) experiments with stable nitroxide biradical in a frozen solution a sequence ps–T–p/2 ns–s–pns–s–refocused FID was used [101]. Such a sequence allowed one to eliminate the major unwanted contributions to the FID modulation originating from spectral diffusion and from excitation of forbidden electron-nuclear spin transitions by mw pulses. The proposed method can be employed to measure interspin distances in the range of 13 Å $\leq$ r $\leq$ 25 Å.

6.11   ESR Nutation Spectroscopy

Nutation is a rocking, swaying, or nodding motion in the axis of rotation of a largely axially symmetric object 1. The nutation frequency of the moment on resonance (Rabi frequency) is described by the transition moment and transition probability between the spin sublevels involved in the transition [102–109]. Electron spin nutation ESN is based on general phenomena in which an electron-spin magnetic moment/magnetisation precesses around an effective magnetic field in an oscillating microwave irradiation field. Determination of relatively short distances between weakly exchange-coupled electron spins is a main goal of application of the pulse- ESR based electron spin ESN spectroscopy The experimentally detected nutation frequency as a function of the microwave irradiation strength ω_1 (angular frequency) for cases of weakly exchange-coupled systems can be classified into three categories; D_{12} (spin dipolar interaction)-driven, Δg-driven and ω_1-driven nutation behaviour, where ω_1-driven nutation depends on the strength of ω_1.

The influence of anisotropic hyperfine interactions on transient nutation electron paramagnetic resonance (EPR) of light-induced spin-correlated radical pairs was studied theoretically using the density operator formalism. Dzuba and Kawamori [101] for On the basis of a model system comprised of a weakly coupled radical pair and one hyperfine-coupled nucleus with $I = 1/2$, analytical expressions for the time evolution of the transient EPR signal during selective microwave excitation of single transitions were given. In the frame of accepted model, the sudden light-induced generation of a triplet radical pair state from a singlet-state precursor. Two mechanisms of coherent nuclear modulations in the time-resolved EPR signals of doublets and radical pairs were presented for various microwave magnetic field strengths. The first modulation scheme is induced by electron and nuclear coherences initiated by the laser excitation pulse as monitored by a weak microwave magnetic field. In a second mechanism the modulations are both created and detected by microwave radiation. The goal of synchronous-nutation method based on behavior of the associated nuclear spin system is to measure a cross-relaxation or slow-exchange rate constant between two spins avoiding spin diffusion [103]. A three spin system in which two of the spins are forced to nutate synchronously has been considered using a master equation which described the evolution of density operator representing the state of the spin system. The 2-D ESN spectroscopy was used to study Eu^{2+} ion ($S = 7/2, I = 5/2$), which has two isotopes (^{151}Eu [47.9 %] and ^{153}Eu [52.1 %]), in a CaF_2 single crystal [105]. The complicated fine-structure hyperfine ESR spectra, related to the allowed and forbidden transitions in high-spin systems, was identified by invoking the spectral simulation of the 2-D ESN spectra on the basis of transition moment analyses. The analyses were based on exact numerical calculations of the transition moments as well as a perturbation-based analytical approach combined with reduced rotation matrices for the nuclear part of the transition moment. Figure 6.20 llustrates result of the analysis showing the contour plot of the X-band 2-D ESN spectra of the Eu^{2+} ($^8S_{7/2}$), as shown in Fig. 6.20. In work [107], two-dimentional (2D-ESN) spectroscopy was used to determine spin dipolar interactions for spin distances within 2.0 nm in a system of weakly exchange-coupled molecular spins in non-oriented media. In the extreme weak irradiation limit, the nutation frequency ω_n for the $|MS\rangle \rightarrow |MS + 1\rangle$ allowed transition of the isolated spin state S is described by an expression,

$$\omega_n = |G|\sqrt{S(S+1) - M_S(M_S + 1)}|\langle M_1(M_s)|M'_I(M_s + 1)\rangle|\omega_1 \qquad (6.21)$$

where ω_1 denotes the irradiation strength B1, in units of angular frequency, of the microwave field and G is the anisotropy of the g-factor. The third term of (6.21) shows the contribution of the nuclear spin transitions involved during the ESR transition. In order to analyse quantitatively the 2D-ESN behaviour of a biradical, the frequency ω_n was measured as a function of ω_1 over a wide range. It was shown that for hetero-spin biradicals, such as Biradical 6.1, Δg effects can be a dominating feature in biradical nutation spectroscopy.

Thus ESN spectroscopy has been proved to be a useful complement to other modern pulse ESR methods.

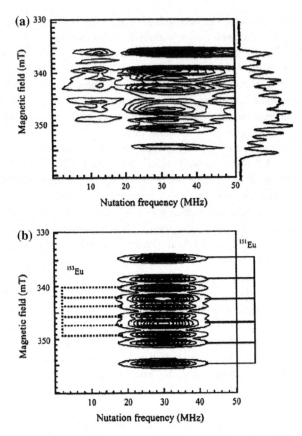

Fig. 6.20 Contour plot of the X-band 2-D ESN spectra of the Eu2+(8S7/2) ion with B 0 parallel to the [105] axis in the magnetic field range of 330–360 mT, where only the M s = 1/2) → M s = −1/2) transition is involved. On the right-hand side of a the ESE-detected field-swept fine-structure hyperfine spectrum through the magnetic field window is displayed, **a** The 2-D ESN spectra observed at 3.5 K. **b** The 2-D ESN spectra calculated by invoking the theoretical transition moment analysis. The *dotted* and *solid lines* stand for the hyperfine transitions from ∼5∼Eu and mEu, respectively

Biradical 6.1 .

References

1. S.A. Dikanov, Yu. D, *Tsvetkov Electron spin echo envelope modulation (ESEEM) spectroscopy* (CRC Press, 1992)
2. O. Grinberga, L. Berliner (eds.), *Very High Frequency (VHF) ESR/EPR* (Springer, 2011)
3. A. Lund, M. Shiotani, S. Shimada (eds.), *Principles and Applications of ESR Spectroscopy* (Springer, 2011)
4. S.K. Misra (ed.), *Multifrequency Electron Paramagnetic Resonance: Theory and Applications* (Wiley-VCH, 2011)
5. G.R. Eaton, S.S. Eaton, D.P. Barr, R.T. Weber, *Quantitative EPR* (Springer, 2010)
6. G.M. Smith, D.J. Keeble, Schiemann O (eds.), *Introduction to Modern Epr Spectroscopy* (CRC, 2010)
7. K. Moebius, A.N. Savitsky, *High-Field EPR Spectroscopy of Proteins and their Model Systems: Characterization of Transient Paramagnetic States* (Royal Society of Chemistry, 2009)
8. W.R. Hagen, *Biomolecular EPR Spectroscopy* (CRC, 2008)
9. S.S. Eaton, G.R. Eaton, L.J Berliner (eds.), Biomedical EPR —Part B: Methodology, Instrumentation, and Dynamics, in: *Biological Magnetic Resonance*, vol. 24 (Springer, 2005)
10. L.J. Berliner, S.S. Eaton, G.R. Eaton (eds.), Distance measurements in biological systems by EPR in *Biological Magnetic Resonance*, vol. 19 (Springer, 2001)
11. G.R Eaton, SS.Eaton, K.M. Salikhov, *Foundations of Modern EPR* (World Scientific, 1998)
12. M.H. Levitt, *Spin Dynamics: Basics of Nuclear Magnetic Resonance* (Wiley, 2008)
13. A. Schweiger, G. Jeschke, *Principles of Pulse Electron Paramagnetic Resonance* (Oxford University Press, 2001)
14. B.C. Gilbert, D.M. Murphy, V. Chechnik (eds.), *Electron Paramagnetic Resonance, Specialist Periodical Report*, vol. 23 (Royal Society of Chemistry, 2012)
15. D. Jeschke, Annu. Rev. Phys. Chem. **63**, 419 (2012)
16. I. Kaminker, H. Yagi, T. Huber, A. Feintuch, G. Ottin, D. Goldfarb, Phys. Chem. Chem. Phys. **14**, 4355 (2012)
17. A. Savitsky, A.A. Dubinskii, H. Zimmermann, W. Lubitz, K. Möbius, J. Phys. Chem. B **115**, 11950 (2011)
18. Y.D. Tsvetkov, A.D. Milov, A.G. Maryasov, Russ. Chem. Rev. **77**, 487 (2008)
19. P.P. Borbat, J.H. Freed, (eds.), Distance measurements in biological systems by EPR, in *Biological Magnetic Resonance*, ed. by L.J. Berliner, S.S. Eaton, G.R, Eaton vol. 19 (Springer, 2001), p. 383
20. B. Kasumaj, S. Stoll, J. Magn. Reson. **190**, 233 (2008); B. Kasumaj, H. Dube, N. Zoelch, F. Diederich, G. Jeschke, J. Magn. Reson. **223**,187 (2012)
21. L.V. Kulik, I.A. Grigor'ev, E.S. Salnikov, S.A. Dzuba, Y.D. Tsvetkov, J. Phys. Chem. A **107**, 3692 (2003)
22. M. Di Valentin, E. Salvadori, E.; S. Ceola, S. and D. Carbonera. Appl. Magn. Reson. **37**, 191 (2009)
23. S.A. Dikanov, in *Electron Paramagnetic Resonance*, ed by. B.C. Gilbert, D.M. Murphy, V. Chechnik (RCS Publishing, 2013) p. 103
24. L. Kulik, W. Lubitz, Photosynth. Res. **102**, 391 (2009)
25. K.M Salikhov, Yu.D. Tsvetkov, in *Time Domain Electron Spin Resonance*, ed by. L. Kevan, R.N. Schwartz (Wiley, New York, 1979) p. 231
26. D.A. Erilov, R. Bartucci, R. Guzzi, D. Marsh, S.A. Dzuba, L. Sportelli, J. Phys. Chem. B **108**, 4501 (2004)
27. A.D. Milov, YuD Tsvetkov, A.G. Maryasov, M. Gobbo, C. Prinzivalli, M. De Zotti, F. Formaggio, C. Toniolo, Appl. Magn. Res. **44**, 495 (2013)
28. P.P. Borbat, E.R. Georgieva, J.H. Freed, J. Phys, Chem. Lett. **4**, 170 (2013)
29. E. Wong, BE208, UCSD Winter 2008 Bioengineering 208 Magnetic Resonance Imaging Winter 2008 Lecture

30. P.P. Borbat, H.S. Mchaourab, J.H. Freed, J. Am. Chem. Soc. **124**, 5304 (2002)
31. E. Hahn, Phys. Rev. **80**, 580 (1950)
32. S.A. Dzuba, E.S. Salnikov, L.V. Kulik, Appl. Magn. Resn. **30**, 637 (2006)
33. M.N. Uvarov, L.V. Kulik, T.I. Pichugina, S.A. Dzuba, Spectrochim. Acta, Part A: Mol. Biomol. Spectrosc. **78A**, 1548 (2011)
34. A.D. Milov, K.M. Salikhov, Yu.D. Tsvetkov, Sov Phy.-Solid State **15**, 802 (1973)
35. K. Milov, M. Salikhov, Yu.D. Tsvetkov, M.D. Shchirov, Fiz. Tverd. Tela (Leningrad) **23**, 975 (1981)
36. A.D. Milov, R.I. Samoilova, YuD Tsvetkov, F. Formaggio, C. Toniolo, J. Raap, Appl. Magn. Reson. **29**, 703 (2005)
37. J.R. Klauder, P.W. Anderson, Phys. Rev. **125**, 912 (1962)
38. A.D. Milov, A.G. Maryasov, YuD Tsvetkov, Appl. Magn. Reson. **15**, 107 (1998)
39. G. E Pake, J. Chem. Phys. **16**, 327 (1948)
40. M-T. Türke, M. Bennati, Phys. Chem. Chem. Phys. **13**, 3630 (2011)
41. A. Kawamori, N. Katsuta, H. Mino, A. Ishii, J. Minagawa, T. Ono, J. Biol. Phys. **28**, 413 (2002)
42. A.D. Milov, Yu.D. Tsvetkov, A,G. Maryasov, M. Gobbo, C. Prinzivalli, M. De Zotti, F. Formaggio F, C. Toniolo, App. l Magn. Res. **44**, 495 (2013)
43. G.Yu. Shevelev, O.A. Krumkacheva, A.A. Lomzov, A.A.D.V. Kuzhelev, D.V. Trukhin, O.Yu. Rogozhnikova, V.M. Tormyshev, D.V. Pyshnyi, M.V. Fedin, E.G. Bagryanskaya, J. Phys. Chem. **B** (2015) (Ahead of Print)
44. B. Paulus, B. Illarionov, D. Nohr, G. Roellinger, S. Kacprzak,†. M. Fischer, S. Weber, A. Bacher, and E. Schleicher, J. Phys. Chem. **B 118**, 13092 (2014)
45. L.V. Kulik, S.A. Dzuba, I.A. Grigoryev, Y.D. Tsvetkov, Chem. Phys. Lett. **343**, 315 (2001)
46. D. Abdullin, F. Duthie, Fraser; A. Meyer, E.S. Mueller, Elisa S., G. Hagelueken, O. Schiemann, J. Phys. Chem. **B** (2015). (Ahead of Print?)
47. A.D. Milov, R.I. Samoilova, Y.D. Tsvetkov, C. Peggion., F. Formaggio, C. Toniolo, J. Phys. Chem. **B118**, 7085 (2014)
48. K.M. Salikhov, I.T. Khairuzhdinov, R.B. Zaripov, App. Magn. Res. **45**, 573 (2014)
49. K.M. Salikhov, I.T. Khairuzhdinov, App. Magn. Res. **46**, 67 (2015)
50. V.V. Kurshev, A.M. Raitsimring, YuD Tsvetkov, J. Magn. Reson. **81**, 441 (1989)
51. A.V. Astashkin, H. Hara, A. Kawamori, The pulsed electron–electron double resonance and 2 + 1 electron spin echo study of the oriented oxygen-evolving and Mn-depleted preparations of photosystem II. J. Chem. Phys. **108**, 3805–3812 (1998)
52. G. Jeschke, I.P. Chechik, A. Godt, H. Zimmermann, J. Banham, C.R. Timmel, H. Jung, Deer analysis: a comprehensive software package a for analyzing pulsed ELDOR data. Appl. Magn. Reson. **30**, 473–498 (2006)
53. A.D. Milov, A.B. Ponomarev, Y.D. Tsvetkov, Chem. Phys. Lett. **110**, 67 (1984)
54. D. Margraf, B.E. Bode, A. Marko, O. Schiemann, T.F. Prisner, Mol. Phys. **105**, 2153 (2007)
55. G. Jeschke, M. Sajid, M. Schulte, N. Ramezanian, A. Volkov, H. Zimmermann, A. Godt, Flexibility of shape-persistent molecular building blocks composed of p-phenylene and ethynylene units. J. Am. Chem. Soc. **132**, 10107 (2010)
56. J.E. Banham, C.M. Baker, S. Ceola, I.J. Daym, G.H. Grant, E.J.J. Groenen, C.T. Rodgers, G. Jeschke, C.R. Timmel, J. Magn. Reson. **191**, 202 (2008)
57. A. Potapov, H. Yagi, T. Huber, S. Jergic, N.E. Dixon, G. Otting, D. Goldfarb, J. Am. Chem. Soc. **132**, 9040–9048 (2010)
58. D. Grohmann, D. Klose, J. P.Klare, C.W.M Kay., H-J. Steinhoff, F. Werner, J. Am. Chem. Soc. **132**, 5954 (2010)
59. A.V. Astashkin, A. Kawamori, Biophys. Techn. Photosynth. **26**, 325 (2008)
60. B.E. Bode, J. Plackmeyer, T.F. Prinser, O. Schiemann, J. Phys, Chem. A **112**, 5064 (2008)
61. V.N. Syryamina, S.A. Dzuba, Chem. Phys. **137**, 145102 (2012)
62. O. Schiemann, N. Piton, Y. Mu, G. Stock, W.E. Engels, T.F. Prisner, J. Am. Chem. Soc. **126**, 5722 (2004)

63. N.A. Kuznetsov, A.D. Milov, V.V. Koval, R. I.Samoilov, Y A, Grishin, D. G Knorre., Y. D. Tsvetkov, O. S. Fedorova, and S. A Dzuba. Phys. Chem. Chem. Phys. **11**, 6826 (2009)
64. D. Grohmann, D. Klose, J.P. Klare, C.W.M. Kay, H-J. Steinhoff, F. Werner, RNA-binding to archaeal RNA polymerase subunits F/E: a DEER and FRET study. J. Am. Chem. Soc. **132**, 5954–5955 (2010)
65. B. Endeward, J.A. Butterwick, R. MacKinnon, T.F. Prisner, J. Am. Chem. Soc. **131**, 15246 (2009)
66. G. Jeschke, Y. Polyhach, Phys. Chem. Chem. Phys. **9**, 1895 (2007)
67. S. Lee, S, D. Budil, and J. H. Freed. J. Chem. Phys. **101**, 5529 (1994)
68. Y.-W. Chiang, A.J. Costa-Filho, B. Baird, J.H. Freed, J. Phys. Chem. B **115**, 10462 (2011)
69. B. Fresch, D. Frezzato, G.J. Moro, G. Kothe, J.H. Freed, J. Phys. Chem. B **110**, 24238 (2006)
70. R.R. Ernst, G. Bodenhausen, A. Wokaun, *Principles of Nuclear Magnetic Resonance in One and Two Dimentions* (Clerandon Press, Oxford, 1987)
71. J. Tang, J.R, Noris. Chem. Phys. Lett. **233**, 192 (1995)
72. S.A. Dzuba, M.K. Bosh, A.Y. Hoff, Chem. Phys. Lett. **248**, 427 (1996)
73. S. Saxena, J.H. Freed, J. Chem. Phys. **107**, 1317 (1997)
74. P.P. Borbat, J.H. Freed, Chem. Phys. Lett. **313**, 145 (1999)
75. P.P. Borbat, H.S. Mchaourab, J.H. Freed, J. Am. Chem. Soc. **124**, 5304 (2002)
76. S. Ruthstein, M. Ji, P. Mehta, L. Jen-Jacobson, S. Saxena, J. Phys. Chem. B **117**, 6227 (2013)
77. W.B. Mims, Phys. Rev. B. **6**, 3543 (1973)
78. B. Kasumaj, H. Dube, N. Zoelch, F. Diederich, G. Jeschke, J. Magn. Reson. **223**, 187 (2012)
79. Y. Deligiannakis, M. Louloudi, N. Hadjiliadis, Coord. Chem. Rev. **204**, 1 (2000)
80. A. Schweiger, Pure Appl. Chem. **69**, 809 (1992)
81. N. Manukovsky, V. Frydman, Goldfarb, J. Phys. Chem. B (2015) (Ahead of Print)
82. P. Hofer, A. Grupp, H. Nedenfuhr, P. Mehring, Chem. Phys. Lett. **132**, 279 (1986)
83. M. Albertini, P. Berto, F. Vallese, M. Di Valentin, P. Costantini, D. Carbonera, J. Phys. Chem. B (2015) (Ahead of Print)
84. D. Goldfarb, V. Kofman, J. Libman, A. Shanzer, R.S. Rahmatouline, A. Schweiger, J. Am. Chem. Soc. **120**, 7020 (1998)
85. A. Savitsky, A.A. Dubinskii, M. Plato, Y.A. Grishin, H. Zimmermann, K. Mobius, J. Phys Chem. B **112**, 9079 (2008)
86. W.B. Mims, Proc. R. Soc. Lond. Ser. A-Mathe. Phys. Sci. **283**, 452 (1965)
87. M. Mehring, P. Hofer, A. Grupp, Ber. Bunsenges. Phys. Chem. **91**, 1132 (1987)
88. J. Niklas, T. Schulte, S. Prakash, M. van Gastel, E. Hofmann, W. Lubitz, J. Am. Chem. Soc. **129**, 15442 (2007)
89. J. Telser, Electron-Nuclear Double Resonance (ENDOR) Spectroscopy. Encyclopedia of Inorganic and Bioinorganic Chemistry (John Wiley and Sons, 2011)
90. M. Houisjen, J.S. Hyde, J. Chem. Phys. **60**, 1682 (1974)
91. J.H. Freed, S. Wang, J. Phys. Chem. **78**, 1155 (1974)
92. S.S. Eaton, Eaton, in ed by. L.J. Berliner, G.R. Eaton, S.S. Eaton, in *Magnetic Resonance in Biology. Distance Measurement in Biological Systems by ESR*, vol. 19 (Kluwer Academic: New York, 2000), p 29
93. J.R. Biller, V.M. Meyer, H. Elajaili, G.M. Rosen, S.S. Eaton, G.R. Eaton, J. Magn. Reson. **225**, 52 (2012)
94. D.J. Francis, W.L. Hubbell, C.S. Klug, J. Magn. Reson. **43**, 405 (2012)
95. A.H. Beth, B.H. Robinson, in *Biological Magnetic Resonance*, ed by. L.J. Berliner, J. Reubin, vol. 8 (Plenum Press, 1989), p. 179
96. H. Sato, B.A. Filas, S.S. Eaton, G.R. Eaton, A.A. Romanyukha, R. Hayes, A.M. Rossi, Radiat. Meas. **42**, 997 (2007)
97. F. Mamedov, P.J. Smith, S. Styring, R.J. Pace, Phys. Chem. Chem. Phys. **6**, 4890 (2004)
98. L.V. Kulik, W. Lubitz, J. Messinger, Biochemistry **44**, 9368 (2005)
99. S.A. Dzuba, Y. Kodera, H. Hara, A. Kawamori, Appl. Magn. Reson. **6**, 391 (1994)
100. A. Schweiger, C. Gemperle, R.R. Ernst, J. Magn. Reson. **86**, 70 (1990)

101. S.A. Dzuba, A. Kawamori, Concepts Magnetic Reson. **8**, 49 (1996)
102. K.B. Konov, A.A. Knyazev, YuG Galyametdinov, N.P. Isaev, L.V. Kulik, Appl. Magn. Res. **44**, 949 (2013)
103. S. Weber, G. Kothe, J.R. Norris. J. Chem. Phys. **106**, 6248 (1997)
104. B. Boulat, I. Najfeld, M. Rance, J. Magn. Reson., Ser. A **120**, 223 (1996)
105. H. Matsuoka, K. Sato, D. Shiomi, T. Takui, Appl. Magn. Reson. **23**, 517 (2003)
106. V. Maurel, L. Skorka, N. Onofrio, E. Szewczyk, D. Djurado, L. Dubois, J–M. Mouesca, Kulszewicz-Bajer, J. Phys. Chem. B **118**, 7657 (2014)
107. K. Ayabe, K. Sato, S. Nishida, T. Ise, S. Nakazawa, K. Sugisaki, Y. Morita, K. Toyota, D. Shiomi, M. Kitagawa, T. Takui, Phys. Chem. Chem. Phys. **14**, 9137 (2012)
108. A.P. Saiko, G.G. Fedoruk, S.A. Markevich, Multiphoton transitions in a spin system driven by strong bichromatic field. J. Exp. Theor. Phys. **105**, 893–899 (2007)
109. K. Sato, M. Yano, M. Furuichi, D. Shiomi, T. Takui, K. Abe, K. Itoh, A. Higuchi, K. Katsuma, Y. Shirota, J. Am. Chem. Soc. **119**, 6607–6613 (1997)

Chapter 7
Miscellaneous Methods of Investigation of Electron Spin Interactions Based on Optical and Other Techniques

Abstract Although electron spin resonance and nuclear spin resonance experiments provide sensitive and effective information about electron spin effects, other physical methods, optical spectroscopy in particular, have also contributed significantly in his area. In the present chapter a general survey is made on advanced optical, Mössbauer, magnetic, neutron scattering and muon spin spectroscopy which taken in combination open new horizons for deep insights into the mechanism and scale of electron spin effects. This powerful arsenal has been used for detailed investigations of chemical and biological processes of great importance for fundamental and applied research.

7.1 Introduction

Practically speaking, all modern optical methods for investigating electron spin dependent processes are based on laser techniques. Since the development of short pulse laser systems, ultrafast time-domain techniques allow the development of detailed insights into the dynamic of molecular structures. After an ultrashort optical trigger pulse induces a transition to a higher electronic state, the evolution of the molecular system is followed by time resolved femto-and picosecond absorption, fluorescence, IR and Raman spectroscopy. These techniques taring in combination allow to elucidate the evolution of the molecular structure during ultrafast physical and chemical processes related to electron spin effects. Quantitative parameters modeling singlet-triplet transitions, triplet-triplet transfer, singlet and triplet excited state quenching, intermolecular quenching, intersystem crossing, long-distance electron transfer, and spin-polarized processes can be determined from analysis of these experiments.

Intramolecular electron spin nuclear spin interactions can be studied by Mössbauer (gamma-resonance, GR, spectroscopy). Analysis of the superfine structure of the GR spectra makes it possible to measure the intensity of internal magnetic fields and the value of the electron spin magnetic moment of Mössbauer atoms, ^{57}Fe for example. Measurement of the microscopic magnetic susceptibility

© Springer International Publishing Switzerland 2016   159
G. Likhtenshtein, *Electron Spin Interactions in Chemistry and Biology*,
Biological and Medical Physics, Biomedical Engineering,
DOI 10.1007/978-3-319-33927-6_7

permits the study of electron spin effects in magnetic materials. Individual magnetic moment and exchange integral contributions can be measured by this technique.

In addition to optical methods, neutron scattering and muon spin spectroscopy offer useful methods for the quantitative investigation of static and dynamical electron spin effects in solids and liquids.

7.2   Optical Methods

7.2.1   *Intramolecular Excited Triplet States Decay Processes*

The triplet-state population may be produced via endogenous or exogenous photosensitization from other triplet species or by efficient intersystem crossing (ISC) from the initially photoexcited singlet state (Section 2.X and relative references). There are three couplings enabling the spin flip mechanism, which enhances the efficiency of the ISC process: media-induced spin or spin–lattice relaxation, hyperfine coupling, and spin–orbit coupling (SOC) [1–4]. Efficiency of optical methods can be illustrated by several examples.

The properties of the isolated nucleo bases uracil, thymine, and 1-methylthymine in their ground and low-lying excited states have been studied by optical methods [5]. For uracil, the transition rates for three different nonradiative singlet-triplet pathways were calculated:

$$\text{(a)}\ S_1\left(^1n \rightarrow \pi^*\right) \rightsquigarrow T_1\left(^3\pi \rightarrow \pi^*\right)$$
$$\text{(b)}\ S_2\left(^1\pi \rightarrow \pi^*\right) \rightsquigarrow T_2\left(^3n \rightarrow \pi^*\right)$$
$$\text{(c)}\ S_2\left(^1\pi \rightarrow \pi^*\right) \rightsquigarrow T_3\left(^3\pi \rightarrow \pi^*\right)$$

The observed high quantum yield of phosphorescence of BP was consistent with an efficient ISC occurring via spin–orbit coupling from $S_1(n\pi^*)$ to $T_2(\pi\pi^*)$, followed by rapid ISC from $T_2(\pi\pi^*)$ to $T_1(n\pi^*)$.

The decarboxylation reaction of ketoprofen (KP) in different acetonitrile-water mixtures producing a carbanion or biradical intermediate was investigated by using femtosecond transient absorption and nanosecond time-resolved resonance Raman spectroscopies [6]. A global fit of the fs-TA transient absorption obtained in a pH 0 acidic solution yielded a time constant of 8.7 ps for ISC and a 1.73 ns time constant for the decarboxylation reaction. The photophysical properties of a SnIV phthalocyanine which coordinates 2 myristate groups (*cis*-PcSn(Myr)$_2$.) through their carboxylate functionalities in a *cis* disposition at the Sn center were studied [7]. Role of caging in the thiol-activated triplet-triplet annihilation (TTA) upconversion in thiol-activated triplet-triplet annihilation (TTA) upconversion was studied with steady-state UV-vis absorption spectroscopy, fluorescence spectroscopy, electrochemical characterization, nanosecond transient absorption spectroscopy, and DFT/TDDFT computations [8]. The results indicated that 2,4-dinitrobenzenenesulfonyl-caged triplet photosensitizer shows a shorter triplet state lifetime (24.7 μs) than the uncaged triplet photosensitizer (86.0 μs).

The photophysical properties of the triad

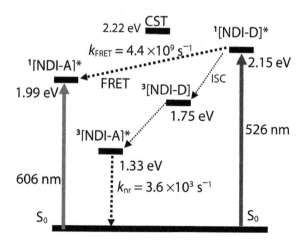

were studied with steady-state UV-vis absorption spectra, fluorescence spectra, nanosecond transien tabsorption spectra, cyclic voltammetry, and DFT/TDDFT calculation. Figure 7.1 illustrates the competing ISC and FRET (fluorescence resonance energy transfer) processes [9].

Along with experimental measurements, advanced theoretical calculations provide deep insight into intramolecular electron spin interactions. Ab initio molecular dynamics including nonadiabatic and spin–orbit couplings on equal footing is used to unravel the deactivation of cytosine after UV light absorption [10]. Intersystem crossing Intersystem crossing was found to compete directly with internal conversion in tens of femtoseconds **and** close degeneracy between singlet and triplet states can compensate for very small spin–orbit couplings, leading to efficient ISC. Deactivation pathways of keto-cytosine including IC and ISC is shown in Fig. 7.2.

Fig. 7.1 Photophysical processes involved in N − 1 upon [9]

Fig. 7.2 Deactivation pathways of keto-cytosine including IC (*in black*) and ISC (*in gray*). The propensity of each path is sketched by the thickness of the *arrows*. The *dotted line* indicates the deactivation pathway of T_1 [10]

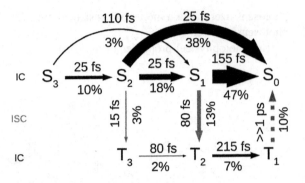

The information on the spin-orbit interaction and intersystem crossing in various molecules and their electronic structure was acquired in [11–13].

7.2.2   Intermolecular Singlet and Triplet Excited States Quenching

7.2.2.1   Dioxygen as a Quencher

The phosphorescence (excited triplet state) quenching technique relies on changes in the phosphorescence life time of a molecular probe when it is quenched by oxygen [14–23]. The quenching process can be illustrated by using a Jablonski diagram (Section X) and it can be described by the Stern–Volmer equation [4].

$$\frac{I_f^0}{I_f} = 1 + k_q \tau_0 \cdot [Q] \tag{7.1}$$

where I_f^0 is the intensity, or rate of fluorescence, without a quencher, I_f is the intensity, or rate of fluorescence, with a quencher, k_q is the quencher rate coefficient, τ_0 is the lifetime of the emissive excited state of A, without a quencher present and $[Q]$ is the concentration of the quencher

$$\frac{I_0}{I} = \frac{\tau_0}{\tau} = 1 + K_{sv}[O_2] = 1 + k_q \tau_0 [O_2] \tag{7.2}$$

where I and I_0 are the luminescence intensities with and without quencher (dioxygen), respectively, τ_0 and τ are the corresponding luminescence lifetimes, K_{SV} is the Stern–Volmer constant and kq is the kinetic quenching constant that reflects the efficiency of the quenching process and can be used to compare luminophores with different lifetimes.

The study of oxygen quenching of the excited state of fluorophores like porphyrins and phthalocyanines with the consequent generation of the cytotoxic singlet oxygen, has demonstrated potential applications in so-called PhotoDynamic Therapy (PDT) [16, 17]. These types of probes have long lifetimes (100 µs) and quenching constant values that make them well suited for measurement of physiological pO_2.

Quenching of excited triplet states T_1 of many substances by ground state molecular oxygen O_2 produces singlet oxygen [18]

$$^*F + O_2(X^3\sum_g) \rightarrow F + O_2(^1\Delta_g) \quad \text{or} \quad O_2(^1\sum_g) \quad (7.3)$$

The two possible states of singlet oxygen differ only in the spin and occupancy of the two degenerate antibonding π_g^* orbitals: in the case of $O_2(^1\Delta_g)$ (the excited state with lowest energy) two electrons are paired in the same π_g^* orbital, whilst in the case of $O_2(^1\Sigma_g)$ two electrons with opposite spin occupy the two degenerate antibonding π_g^* orbitals. As an example, on the base of measurement of rate constants for a series of 10 biphenyl sensitizers of very different oxidation potential, E_{ox}, and constant and large E_T it was demonstrated that the quenching of triplet states by O_2 producing $O_2(^1\Sigma_g^+)$, $O_2(^1\Delta_g)$, and $O_2(^3\Sigma_g^-)$, proceeds via two different channels [19]. One quenching channel originates from excited $^{1,3}(T_1 \cdot ^3\Sigma)$ complexes with no CT character (nCT); the other originates from $^{1,3}(T_1 \cdot ^3\Sigma)$ exciplexes with partial charge transfer character (pCT). A fast intersystem crossing equilibrium between $^1(T_1 \cdot ^3\Sigma)$ and $^3(T_1 \cdot ^3\Sigma)$ is only observed in the nCT but not in the pCT channel.

7.2.2.2 Quenching by Organic molecules

Laser-flash photolysis was employed to determine absolute rate constants for quenching of the 5-deazariboflavin (^{3}5DRF*) triplet excited state by amines and phenols in methanol solution [20]. The observed reactivity order for triplet quenching by aliphatic amines shows analogous behavior to the corresponding trend in oxidation potentials for these reducing agents; quenching by aromatic amines is significantly more efficient. In the case of phenols, the quenching rate constants render a Hammett reaction constant (ρ) value of -1.2 ± 0.1. This negative ρ value is in accord with a mechanism in which hydrogen-atom transfer from phenols to the 5-deazariboflavin triplet excited state takes place via an electron–proton transfer sequence.

To validate mechanism of quenching of protein phosphorescence, the rate of quenching of the buried Trp residues of RNase T1 and parvalbumin by acrylamide and the bigger double-headed derivative bisacrylamide was experimentally compared [21]. This study demonstrated that acrylamide quenching of superficially buried Trp residues of RNase T1 and parvalbumin is due to long-range interactions with the quencher in the solvent. It was suggested that the most probable mechanism of the quenching is electron transfer from the tryptophane segment and the quenchers.

In work [22] time-resolved laser spectroscopy was applied to measure the excited triplet state quenching by Dissolved Natural Organic Matter (DOM) for the anionic triplet sensitizers 4-carboxybenzophenone (CBBP) and 9,10-anthraquinone-2,6-disulfonic acid (2,6-AQDS) and for the triplets of 2-acetonaphthone (2AN) and 3-methoxyacetophenone (3MAP). No quenching was detected for CBBP and 2,6-AQDS, while second-order quenching rate constants with DOM for 2AN and 3MAP in the range of 1.30–3.85 $\times$ 10^7 L mol c^{-1} s^{-1} were measured.

7.2.3 Nitroxides as Quenchers

7.2.3.1 Intermolecular Quenching

The excited-state quenching of aromatic and other molecules by nitroxide radicals has been extensively investigated by optical measurements [23–27]. Quenching of triplet states of molecules by nitroxyl radicals can occur by charge transfer, intersystem crossing and energy transfer from the triplet molecule to the radical. The contributions of these mechanisms depends on the triplet energy of chromophors and the electron donor ability of the radical. Quenching of triplet states of ketones by nitroxyl radicals occurs by charge transfer and by energy transfer from the triplet molecule to the radical.

By flash photolysis and fluorimetric measurements, it was established that quenching of triplet states of ketones by nitroxyl radicals occurs by charge transfer and by energy transfer from the triplet molecule to the radical [23]. The relative contributions of of these processes depends on the triplet energy of the ketone and the electron donor ability of the radical. In contrast, quenching of the triplet state of anthanthrone by nitroxyl radicals occurs by charge transfer from the radical to the triplet molecule. The quenching mechanism for pyrene fluorescence by 2,2,6,6-tetramethylpiperidine-1-oxyl (TEMPO) and galvinoxyl radicals has been studied by measuring the time evolution of the transient absorption [24]. It was demonstrated from an analysis of the dual exponential time profiles of the transient absorption due to $Sm \leftarrow S_1$ and $Tn \rightarrow T_1$ transitions that radicals largely enhance the intersystem crossing from the S_1 state to triplet manifolds. Quenching mechanisms of S_1 and T_1 coronene by the (TEMPO) radical were investigated by probing absorptive (abs) and emissive (em) chemically induced dynamic electron polarizations (CIDEP), which are generated by interactions of S_1 and T_1 coronene with TEMPO radicals, respectively [25] Coronene has a relatively long fluorescence lifetime, and thus, it is easy to study both S_1 and T_1 quenching processes by using the Cor–TEMPO system as a model system. The time-resolved electron spin resonance (TR-ESR) method, TR Thermal Lensing, Transient Absorption, and Fluorescence Measurements were used in combination. It was shown that that quenching of T_1 coronene by TEMPO in benzene occurs through an electron-exchange interaction in an RT encounter complex.

The quenching mechanism of interactions between dendrimer-entrapped-gold nanoparticles clusters and nitroxide radical was explored by combined fluorescence and ESR spectroscopies—[26]. This system was suggested to be the fluorescence probe for sensing ascorbic acid (AA) in living cells.

7.2.3.2 Quenching in Donor-Acceptor Pairs

Starting from the first publication by the Likhtenshtein group [28] on quenching of excited states of a fluorophore by covalently linked radical in the supermoleculer molecule, this field has attracted much attention because of the growing interest of high-spin excited states of organic molecules such as quartet and quintet states [29–40]. The property of stable nitroxide radical moiety to quench the excited state of the chromophore segment was exploited for developing new molecular probes and also for modeling intramolecular photochemical and photophysical processes in the course of the light energy conversion and construction of new photoswitching magnetic materials [29–41] and references therein.

Steady state and time-resolved absorption and fluorescence measurements were used to probe the mechanisms of excited singlet state quenching by stable nitroxide radicals in a series of fluorescamine-derivatized nitroxides with direct connection between the donor and acceptor segments [41]. It was suggested that the rapid singlet state quenching in this series of compounds thus appears to arise from enhanced internal conversion. The rate constants of intramolecular fluorescence quencing (k_q) measured by the time-domain picosecond fluorescence technique, fluorescence quantum yields of paramagnetic and diamagnetic forms and the rate constant of photoreduction (k_{red}) in various solvents for a series of dansyl compound containing various nitroxides were measured [37–39]. A positive correlation between the rate constant of the nitroxide fragment photoreduction k_{red} and the equilibrium constant K_{eq} for the chemical exchange reaction between different nitroxides depending on the nitroxide redox potential. was established Nevertheless, the kq values were found not to be dependent on K_{eq}. On the basis of these and other available data, two mechanisms of IFQ were proposed: the dominant mechanism is intersystem crossing and the ancillary mechanism is irreversible intramolecular electron transfer (ET) from the excited singlet of the fluorophore (donor D) to nitroxide (acceptor A) followed by fluorophore segment regeneration and hydroxyl formation. The latter mechanism is responsible for photoreduction.

7.3 Fluorescence Correlation Spectroscopy

Fluorescence correlation spectroscopy (FCS) is a correlation analysis of fluctuations of the fluoresence intensity [42, 43]. The analysis provides model parameters of the physics underlying the fluctuations. One of the applications of these methods is an analysis of the concentration fluctuations of fluorescent molecules in solution. In this application, the fluorescence emitted from a very tiny space in solution containing a small number of fluorescent particles (molecules) is observed. The fluorescence intensity fluctuates due to Brownian motion of the particles.

The transient dark state of the probe thiazole orange, $5'$-$d(T_6D_{514}T_6)$-$3'$, attached to a specific DNA sequence, was investigated using FCS [42]. A laser technique was employed to measure the fluorescence cross-correlation of the solution sample. The transient dark state that was measured in a time scale of a few microseconds for a triplet state. This proceess was found be dependent on photoisomerization, viscosity, oxygen concentration, and hybridization. The photophysical and photochemical behaviors of the thermodynamically stable conformation of the dye, At analysis of experimental data, the energy levels for the ground singlet (S_0), first excited singlet (S_1), and triplet states (T_1), of the thermodynamically stable conformation of the dye were taking into consideration. On the basis of a detailed study of the above mentioned processes the effective intersystem crossing rate (k_{isc}') and triplet state deactivation rate (k_T) ware determined.

An approach to study bimolecular interactions in model lipid bilayers and biological membranes, exploiting the influence of membrane-associated electron spin resonance labels on the triplet state kinetics of membrane-bound fluorophores has been introduced [43]. Singlet-triplet state transitions within the dye Lissamine Rhodamine B (LRB) were studied, in aqueous solutions, with LRB bound to a lipid in a liposome, and in the presence of different local concentrations of the electron spin resonance label TEMPO. Both in solution and in membranes, the measured relative changes in the singlet-triplet transitions rates were found to reflect well the expected collisional frequencies between the LRB and TEMPO molecules.

7.4   Hole Burning Optical Spectroscopy

Spectral hole burning is the frequency selective bleaching of the absorption or fluorescence spectrum of a material, which leads to an increased transmission (a "spectral hole") at the selected frequency [44]. Two basic requirements, were formulated for the observation of this phenomenon, are: (1) The spectrum is inhomogeneously broadened and (2) the material undergoes, subsequent to light absorption, a modification which changes its spectrum. The frequency selective irradiation is usually accomplished by using an extremely narrow band monochromatic laser. The electronic ground and lowest excited triplet states of 2,2-dinaphthylcarbene were explored using hole-burning spectroscopy [45]. The hole-burning experiment on the 0,0 vibronic line of a $T_0 \rightarrow T_1$ fluorescence excitation spectrum was performed on the substrate dispersed in n-hexane and n-heptane at 1.7 K. The zero-field splitting (ZFS) of the ground T_0 and excited T_1 states was suggested to originate from the spin-spin interaction between two unpaired electrons was suggested as a cause of the zero-field splitting (ZFS) of the ground T_0 and excited T_1 states. In the frame of the accepted model, the holes and antiholes located far from the central hole (13–15 GHz away) should reflect the

ZFS of the ground T_0 state, whereas the holes located in the neighborhood of the central hole (within the (2 GHz range) should reflect the ZFS of the excited T_1 state. It was also suggested that intersystem crossing and both intersystem crossing steps, $T_1 \rightarrow Sn$ and $S_1 \rightarrow T_0$, which contribute to the shape of the hole-burning spectrum, provide main depopulation channel of the excited T_1 state.

The spectral position of the deep holes located far from the central hole provide direct information about the energy separation between the ZFC triplet sublevels [45].

$$\Delta E(T_{0z}, T_{0x}) = D_0 - E_0 = 13.65 \pm 0.05 \, \text{GHz}$$
$$\Delta E(T_{0x}, T_{0y}) = 2E_0 = 1.3 \pm 0.05 \, \text{GHz}$$

(7.4)

where D_0 and E_0 are associated with anisotropic interaction.

Spectral hole-burning absorption spectroscopy on the 879 nm, $^4I_{9/2} \rightarrow {}^4F_{3/2}$ transition in $Nd3^+:YVO_4$ was performed [46]. Under the application of a small magnetic field of 300 mT, the strong spectral holes with a homogeneous linewidth of 63 kHz at 2.1 K were observed at an applied field The holes were found to be long lived compared to the excited-state lifetime, indicating population trapping in the Zeeman sublevels. The authors concluded that these results indicate that the dominant relaxation process between the Zeeman levels at low magnetic fields is not spin lattice relaxation but rather a magnetic spin interaction.

Thus, the hole-burning method has been demonstrated to be a useful technique for studies of triplet-triplet transitions and can be extended to other systems with triplet ground states.

7.5 Organic Light-Emitting Diode (OLED) Spectroscopy

Organic light-emitting diodes OLED, a device, which emits light in response to an electric current is expected to be the next-generation solid-state lighting technology. Since the pioneering work by Tang and Vanslyke in 1987 [47]. OLEDs have attracted considerable attention in both basic and applied areas and is expected to be the next-generation solid-state lighting technology. The organic light-emitting diode (OLED) method was used for understanding the triplet state properties of organo-transition metal compounds [47–53]. The lowest triplet state T_1 of organo-transition metal complexes plays a pivotal role in organic light-emitting diodes. In a comprehensive review [50], the properties of the emitters were discussed with respect to the harvesting of singlet and triplet excitons that are generated in the course of the electroluminescence process.

In organic molecules, without heavy atoms, internal conversion, IC, from the S_1 to the S_0 state with the same multiplicity, takes place in about 10^{-12} s, while intersystem crossing, ISC times can be as long as 0.1–1 s. On the other hand, in organo-transition metal compounds, due to the small energy separation DE

($S_1 - T_1$), upward processes from T_1 to S_1 are effective at ambient temperature and emission features from both states, triplet and singlet, can occur. As a result, the emission is governed by a Boltzmann distribution [50]:

$$\frac{Int(S_1 \to S_0)}{Int(T_1 \to S_0)} = \frac{k^r(S_1 \to S_0)}{k^r(T_1 \to S_0)} \cdot \exp\left(-\frac{\Delta E(S_1 - T_1)}{k_B T}\right) \qquad (7.5)$$

Here Int ($S_1 \to S_0$) and Int($T_1 \to S_0$), k^r ($S_1 \to S_0$) and $k^r(T_1 \to S_0$) represent the fluorescence and phosphorescence intensities and the corresponding radiative rates, respectively. k_B is the Boltzmann constant and T the absolute temperature. Equation 7.5 allows the determination of the DE value, which is the energy of the Zero Field Splitting (ZFS).

In electro-luminescence processes in the context of the OLED method, the population of excited states occurs via a recombination of negatively charged electrons and positively charged holes (Fig. 7.3) forming sinlet and triplet excitons excitons in the emission layer of an OLED [50]. The intersystem crossing time between these two states is expected to be significantly longer than relaxation times to lower state because of the small SOC constants of the host molecules and the very small energy splitting between the singlet (^{1}CT) and triplet (^{3}CT) charge transfer states As a result of fast and separate relaxations without spin flips, the lowest excited singlet state S_1 and the triplet state of the emitter complex are populated and ISC from both the S_1 to the T_1 state can occur.

As an example, the electroluminescent (EL) properties of, multilayered organic light-emitting diodes (OLEDs) devices two emitting materials, 1,3-bis (10-phenylanthracen-9-yl)benzene (**1**) and 2,6-bis(10-phenylanthracen-9-yl)pyridine (**2**), fabricated in the following sequence: indium-tin-oxide (ITO) were

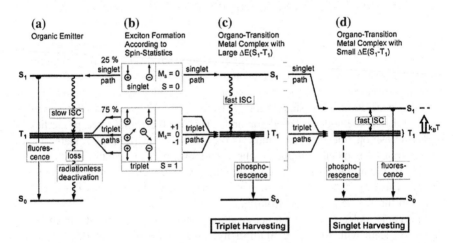

Fig. 7.3 The diagram displays electro-luminescence excitation processes for organic and organo-transition metal emitters, respectively, and explains the effects of triplet harvesting and singlet harvesting [50]

investigated [53]. A device using **1** as an emitting material showed a white emission with maximum luminance, luminous, power, and external quantum efficiency values of 1727 cd/m^2, while device using **2** exhibited a sky-blue emission with maximum luminance, luminous, power, and external quantum efficiency values of 2279 cd/m^2.

7.6 Mössbauer Spectroscopy

Mössbauer or gamma-resonance (GR) spectroscopy is a spetroscopic technique based on the Mössbauer effect [54–57]. This effect, discovered by Rudolf Mössbauer in 1957, consists of the recoil-free, resonant absorption and emission of gamma rays in solids [54]. The principle of conservation of momentum requires a nucleus to recoil during emission or absorption of a gamma ray. Nuclei in a solid crystal, are not free to recoil because they are bound in place in the crystal lattice. When a nucleus in a solid emits or absorbs a gamma ray, some energy can still be lost as recoil energy of phonons. In Mössbauer spectroscopy, when gamma rays emitted by one nucleus can be resonantly absorbed by a sample containing nuclei of the same isotope, this absorption is measured. Three types of nuclear interaction may be observed: an isomer shift (a chemical shift); quadrupole splitting; and magnetic or hyperfine splitting (Zeeman effect).The latter is a source of information of electron spin nuclear spins interaction. Due to the high energy and extremely narrow line widths of gamma rays, Mössbauer spectroscopy is a very sensitive technique in terms of energy resolution, capable of detecting changes in just a few parts per 10^{11}!. Therefore Mössbauer spectroscopy probes tiny changes in the energy levels of an atomic nucleus in response to its environment.

Electron spin—nuclear spin effects are revealed in hyperfine splittings which result from the interaction between the nucleus and the local magnetic field. A nucleus with spin, I, splits into 2I + 1 sub-energy levels in the presence of magnetic field. For example, a nucleus in ^{57}Fe with spin state I = 3/2 will have splitting is of the order of 10^{-7} eV into 4 non-degenerate sub-states with m$_I$ values of +3/2, +1/2, −1/2 and −3/2 t. Only six possible transitions for a 3/2 to 1/2 transition are allowed because he selection rules for magnetic dipoles ensures that transitions between the excited state and ground state can only occur where m$_I$ changes by 0 or 1 The magnitude of the internal magnetic field in paramagnetic compound can reach 1–50 T and therefore in a number of cases the splitting is observed in the absence of a magnetic field. Application of a high-strength magnetic field can also reveal the hyperfine structure of GR spectra. Fast spin electron relaxation and antiferromagnetic interaction between atoms would average the magnetic field at the nucleus and the magnetic hyperfine structure would disappear.

Below a few typical examples of the Mössbauer spectroscopy application in investigation of hyperfine interactions. EPR and Mössbauer Spectroscopy and Density Functional Theory analysis of a High-Spin Fe^{IV}–Oxo Complex were reported [58]. The work gave a detailed characterization of the structurally similar complexes $[Fe^{IV}H_3buea(O)]^-$, $[Fe^{III}H_3buea(O)]^{2-}$, and $[Fe^{III}H_3buea(OH)]^-$ (H_3buea = tris[(N'-$tert$-butylureaylato)-N-ethylene]aminato):

$[Fe^{III}H_3buea(O)]^{2-}$, **2**　　　　$[Fe^{III}H_3buea(OH)]^-$, **2-OH**　　　　$[Fe^{IV}H_3buea(O)]^-$, **1**

Using Mössbauer and dual-frequency/dual-mode electron paramagnetic resonance (EPR) spectroscopies. Isomer shifts are reported relative to Fe metal at 298 K. High-field Mössbauer spectroscopy showed an ^{57}Fe A_{dip} tensor of (+5.6, +5.3, −10.9) MHz and A_{iso} = −25.9 MHz for the $[Fe^{IV}H_3buea(O)]^-$ complex, and the results of DFT calculations were in agreement with the nuclear parameters of the complex.

Simulations of the Mössbauer spectra were calculated with least-squares fitting using the program *SpinCount* and the standard spin Hamiltonian [58]

$$H = \beta_e \mathbf{B}.\mathbf{g}.\mathbf{S} + \mathbf{S}.\mathbf{D}.\mathbf{S} + \mathbf{S}.\mathbf{A}.\mathbf{I} - g_n b_n \mathbf{B}.\mathbf{I} + \frac{eQV_{zz}}{12}\left[3I_z^2 - I(I+1) + \eta(I_x^2 - I_y^2)\right]$$

(7.6)

^{57}Fe Mössbauer spectroscopy was applied to investigate the superconductor parent compound $Fe_{1+x}Te$ for x = 0.06, 0.10, 0.14, 0.18 within the temperature range 4.2–300 K [59]. A transmission integral has been applied to account for the absorption profile of a spin density wave (SDW) component by supposing that there is a *quasi*-continuous distribution of the hyperfine field following the shape of SDW Magneto-electric $GaFeO_3$ was characterized by x-ray diffraction, dc-magnetization, ac-susceptibility, low temperature and high field ^{57}Fe Mössbauer spectroscopy and dielectric constant measurements [60]. The zero-field Mössbauer spectra of 280 and 5 K were analyzed with the NORMOS-SITE program by considering one Fe site for the estimation of the hyperfine parameters. A hyperfine field (B_{INT}) of 49.2 ± 0.1 T is observed from the 5 K data. The values of the effective hyperfine field H_{eff} (vector sum of the applied and internal hyperfine fields external field (5 T) for different $GaFeO_3$ samples) were found to vary from 44.5 to 55.7 T. Advances in Mössbauer spectroscopy with a high velocity resolution in biomedical, pharmaceutical, cosmochemical and nanotechnological research were revewed in.

The detailed information regarding interstitial site and hyperfine interaction in non-magnetic Zn ion doped $Ba_3CO_2Fe_{24}O_{41}$ external-field Mössbauer spectroscopy has been provided in [61]. Figure 7.4 shows the Mössbauer spectra of Ba2Co2–xZnx Fe24O41 (x = 0.0, 0.5, 1.0, 1.5, and 2.0) at 4.2 K with applied field ranging

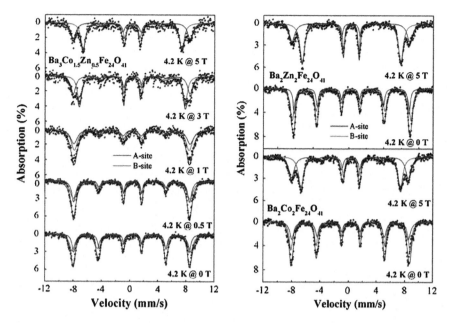

Fig. 7.4 Mössbauer spectra of Ba3Co2-xZnxFe24O41 (x = 0.0, 0.5, and 2.0) at 4.2K with applied field ranging from 0 to 50 kOe [61]

from 0 to 50 kOe. The analysis of the spectra indicate that the presence of strong external field affect the hyperfine structure, hyperfine structure of Fe ions has two opposite effects to external field and all the spectra were least-squares fitted with two-sextets.

Further examples of the Mossbauer spectroscopy application can be find in references [62, 63].

7.6.1 Magnetic Measurements

Magnetic susceptibilitymeasurements for the investigation of electron spin effects in magnetic materials have been reviewed in a variety of recent publications [64–71]. The magnetic susceptibility χ is a dimensionless proportionality constant that indicates the degree of magnetization of a material in response to an applied magnetic field, The *volume magnetic susceptibility*, represented by the symbol χ_v is defined in the International System of Units.

$$M = \chi_v H \qquad (7.7)$$

where M is the magnetization of the material (the magnetic dipole moment per unit volume), measured in amperes per meter, and H is the magnetic field strength, also measured in amperes per meter. The Curie constant is a material-dependent property that relates a material's magnetic susceptibility to its temperature. For a two-level system with magnetic moment μ, the formula reduces to,

$$C = \frac{1}{k_B} N\mu^2 \tag{7.8}$$

where N is the number of magnetic atoms (or molecules) per unit volume, μ is the Bohr magneton, and k_B is the Boltzman constant. The constant C is used in Curie's Law, which states that for a fixed value of a magnetic field, the magnetization of a material is inversely proportional to the temperature (7.8).

In a system with a zero field splitting D when the magnetic field is perpendicular to the axial direction [67]:

$$\chi_\perp = \frac{2Ng_\perp^2\mu_B^2}{D}\frac{1 - \exp(-D/kT)}{1 + 2\exp(-D/kT)} \tag{7.9}$$

For triangle with the exchange integral $|J|$

$$\chi = \frac{Ng^2\mu_B^2}{4kT}\frac{5 + \exp(-3J/kT)}{1 + \exp(-3J/kT)} \tag{7.10}$$

The Ru-Ru spin-singlet formation in $La_{2-x}Ln\,x\,RuO_5$ (Ln = Pr, Nd, Sm, Gd, Dy) was investigated by measurements of the specific heat and magnetic susceptibility [69]. A field under cooled conditions of H = 1000 Oe was applied for all measurements. It was found that the effective magnetic moments μ_{eff} (in μB) of the $Ln3$ + ions in selected $La_2 - x\,LnxRuO_5$ compounds vary from 0.27 to 3.05 and an intradimer exchange of $J_0/k_B \approx$ 260 K (23.5 meV) was reported. The mixed-valent $Fe^{II}Fe^{III}$ imido complexes $Fe(^iPrNPPh_2)_3Fe \equiv NR$ (R = ^tBu Ad 2,4,6-trimethylphenyl) have been characterized via X-ray crystallography, Mössbauer spectroscopy, cyclic voltammetry, and SQUID magnetometry, and computational methods [70]. The effect of the metal–metal interaction on the electronic structure and bonding in diiron imido complexes was discussed. Solid state dc and ac magnetic susceptibility measurements in $Mn_3O(O_2CMe)_3(dpd)_{3/2})]_2(I_3)_2)$ revealed that each Mn_3 subunit of the dimer is a single-molecule magnet (SMM) with an S = 6 ground state [71]. In this structure the two units are very weakly ferromagnetically exchange coupled. High-frequency EPR spectroscopy on a single crystal displays signal splittings indicated quantum superposition/entanglement of the two SMMs. The experimental data were found to fit to following parameters: $J = +6.8(1)$ cm^{-1}, $J = +26.3(4)$ cm^{-1}, and $g = 1.95(1)$, with temperature-independent paramagnetism (TIP) held constant at 600×10^{-6} cm^3 mol^{-1}.

7.7 Neutron Scattering and Muon Spin Spectroscopy

7.7.1 Inelastic Neutron Scattering

The neutron is a subatomic particle, with no net electric charge and with magnetic moment of the neutron to be $\mu_n = -1.93(2)$ μ_N, where μ_N is the nuclear magneton. (https://en.wikipedia.org/wiki/Neutron) Inelastic neutron scattering (INS) observes the change in the energy of the neutron as it scatters from a sample and can be used to probe magnetic and quantum excitations or electronic transitions [72–74]. Inelastic neutron scattering echo techniques (neutron spin echo, NSE, and neutron resonance spin echo), has a strong analogy to the Hahn spin echo because using the quantum mechanical phase of the neutrons in addition to their amplitudes. In both cases the loss of polarization (magnetization) due to dephasing of the spins in time is restored by an effective time reversal operation, that leads to a rephasing of polarization). In NSE, different neutron velocities in the incoming neutron beam is a cause of the dephasing. Inelastic light scattering is an intensively used tool in the study of electronic properties of solids. In work [72], the spectroscopic intensities of the transitions within the ground-state multiplet were analyzed. The effects of a mixing with higher-lying spin multiplets were investigated using the experimental techniques of inelastic neutron scattering and electron paramagnetic resonance. Three main effects of spin mixing are observed: (1) a pronounced dependence of the INS intensities on the momentum transfer Q emerges in first order, signaling the many-spin nature of the wave functions in exchange-coupled clusters; (2) the intensities of the transitions within the spin multiplet are affected differently by spin mixing; (3) the higher-order contributions to the cluster magnetic anisotropy which come from the single-ion ligand-field terms and spin mixing, respectively. The analytical results were illustrated by means of three examples: an antiferromagnetic heteronuclear dimer; a Mn−[3 × 3] grid molecule; and the single-molecule magnet Mn$_{12}$.

A review [74], described the static and dynamic magnetic properties of hole-doped cuprate superconductors measured with neutron scattering. Experiments on the monolayer La$_{2-x}$(Sr,Ba)$_x$CuO$_4$ and bilayer YBa$_2$Cu$_3$O$_{6+x}$ cuprates were described in deteils. It was found that, at zero hole doping, both investigated classes of materials are antiferromagnetic insulators with large superexchange constants of $J > 100$ meV. The spin fluctuations in La_{2 − x}Sr_{x}CuO_{4} and YBa_{2}Cu_{3}O_{7 − x} were the focus of a recent review [75].

7.7.2 Spin Muon Spectroscopy

The muon is an elementary particle similar to the electron, with electric charge of
-1 e and a spin of $\frac{1}{2}$, but with a much greater mass (https://en.wikipedia.org/wiki/
Muon). Muon spin spectroscopy is an experimental technique based on the
implantation of spin-polarized muons in matter and on the detection of the influence
of the atomic, molecular or crystalline surroundings on their spin motion [76–78].
The motion of the muon spin is due to the magnetic field experienced by the particle
and may provide information on its local environment in a very similar way to other
magnetic resonance techniques, such as electron spin resonance (ESR) and, more
closely, nuclear magnetic resonance (NMR). If an external magnetic field is applied
in the direction of the initial muon spin, as it is in so called longitudinal field
experiments, the Hamiltonian for muonium can be written as [76]

$$H = -\gamma_\mu h I \times B + \gamma_e h S \times B + h S \times A \times I \qquad (7.11)$$

where γ_m and γe are the muon and electron gyromagnetic ratios, B is the applied
longitudinal field, I and S are the muons and electrons spins and A is the hyperfine
tensor The energy levels as a function of the applied magnetic field that originates
from this Hamiltonian for an isotropic hyperfine interaction are were also shown.
    To generate radicals a light isotope of hydrogen called muonium (Mu = [μ+ , e−])
was used. These radicals were investigated by so called avoided level crossing muon
spin resonance (ALC-μSR) techniques, is based on measuring the asymmetry of the
muon decay as a function of a magnetic field applied parallel to the initial direction of
the muon spin [77]. The techniques techniques involve injecting a beam of
spin-polarized positive muons into a sample and detecting the positron produced by
the decay of each muon. Muons can react with a radiolytic electron forming Mu,
which can react with unsaturated molecules to produce a muoniated radical. where
The ensemble of muons appeared to be ~ 100 % polarized spin labels. The decay of
the muon occurs with lifetime (2.2 μs) which is comparable to many molecular
processes. An example of typical ALC-μSR spectrum of p-xylene is shown in
Fig. 7.5.

• The resulting radicals were identified by comparing the experimental proton
  hyperfine coupling constants (A^{exp}) with values (A^{cal}) calculated for the possible
  structures using density function theory (A^{cal}). For example for compound 2 the
  A values (in MHz) were found as A^{exp} (benzene CH) == −11.8), A^{cal} (benzene
  CH) = −8.27 and A^{exp} (bridging CH_2) == 2.86, A^{cal} (bridging CH_2) = 2.89.
• In published work [78] reactivity of the multifunctional cyclic silylene **4** and its
  carbene complex **5** (Fig. 7.6) have been investigated by a combination of muon
  spin spectroscopy and computation. Reaction of muonium with **5** produced two
  radicals, but with markedly different hyperfine constants.

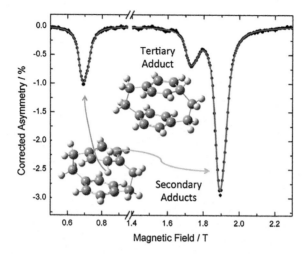

Fig. 7.5 Structure and background subtracted ALC-μSR spectrum of *p*-xylene at 280 K [77]

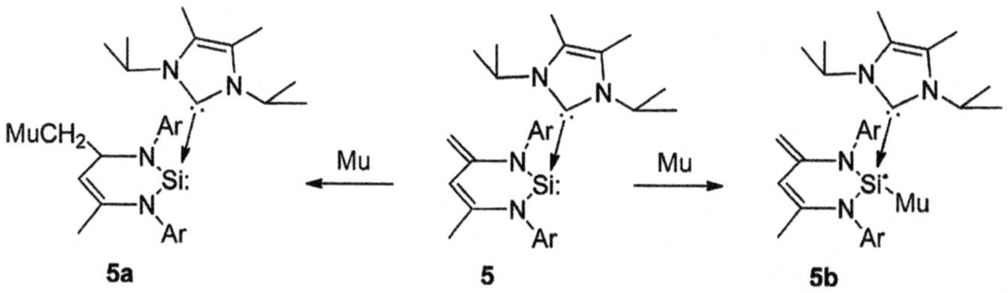

Fig. 7.6 Reaction of zwitterionic N-heterocyclic silylene with muon [78]

References

1. N.J. Turro, J.C. Scaiano, *Modern Molecular Photochemistry of Organic Molecules.* (University Science Books, 2010)
2. G.I. Likhtenshtein, *Stilbenes: Application in Chemistry* (Life Science and Material Science WILEY-VCH, Weinhem, 2009)
3. O. S. Wolfbeis, *Fluorescence Spectroscopy: New Methods and Applications* (Springer Verlag, 1993)
4. J.R. Lakowicz, *Principles of Fluorescence Spectroscopy* (Springer, US., 2013)
5. M. Etinski, T. Fleig, C.M. Marian, J. Phys. Chem. A **113**, 11809 (2009)
6. M.D. Li, J. Ma, T. Su, M.L. Liu, L. Yu, D.L. Phillips, J. Phys Chem. B **116**, 5882 (2012)
7. E. Gutierrez-Meza, R. Noria, G. Granados, V. Gomez-Vidales, J.Z. Ramirez, H.I. Beltran, J. Peon, J. Phys. Chem. B **116**, 14107 (2012)
8. C. Zhang, J. Zhao; X. Cui, X. Wu, J. Org. Chem. **80**, 5674 (2015)
9. S. Wu, F. Zhong, J. Zhao, S. Guo, W. Yang, T. Fyles, Broadband Visible Light-Harvesting Naphthalenediimide (NDI) Triad: Study of the Intra-Intermolecular Energy/Electron Transfer and the Triplet Excited State. J. Phys. Chem. A **119**, 4787 (2015)

10. M. Richter, P. Marquetand, J. Gonzalez-Vazquez, I. Sola, L. Gonzalez, J. Phys. Chem. Lett. **3**, 3090 (2012)
11. G. Cui, W. Fang, J. Chem. Phys. **138**, 044315/1 (2013)
12. M. Bialkowska, A. Makarewicz, M. Banasiewicz, B. Kozankiewicz, Chem. Phys. Lett. **555**, 131 (2013)
13. S. Kihara, D.A. Hartzler, G.S. Orf, R.E. Blankenship, S. Savikhin, J. Phys. Chem. B **119**, 5765 (2015)
14. A. Ruggi, F.W.B. van Leeuwen, A.H. Velders, Coord. Rev. **255**, 2542 (2011)
15. A.D. Estrada, A. Ponticorvo, T.N. Ford, A.K. Dunn, Optics Lett. **33**, 1038 (2008)
16. M.H. Abdel-Kader, *Photodynamic Therapy. From Theory to Application* (Springer, Berlin Heidelberg, 2014)
17. J.M. Dabrowski, L.G. Arnaut, Photochem. Photobiol. Sci. (2015), Ahead of Print
18. L.J. Andrews, E.W. Abrahamson, Chem. Phys. Lett. **10**, 113 (1971)
19. R. Schmidt, F. Shafii, J. Phys. Chem. A **105**, 8871 (2001)
20. M. Barra, B. VanVeller, C. Verberne, J. Photochem. Photobiol. A Chem. **225**, 113 (2011)
21. G.B. Strambini, M. Gonnelli, J. Phys. Chem. B **114**, 9691 (2010)
22. J. Wenk, S.N. Eustis, K. McNeill, S. Canonica, Environ. Sci. Technol. **47**, 12802 (2013)
23. A.S. Tatikolov, V.A. Kuz'min, Dokl. Phys. Chem. (Engl. Transl.) **223**, 742 (1975)/Λ Karpiuk J, Grabowski ZR, Mechanism and kinetics of fluorescence quenching of aromatic hydrocarbons by a stable nitroxyl radical. Chem. Phys. Lett. **160**, 451–466 (1989)
24. T. Suzuki, K. Ob, Chem. Phys. Lett. **246**, 130 (1995)
25. M. Mitsui, Y. Kobori, A. Kawai, K. Obi, J. Phys. Chem. A **108**, 524 (2004)
26. C.-P. Liu, T.-H. Wu, C.-Y. Liu, H.-J. Cheng, S.-Y. Lin, J. Mater. Chem. B: Mater. Biol. Med. **3**, 191 (2015)
27. K. Zamojc, W. Wiczk, B. Zaborowski, D. Jacewicz, L. Chmurzynski,. Spectrochimica Acta, Part A: Molecular and Biomolecular Spectroscopy **136** (Part_C), 1875 (2015)
28. I.M. Bystryak, G.I. Likhtenshtein, A.I. Kotelnikov, O.H. Hankovsky, K. Hideg, Russ. J. Phys. Chem. **60**, 1679 (1986)
29. N.V. Blough, D.J. Simpson, J. Am. Chem. Soc. **110**, 1915 (1988)
30. G.I. Likhtenshtein, *New Trends In Enzyme Catalysis and Mimicking Chemical Reactions* (N. Y., Kluwer Academic/ Plenum Publishers, 2003)
31. G.I. Likhtenshtein, J. Yamauchi, S. Nakatsuji, A, Smirnov, R. Tamura, *Nitroxides: Application in Chemistry, Biomedicine, and Materials Science* (WILEY-VCH, Weinhem, 2008)
32. G.I. Likhtenshtein *Solar Energy Conversion. Chemical Aspects* (WILEY-VCH, Weinhem, 2012)
33. E. Lozinsky, V.V Martin, T.A. Berezina, A. Shames, A.L, Weis, G.I. Likhtenshtein, J. Biochem. Biophys. Meth. **38**, 29 (1999)
34. P. Parkhomyuk-Ben Arye, N. Strashnikova, G.I. Likhtenshtein, J Biochem. Biophys. Meth. 51:1–15 (2002)
35. N. Medvedeva, V.V., Martin and. G. I. Likhtenshten Dual fluorophore-nitronyl probe for investigation of superoxide dynamics and antioxidant status of biological systems. J. Photochem. Photobiol. A: Chem. **163**, 45 (2004)
36. E.M. Lozinsky, L.V. Martina, A.L. Shames, N. Uzlaner, A. Masarwa, G.I. Likhtenshtein, D. Meyerstein, V. V. Martin, and Z
37. G.I. Likhtenstein, K. Ishii, S. Nakatsuji, Photochem. Photobiol. **83**, 871 (2007)
38. G.I. Likhtenshtein, Pure Appl. Chem. **80**, 2125 (2008)
39. G.I. Likhtenshtein, D. Pines, E. Pines, V. Khutorsky, Appl. Magn. Reson. **35**, 459 (2009)
40. G.I. Likhtenshtein, App. Biochem. Biotechn. **152**, 135 (2009)
41. S.E. Herbelin, N.V. Blough, J. Phys. Chem. B **102**, 8170 (1998)
42. H.S. Shin, A. Okamoto, Y. Sako, S.W Kim, S.Y. Kim, C.G. Pack, J. Phys. Chem. A **117**, 27 (2013)

43. A. Chmyrov, S. Johansson, A. Andersson, L. Mäler, J. Widengren, Biophys. J. **99**, 3821 (2010)
44. S. Volker, Hole-burning spectroscopy. Annu. Rev. Phys. Chem. **40**, 499 (1989)
45. B. Kozankiewicz, M. Aloshyna, A.D. Gudmundsdottir, M.S. Platz, M. Orrit, P. Tamarat, J. Phys. Chem. A **103**, 3155 (1999)
46. S.R. Hastings-Simon, M. Afzelius, J. Minář, M. Bettinelli, M.M.U. Staudt, B. Lauritzen, H. de Riedmatten, N. Gisin, A. Amari, A. Walthe, S. Kröll, Phys. Rev. B **77**, 12511 (2008)
47. C.W. Tang, S.A. Vanslyke, Appl. Phys. Lett. **51**, 913 (1987)
48. J.W. Kim, S.I. You, N.H. Kim, J.-A. Yoon, K.W. Cheah, F.R. Zhu, W.Y. Kim, Scientific reports (2014) (Ahead of Print)
49. G.L. Ingram, Z.-H. Lu, J. Photonics Energ. **4**, 40993/1 (2014)
50. H. Yersin, A.F. Rausch, R. Czerwieniec, T. Hofbeck, Coord. Chem. Rev. **255**, 2622 (2011)
51. H. Yersin (ed.), *Highly efficient OLEDs with phosphorescent materials* (Wiley-VCH, Weinheim, 2008)
52. H. Jeong, H. Shin, J. Lee, B. Kim, Y.-I. Park, K.S. Yook, B.-K. An, J. Park, J. Photonics Energ. **5**, 1 (2015)
53. Y.S. Kim, J.-Y. Yoon, H.W. Lee, J. Kim, S.E. Lee, H.W. Lee, Y.K. Kim, S.S. Yoon, J. Photochem. Photobiol. A: Chemistry (2015) (Ahead of Print)
54. R.L. Mössbauer, Zeitschrift für Physik A **151**, 124 (1958)
55. D-P. Yang, Moessbauer spectroscopy, in *Encyclopedia of Applied Spectroscopy*, edited by D. L. Andrews 51–85 (2009)
56. Y. Yoshida, G. Langouche, *Mössbauer Spectroscopy. Tutorial Book* (Springer, 2013)
57. D. Dickson, F.J. Berry, *Mössbauer Spectroscopy* (Cambridge University Press, 2005)
58. R. Gupta, D.C. Lacy, E.L. Bominaar, A.S. Borovik, M.P. Hendrich, J. Am. Chem. Soc. **134**, 9775 (2012)
59. A. Błachowski, K. Ruebenbauer, P. Zajdel, E.E. Rodriguez, M.A Green, J. Phys. Condens. Matter **24**, 386006
60. K. Sharma, V. Raghavendra Reddy, A. Gupta, A. Banerjee, A.M. Awasthi, J. Phys. Condens. Matter **25**, 076002 (2013)
61. J.T. Lim, C.S. Kim, J. Appl. Phys. **117**, 17B743/1 (2015)
62. A. Hanc-Kuczkowska, J. Kansy, G. Dercz, L. Pajak, J. Lelatko, J. Deniszczyk, Solid State Phenomena 203–204 (Applied Crystallography XXII), 137–141, 6 p. (2013)
63. M. Budzynski, V.I. Valkov, A.V. Golovchan, V.I. Mitsiuk, A.P. Sivachenko, Z. Surowiec, T. M. Tkachenka, Physica B (Amsterdam, Neth.) **452**, 37 (2014)
64. S.M. Walker, *Magnetism* (Lerner Publishing Group 2008)
65. N.A. Spaldin, *Magnetic Materials: Fundamentals and Applications* (Cambridge University Press, 2010)
66. S. Kuppuswamy, T.M. Powers, B.M Johnson, M.W. Bezpalko, C.K. Brozek, M. Foxman, L. A. Berben, C.M. Thomas, Inorganic Chemistry (Ahead of Print)
67. J. Yamauchi, Fundamentals of magnetism, in ed. by G.I. Likhtenshtein, J. Yamauchi, S. Nakatsuji, A. Smirnov, R. Tamura, *Nitroxides: Application in Chemistry, Biomedicine, and Materials Science* (WILEY-VCH, Weinhem, 2008)
68. S. Kuppuswamy†, T.M. Powers, B.M. Johnson, M.W. Bezpalko, C.K. Brozek, B.M. Foxman, L.A. Berben, C.M. Thomas*†, Inorg. Chem. Article ASAP
69. S. Riegg, A. Guenther, H.-A. Krug von Nidda, M.V. Eremin, M.V.A. Reller, A. Loidl, S.G. Ebbinghaus, Eur. Phys. J. B: Condens Matter Complex Syst. **85**, 413 (2012)
70. S. Kuppuswamy, T.M. Powers, B.M. Johnson, M.W. Bezpalko, C.K. Brozek, B.M. Foxman, L.A. Berben, L.A. Thomas, Inorg. Chem. **52**, 4802–4811 (2013)
71. T.N. Nguyen, M. Shiddiq, T. Ghosh, K.A. Abboud, S. Hill, G. Christou, J. Am. Chem. Soc. **137**, 7160 (2015)
72. O. Waldmann, H.U. Güdel, Phys. Rev. B **72**, 094422 (2005)
73. T.P. Devereaux, R. Hackl, Rev. Mod. Phys. **79**, 175 (2007)

74. R.J. Birgeneau, C. Stock, J.M. Tranquada, K. Yamada, J. Phys. Soc. Japan **75**, 111003/1 (2006)
75. T.E. Mason, Neutron scattering studies of spin fluctuations in high temperature superconductors. In *Handbook on the Physics and Chemistry of Rare Earths: Vol. 31, High-Temperature Superconductors-II*, eds. by K.A Gschneidner, L. Eyring, M.B. Maple (Elsevier, Amsterdam, 2001), Chap. 198, p. 281
76. S. Han, K. Wang, M. Willis, L. Nuccio, F.L. Pratt, J.S. Lord, K.J. Thorley, J. Anthony, A. J. Drew, S. Zhang, L. Schulz, Synthetic Metals xxx (2015) xxx–xxx
77. I. McKenzie I, R. Scheuermann, and K. Sedlak, J. Phys. Chem. A 116:7765 (2012)
78. P.W. Percival, B.M. McCollum, J.-C. Brodovitch, M. Driess, A. Mitra, M. Mozafari, R. West, Xiong, Y.and Yun; and S. Yao,. Organometallics **31**, 2709 (2012)

Chapter 8
Electron Spin Effects in Chosen Chemical and Physical Processes

Abstract This chapter illustrates the pivotal role of electron spin—electron spin and electron—nuclear interactions in various areas of photochemistry and photophysics. Advanced methods based on these interactions allow to elucidate detail mechanism of radical reactions and also to control them. Long distance exchange interactions govern the process of the photoinduced charge separation in donor-acceptor pairs, which is in fact the act of conversion of light energy to chemical energy. Exchange interaction also form a basis for triplet-triplet energy transfer and annihilation, processes of great basic and applied importance. Analogy between spin exchange, triplet-triplet energy transfer and electron transfer have been stressed.

8.1 Introduction

Exchange electron spin-electron spin interactions play decisive role in radical chemical reactions, electron transfer, triplet-triplet energy transfer, annihihalation, and magnetic field effects. More, molecular geometries, which are decisive in many chemical and physical processes can be established on the basis of measurement of the spin-spin exchange and dipole-dipole interactions. Though traditional experimental physical methods such as continuous wave (CW) ESR and optical methods are successively used, application of advance pulse methods opens new era in this important basic and applied field (Chaps. 5 and 6). The time-domain nano-, pico-, and femtosecond optical spectroscopies and other physical methods can also contribute in solving dynamic and structural problems (Chap. 7).

8.2 Reactions Involving Radicals

Chemically induced electron spin polarization (CIDEP) of reactive radicals discovered by Fessenden [1] can arise from a variety of mechanisms such as the triplet mechanism (TM), radical pair mechanism (RPM), spin-correlated radical pair

© Springer International Publishing Switzerland 2016
G. Likhtenshtein, *Electron Spin Interactions in Chemistry and Biology*,
Biological and Medical Physics, Biomedical Engineering,
DOI 10.1007/978-3-319-33927-6_8

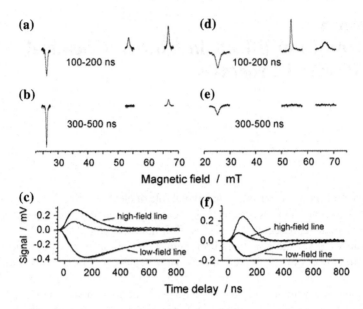

Fig. 8.1 TREPR spectra (**a**, **b**) and kinetic decays (**c**) measured in benzene, and TREPR spectra (**d**, **e**) and kinetic decays (**f**) measured in poly(ethylene glycol) at L-band (1.5 GHz) for radicals **1** and **2** (see Scheme 8.1 for structures). *Numbers* shown in figures (**a**, **b**, **d**) and (**f**) refer to the time delays of the boxcar integration window. The TREPR kinetics of radical **2** are marked by *arrows*, and the kinetics of radical **1** are not marked [5]

mechanism (SCRP), radical triplet pair mechanism (RTPM), or via cross-relaxation (Chap. 4). Experimental measurement of the CIDEP parameters is a direct way to investigation of kinetic and mechanism of radical reaction [1–8] and references cited in Chap. 4.

The mechanism of formation and decay of low field chemically induced electron spin polarization (CIDEP) in spin-correlated radical pairs (SCRPs) generated by Photoexcitation of (2,4,6-trimethylbenzoyl) diphenylphosphine oxide (TMBDPO) in low magnetic fields in viscous environments was investigated employing L-band (1.5 GHz) time-resolved electron paramagnetic resonance (TREPR) spectroscopy [5]. Figure 8.1 shows the TREPR spectra and kinetic decays after the laser flash. The TREPR kinetics measured for the high and low field lines was found to be different. A suggested reaction mechanism is related to the mixing of electron and nuclear spin states of the radical due to S-T transition.

Peroxyl radicals formed from the addition of oxygen to carbon radicals of *N*-acetyl glycine, serine, and diglycine was directly observed at room temperature via TR-EPR spectroscopy [6]. A scheme of the spin-polarized peroxyl radical adducts formed from the addition of oxygen to amino acid radicals is shown in Fig. 8.2. The peroxyl radicals was shown to exhibit unusually strong chemically induced electron spin polarization which was discussed in terms of the radical pair mechanism and spin polarization transfer processes.

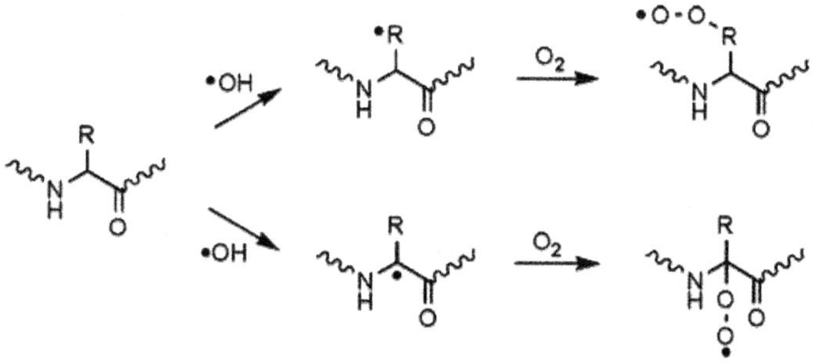

Fig. 8.2 Peroxylation of amino acid residues along the peptide backbone. R = carbon or sulfur atom [6]

Starting from work [8], several publications were related to investigation of such a complicated heterogeneous system as aerosol reverse micelles [9, 10]. X-band time-resolved electron paramagnetic resonance spectra (X-band) of correlated radical pairs created in aerosol AOT reverse micelles and microemulsions were detected using the microreactor model [9]. According to proposed scheme, the radicals are formed inside the water pool using photooxidation of diglycine by the excited triplet states of two different anthraquinone sulfonate salts. The data correlate with the diffusional properties of correlated radical pairs in regular aqueous micelle solutions.

The case, in which spin-spin interactions between them are comparable to the differences in their Zeeman interactions with the external magnetic field was analyzed on example of n-butylamine in work [11]. It was demonstrated that the α-CH$_2$ protons are polarized directly, whereas the β-CH$_2$, γ-CH$_2$ and δ-CH$_3$ protons get polarized only indirectly due to the transfer of polarization from the α-CH$_2$ protons. Results of the measurement of the entire magnetic field dependence of CIDNP are shown in Fig. 8.3.

Examples of recent application of modern ESR pulse techniques were described in publications [14–16] and references therein.

8.3 Electron Transfer in Donor-Acceptor Pairs

8.3.1 Long Distance Electron Transfer and Spin Exchange

For a long time, researchers followed a paradigm that electron transfer between a donor and an acceptor centers can occur only at a short distance (<6–7 Å). The similar idea was accepted for electron spin exchange. Nowadays Long-range Electron Transfer (LRET) up to 30 Å is a wide spread phenomenon of large number chemical in biological reaction, in particular. LRET between donor (D) and acceptor (A) centers can occur by three mechanisms: (1) superexchange via

Fig. 8.3 2 Magnetic field dependence of CIDNP obtained for different protons of n-butylamine (**a**, **b**) and total CIDNP (**c**) of the α-CH$_2$ and β-CH$_2$ protons (*open triangles*) and of all seven protons (*full triangles*). In a CIDNP of different protons is denoted by the following symbols: *full squares*, α-CH$_2$; *open squares* β-CH$_2$; in **b** *full circles*, c-CH$_2$; *open circles*, d-CH$_3$; *asterisk* denotes positions of additional features at 0.15 and 0.7 T [11]

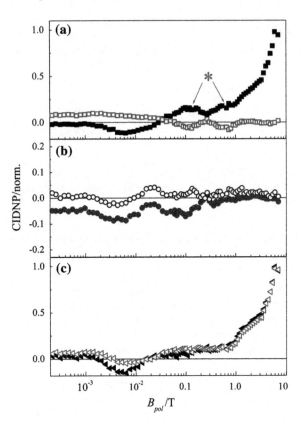

intermediate orbitals, this mechanism involves the direct electronic coupling between adjusting D and A groups (Sect. 1.X), (2) consecutive electron jumps via chemical intermediates and (3) mixing of the both mechanisms [12–25, 17–25].

Investigation the mechanism of efficient photoinduced electron-transfer processes in donor–acceptor systems is essential for developing molecular systems for artificial photosynthesis and for deeper understanding molecular aspects of the natural photosynthesis.

The exchange integral value J is one of the important quantities governing the efficiencies of the spin conversion processes, although its magnitude is usually much smaller (typically $|J| \leq 10^{-2}$ kJ mol^{-1}) than the thermal energy at an ambient temperature. Value of resonance integral (electronic coupling) $|V_{CR}|^2$ is decisive parameter in theory of electron transfer (Sect. 1.4). It was suggested that this parameter relates to the exchange integral J [17–19].

The singlet-triplet (S-T) splitting, ES-T, within the RP is given by the phenomenological parameter, 2 J, the magnitude of the indirect exchange interaction. A connection between spin exchange and electron if similar orbitals are involving in the both processes can be described by (1.9) which is a combination of the Anderson equation for superexchange (X) and the Marcus equation (Y) for electron transfer.

A semi-empirical approach for the quantitative estimation of the effect bridging the group on the long-distance electron transfer LDET was developed by Likhtenshtein [21–24]. An analogy between superexchange in electron transfer (ET) and such electron exchange processes as triplet-triplet energy transfer (TTET) and spin-exchange (SE) was considered. Experimental data on long distance TTET and ET processes were collected. As was shown in [22–26], experimental data on the dependence of k_{TT} and J_{SE} on the distance between the centers (ΔR) lies on two curves, which are approximated by the following equation

$$k_{TT}, J_{SE} \propto \exp(-\beta \Delta R) \qquad (8.1)$$

For systems in which the centers are separated by a "non-conductive" medium (molecules or groups with saturated chemicals bond) $\beta_{TT} = 2.6$ Å^{-1}. For systems in which the radical centers are linked by "conducting" conjugated bonds, β_{SE} is 0.3 Å^{-1}. The data on exchange integrals for paramagnetic center separated by "non conducting" media were obscured.

It was taken in account that TTET is a long distance process and above mentioned analogy can allow to predict corresponding distance dependent for SE and therefore for ET. We suggest that in the first approximation

$$V_{ET}^2, J_{SE}, J_{TT} \propto \exp(-n\beta R) \qquad (8.2)$$

where $(|V_{CR}|^2)$ is the square of the resonance integral, J_{SE} ishe exchange integral (J_{SE}), J_{TT} is the triplet-triplet transfer integral, R_i is the distance between the interacting centers and β_i is a coefficient which characterizes the degree of the integral decay. In the first approximation, n = 2 for the ET and SE processes with the overlap of two orbitals and n = 4 for the TT process in which four orbitals overlap (of ground and triplet states of the donor and ground and triplet states of the acceptor). The spin exchange and TT phenomena may be considered an idealized model of ET without or with only a slight replacement of the nuclear frame. Thus, the experimental dependence of exchange parameters k_{TT} and J_{SE} on the distance between the exchangeable centers and the chemical nature of the bridge connecting the centers may be used for evaluating such dependences for the resonance integral in the ET equations (8.3).

The ratios

$$\gamma = \frac{J_{SE}^0}{J_{SE}^x} \qquad (8.3)$$

where J_{SE}^0 and J_{SE}^x are the exchange integrals for two non-bridged radicals and radicals tethered by a media x, can be considered as parameter of attenuation parameter or factor) of the exchange interaction of SE through the given bridge. Taking into account (8.2) with value n = 2 for SE and ET, an expression for the dependence of the attenuation parameters for SE and ET on the distance between remote donor and acceptor centers $D_R D_A$ can be given as

$$\gamma_{ET} = \gamma_{SE} = \exp(-\beta_i \Delta R) \qquad (8.4)$$

with β_{ET} (nc) = 0.5 β_{TT} = 1.3 Å^{-1} for a "non-conducting" medium and β_{ET} (c) = 0.3 Å^{-1} for a "conducting" bridge. The value of β_{ET} (1.3 Å^{-1}) is found to be close to that obtained by analysis of k_{ET}, which is proportional to $|V_{CR}|^2$, on the distance βR in model and biological systems (Fig. 1.3).

The observed linear dependence of the logarithm of the rate constant of the electron transfer in RCs of purple bacteria and plant photosystem I (log k_{ET}) on the edge-edge distance between the donor and acceptor centers (R) (circles) (Fig. 8.4) corresponds to dependence predicted by (8.4) taken in consideration of similarity of the electron transfer and spin exchange processes characterized by the exchange integral J.

Similar but not identical approach was developed in [25, 26]. The authors also paid attention on similarity of long-range intramolecular electron transfer and positive ions (hole) transfer, and triplet energy transfer. They suggested that three studies are closely related, and the processes can be summarized and designated as (Scheme 8.1).

And that exchange or through saturated bridges, should be 2 times higher than that for ET. The theoretical interpretation relied on a simple fact that the

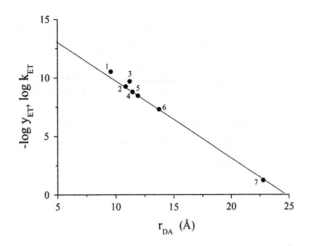

Fig. 8.4 Dependence of maximum rate constant of ET on the edge-edge distance between the donor and acceptor centers. in photosythetic RCs of bacteria and plants: 1. $A_0^{\cdot-} \rightarrow A_1$; 2. H$^{\cdot-} \rightarrow Q_A$; 3.$A_1^{\cdot-} \rightarrow$ Fx; 4. H$^{\cdot-} \rightarrow$ P$^+$; 5. C559$^{\cdot-} \rightarrow$ P$^+$; 6. Fx$^{\cdot-} \rightarrow$ F$_A$; and QA$^{\cdot-} \rightarrow$ P$^+$. The *straight solid line* is related to the dependence of the attenuation parameter for spin exchange in "non-conducting bridges with β_{ET} = 1.3 Å^{-1} [22]

Scheme 8.1 .

D–Sp-A → D$^+$-Sp-A$^-$ (ET)
D$^+$-Sp-A → D$^+$-SpA' (HT)
^{3}D-Sp-A → D-Sp-3A (TT)

$FNO^{\bullet} + h\upsilon \rightarrow F^*NO^{\bullet}$

$F^*NO^{\bullet} \rightarrow FNO^{\bullet}$ (Q[I])

$F^*NO^{\bullet} \rightarrow F^+NO^-$ (CS[I])

$F^+NO^- \rightarrow FNO^{\bullet}$ (CR[I])

$F^+NO^- + RH \rightarrow D\text{-}FNOH + products$ (ChR)

Scheme 8.2 .

Dexter-type TT exchange can be viewed as a concerted transition of one electron and one hole, moving in the same direction. The electron travels from the donor LUMO to the acceptor LUMO, while the hole travels from donor HOMO to the acceptor HOMO. The two transitions were represented by their own electronic coupling matrix elements V_{ET} and V_{HT}, for ET and hole transfer (HT), respectively, which are considered to exhibit very close distance attenuation factors ($\beta_{ET} \approx \beta_{HT}$). Therefore, the coupling matrix element for the TT transfer (V_{TT}) is approximated by the product of the two: $V_{TT} \approx |V_{ET}||V_{HT}| = C \times exp(-2\beta_{ET}R_{DA})$, where C is a system-specific constant, and R_{DA} is the donor–acceptor distance. It follows that: $\beta^{TT}/\beta^{ET} = 2$.

In works [27], it was suggested that $|V_{CR}|^2$ is proportional to 2 J and is highly sensitive to both the radical pairs distance and the structure of the intervening bridge system. For example, for charge recombination (CR) when the reorganization energy for the ET is small:

$$|V_{CR}|^2 = 2J\Delta E_{CR} \qquad (8.5)$$

where $|V_{CR}|$ denotes the electronic coupling and ΔE_{CR} is the vertical energy gap for the charge recombination.

The empirical data on the exchange integral values (J_{ET}) for the spin-spin interactions in systems with known structure, that is, biradicals, transition metal complexes with paramagnetic ligands and monocrystals of nitroxide radicals have served as a basis for the estimation of the value of the attenuation parameter γ_X for the exchange interaction through a given group X [22]. By our definition, the γ_X is

$$\gamma_X = \frac{J_{RYZP}}{J_{RYXZP}} \qquad (8.6)$$

where R is a nitroxide or organic radical, P is a paramagnetic complex or radical and X, Y, and Z are chemical groups in the bridge between R and P.

Table 8.1 shows the results of the calculation of parameter γ_X from empirical data on the spin exchange in the nitroxide biradicals by (8.6)

Data presented in Table 8.1 and (8.4) may be used for the analysis of alternative electron transfer pathways in chemical and biological systems.

Table 8.1 Attenuation parameter γ X calculated from empirical data on the spin exchange in nitroxide biradicals and complexes with paramagnetic ligands: 22 γ_x for individual group, γ_{hb} for hydrogen bond and γ_v for Van der Waals contact for spin exchange processes in nitroxide biradicals and transition metals complexes with paramagnetic ligands [22]

Group, X	γ_x	Group, X	γ_x
C_6H_4	6.00 ± 0.03		
$C = C$	1.7	$-NH-CO-$	55
$C = O$ $\quad$ C	8.4 ± 0.4	γ_v	50
NH	6.5	γ_{hb}	10
O	5	H	12
$S = O$	2.1	SO_2	2.2
	3.5	$RP = O$	2.40 ± 0.03

We suggest the following equations for the spin exchange integral and rate constant of electron transfer in donor-acceptor pairs:

$$J_{SE} = J_0 \rho_{S1} \rho_{S2} \prod^i \gamma_i^{-1} \qquad (8.7)$$

and

$$k_{ET} = k_0 \rho_D \rho_A \prod^i \gamma_i^{-1} \qquad (8.8)$$

where $J_0 = 10^{14}$ s^{-1} exchange integral at Van der Waals contact, $\rho_{S1,2}$—spin density at Van der Waals contact, γ_i—attenuation factors, $k_0 = 5 \times 10^{10}$ s^{-1} rate constant of ET in a notbridged DA pair. Equations (8.7) and (8.8) predicted ratio of values of exchange integral and ET rate constant $J_X/k_X = J_0/k_0 = 200$.

Equation (8.7) with values $J_0 \approx 10^{14}$s^{-1}, $\rho_{S1} \approx \rho_{S2} \approx 0.2$, $\gamma_1 \gamma_2 = 36$ for the donor-acceptor pair [27] and two phenyl bridged groups gave calculated $J_2 = 10^{10}$ s^{-1} as compared with experimental $J_2 = 6 \times 10^9$ s^{-1} [27]. A good correlation between experimental J values (n = 2–5) and that predicted by (8.7) take place (Fig. 8.5, Formula 8.1).

For the rate constant of charge separation in with n = 1, (8.8) predicted $k_{ICS} = 3 \times 10^{10}$s^{-1} versus experimental $k_{ICS} = 5 \times 10^{10}$s^{-1}

As one can see from Fig. 8.6, for compound I with n = 1–3 experimental data of the charge recombination rate constant, k_{CS} versus donor-acceptor distance fit to the dependence predicted by (8.8), while the experimental dependence rate constant of charge separation is steeper that for predicted one. This discrepancy can be explained suggesting that the spin exchange and recombination involve the same triplet orbitals, while charge separation occurs from singlet orbital.

Fig. 8.5 a Logarithmic plot of the exchange integral 2 J versus donor-acceptor distance, r_{DA}. Points experimental data [27]. *Solid line* predicted from (8.8)

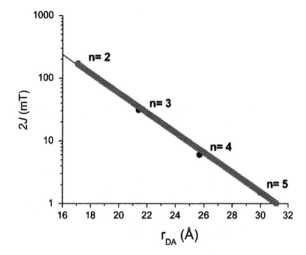

Formula 8.1 .

For rate constant of recombination of charge separated primary and secondary quinnons [28] (Fig. 8.7) in the photosynthetic reaction centers, employment (8.8) led to following calculated value: calculated (this work) $k_{ETqq} = 0.5 \times 10^8$ s^{-1} as compare with $k_{ETqq} = 0.9 \times 10^8$ s^{-1} calculated in Calvo et al. [28] and experimental $k_{ETqq} = 0.3 \times 10^8$ s^{-1} [29].

Aformentioned examples justified that (8.7) and (8.8) can be used of fast estimation of spin exchange integrals and the ET rate constant in certain donor acceptor pairs.

Triplet energy and electron transfer dynamics in Pt porphyrin–rhodamine B dyads comprising platinum(II) *meso*-tetraarylporphyrin (PtP) and Rhodamine B piperazine derivative (pRhB$^+$), linked by oligo-*p*-phenylene bridges (Ph$_n$), upon selective excitation of pRhB$^+$ at a frequency below that of the lowest allowed transition of PtP, $T_1 \rightarrow S_0$ phosphorescence of PtP was observed using Nanosecond Transient Absorption Spectroscopy (NSTA), Femtosecond Transient Absorption Spectroscopy (FSTA) and cyclic voltammetry [30]. The pathway leading to the emissive PtP triplet state includes excitation of pRhB$^+$, ET with formation of the singlet radical pair, intersystem crossing within that pair, and subsequent radical recombination and ET and TT rate distance dependence (Fig. 8.8).

The photochemistry of Zinc(II) phthalocyanine oligophenylene-ethynylene based donor-bride-acceptor dyads: ZnPc-OPE-AuP(+) and ZnPc-OPE-C(60) was

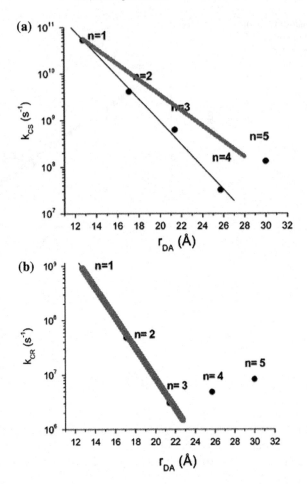

Fig. 8.6 **a** Logarithmic plot of the charge separation rate constant, kCS versus donor-acceptor distance, r_{DA}. **b** Logarithmic plot of the charge recombination rate constant, kCR versus donor-acceptor distance, rDA [27], *solid line* predicted from (8.9)

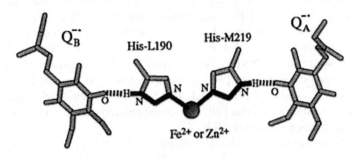

Fig. 8.7 Segment of electron transfer chain of bacterial photosynthetic center [29]

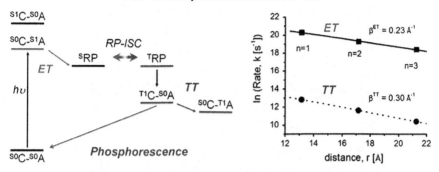

Fig. 8.8 Energy diagram of bichromophoric antenna (A)–core (C) systems, in which the local triplet state of the core ($^C T_1$) is populated via excitation of the antenna, electron transfer (ET) with formation of the singlet radical pair (RP$_S$), intersystem crossing (RP-ISC), and radical recombination (R$_T$) from the resulting triplet radical pair (RP$_T$) ET and TT rate distance dependence [30]

investigated [31]. A gold(III) porphyrin and a fullerene has been used as electron accepting moieties. The results for ZnPc-OPE-AuP(+) indicated a strong electronic coupling over a distance of more than 3 nm. The electronic coupling was manifested in both the absorption spectrum and an ultrafast rate for photoinduced electron transfer ($k_{PET} = 1.0 \times 10^{12}$ s^{-1}. The charge-shifted state in ZnPc-OPE-AuP(+) recombines with a relatively low rate (k_{PET}) $= 1.0 \times 10^9$ s^{-1}. In contrast, the rate for charge transfer in the other dyad, ZnPc-OPE-C(60), is relatively slow ($k_{PET} = 1.1 \times 10^9$ s^{-1}, while the recombination is very fast ($k_{BET} \approx 5 \times 10^{10}$ s^{-1}.

Time-resolved electron paramagnetic resonance (TREPR) spectroscopy at the X-band and Q-band microwave frequencies was used to investigate the spin exchange interaction, J, in serious of biradicaloids [32]. The corresponding spin-correlated radical pairs were formed by photoseparation in covalently bound phenothiazine (PHZ) donor and methylviologen (V) acceptor compounds with polymethylene chain spacers (C8, C10, C12) incorporated in a "through-ring" (rotaxane) fashion to α-cyclodextrin hosts The analysis of TREPR spectra showed that the J values are negative in sign and have absolute values range from 2 to 1000 Gauss.. The results were discussed in terms of through-bond (phenyl ether spacers π-bonds assisting the electronic coupling) versus through-space electronic coupling mechanisms.

Probing protein environment with dual fluorophore-nitroxide (FNO·) molecules in which fluorophore is tethered with nitroxide, a fluorescence quencher opens unique opportunities to study molecular dynamics and micropolarity of the medium, intramolecular fluorescence quenching (IFQ), electron transfer, photoreduction and light energy conversion [33–36]. The following scheme describes photophysical and photochemical processes in the dual (Scheme 8.2).

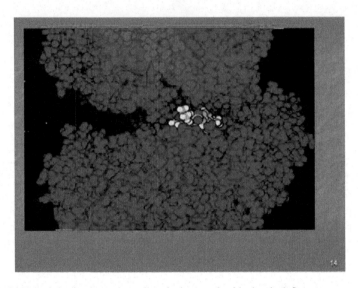

Fig. 8.9 Model of the bovine serum albumin incrusted with the dual fluorescence—nitroxide probe FN1 (Likhtenshtein personal communication)

The fluorophore and nitroxide segments of the probes allow monitoring of molecular dynamics and also make it possible to measure micropolarity of the medium in the vicinity of the donor (by fluorescence technique) and acceptor (by ESR) moieties. Exchange, energetic and molecular dynamic factors affecting photoinduced electron transfer in a donor-acceptor pair (D-A) incorporated into bovine serum albumin were investigated (Fig. 8.9) [33–35]. The kinetics of reversible and irreversible intramolecular electron transfer from the donor to the acceptor were monitored by picoseconds time resolved fluorescent and ESR techniques at temperature range from 70 to 300 K. On the basis of the obtained kinetic data and of the data of the micropolarity in the vicinity of the donor and the acceptor groups, the parameters of the Marcus–Levich theory standard Gibbs energy change ($\Delta G^0 = -1.7$ eV, the reorganization energy ($E_r = 0.9$ eV) the resonance integral ($V_{DA} = 1.5 \times 10^{-3}$ eV) and the Frank—Condon factor for reversible electron transfer were obtained. The electron transfer resonance integral ($V_{DA} = 1.5 \times 10^{-3}$ eV) was estimated using semiempirical methods described in Section X. The role of supper exchange and molecular dynamics in the vicinity of the donor and the acceptor groups in longdistance electron transfer was discussed (Formula 8.2).

In work [37], he charge-transfer process from biphenyl to naphthalene was determined for the radical anions and radical cations of molecules with the structure: (2-naphthyl)-(steroid spacer)-(4-biphenylyl) varied degrees of the spacer unsaturation (Fig. 8.10). The experimental data were explained in framework of a combination of two independent mechanisms: a single-step, superexchange mechanism, and a two-step, sequential charge transfer. The superexchange rate was

Formula 8.2 .

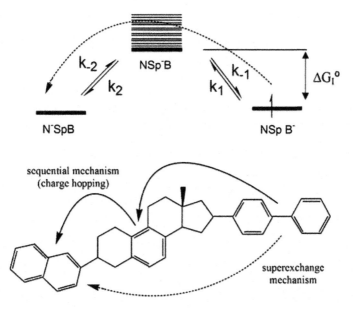

Fig. 8.10 Sequential (charge hopping) and superexchange mechanisms illustrated for electron transfer in an anion having a phenyl group in the spacer. See details in [37]

shown to have a weak dependence on ΔG_I^0, changing 10-fold for a change in ΔG_I^0 of 2 eV, compared to the sequential mechanism in which the rate can change over 10^3 for 0.5 V. It was concluded that superexchange is the dominant mechanism when ΔG_I^0 is large, but the sequential mechanism will dominate at small ΔG_I^0.

The factors affecting electron spin polarization transfer from the photogenerated $D^{+\bullet}$–C–$A^{-\bullet}$ spin-correlated radical pair (RP) to the stable nitroxide in the series of compounds were revealed [38]. OK In the D–C–A–R$^\bullet$ molecules, D is 4-methoxyaniline (MeOAn), 2,3-dihydro-1,4-benzodioxin-6-amine (DioxAn), or benzobisdioxole aniline (BDXAn), C is 4-aminonaphthalene-1,8-dicarboximide, and A is naphthalene-1,8:4,5-bis(dicarboximide or pyromellitimide (Fig. 8.11).

Continuous-wave time-resolved electron paramagnetic resonance (CW-TREPR) spectroscopy using continuous-wave (CW) microwaves at 295 and 85 K and

Fig. 8.11 Schematic illustration polarization transfer in a donor acceptor pair [38]

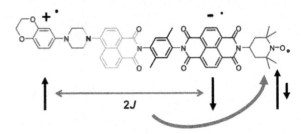

Polarization transfer $\propto |2J|$

electron spin–echo at 85 K were used to probe the initial formation of the spin-polarized radical pair (RP)and the subsequent polarization of the attached $R^{\bullet}$ radical depending on the spin–spin exchange interaction ($2J_{DA}$) of $D^{+\bullet}$–C–$A^{-\bullet}$. The TREPR spectra show that $|2J_{DA}|$ for $D^{+\bullet}$–C–$A^{-\bullet}$ decreases in the order MeOAn$^{+\bullet}$ > DioxAn$^{+\bullet}$ > BDXAn$^{+\bullet}$ as a result of their spin density distributions, whereas the spin–spin dipolar interaction (d_{DA}) remains nearly constant.

The spin–spin interactions in the triradical generated by photoexcitation of the 4-aminonaphthalene-1,8-dicarboximide (ANI) was investigated employing the TR-EPR spectroscopy [39]. Results indicated that in the triradical spin–spin exchange interaction ($2J_{AR}$) negligibly small, whereas their short through-space distance results in a strong dipolar interaction (D_{AR}). Transient nutation experiments (Fig. 8.12) revealed transitions belonged to a species with spin $S = 1$ (Formula 8.3).

Photoinduced electron transfer within a donor–acceptor dyad comprising a zinc porphyrin donor and a tetracationic cyclobis(paraquat-p-phenylene) (CBPQT^{4+}) acceptor was investigated by femtosecond and nanosecond transient absorption spectroscopy, as well as by transient EPR spectroscopy [40]. The experiments showed, that photoexcitation of the dyad produced a weakly coupled ZnP$^{+\bullet}$–CBPQT$^{3+\bullet}$spin-correlated radical-ion pair having a τ = 146 ns lifetime and a spin–spin exchange interaction of only 0.23 mT.

Interatomic currents exchange integral J_{ab} providing full information about the tunneling process related to the rate of electron transfer reaction by the classical Marcus–Levich–Dogonadze equation were calculated using of the Hartree–Fock approximation [41]. The total atomic current given as

$$J_a^{tot} \equiv \frac{1}{2} \sum_b |J_{a,b}| \tag{8.9}$$

is proportional to the probability that the tunneling electron in passing through this atom; as such, it provides a convenient way, The tunneling flux theory to electron transfer reactions was applied to a model system based on the low-potential heme and high-potential heme (heme bL)/(heme bH) redox pair of ubiquinol:cytochrome c oxidoreductase complex (Fig. 8.13).

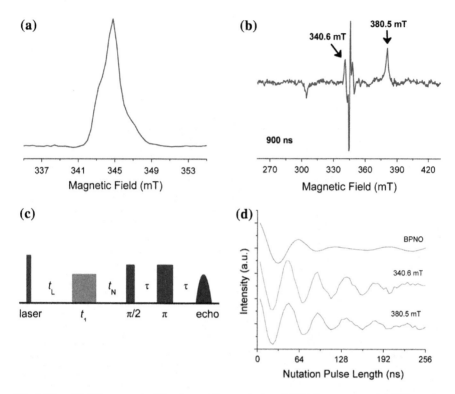

Fig. 8.12 **a** FS-ED spectrum of **1** prior to photoexcitation. **b** FS-ED spectrum of **1** 900 ns after photoexcitation. **c** Schematic illustration of the transient nutation pulse sequence. **d** Representative comparison of the transient nutation spectrum of (*black*) dark state BPNO˙ radical, (*red*) 340.6 mT, and (*blue*) 380.5 mT (see details in [39])

Formula 8.3 .

    In spite of remarkable progress in study elementary processes in donor acceptor systems, the most important problems essential for understanding photosynthesis and invention effective artificial chemical system of light energy transfer in which photoseparated charges can be used for producing energy reached stable products remains to be solved.

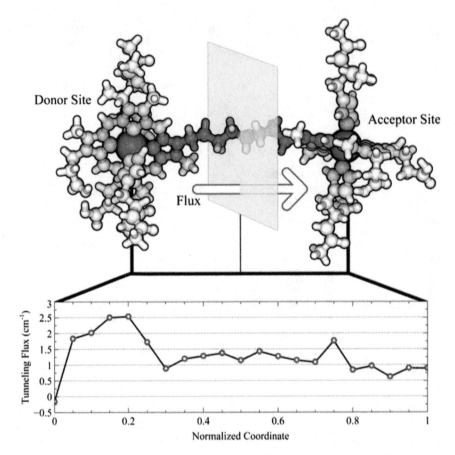

Fig. 8.13 Total tunneling flux in model M1 across the dividing surface between the donor and acceptor redox sites. The *red color* intensity indicates the relative importance of the atoms in the tunneling process. The tunneling flux is expected to be relatively constant in the region between the two redox site [41]

8.4   Triplet-Triplet Energy Transfer and Annihilation

8.4.1   General

Control over emissivity and dynamics of triplet excited states of molecules is central to practical utilization of these states in important areas of technology and medicine. Triplet electronic spin states encompass applications medical photodynamic therapy [42], energy up-conversion by triplet–triplet annihilation [43, 44], biological oxygen sensors [45] and molecular probing [46].

Triplet-triplet energy transfer (TTET) is involved in many biological (e.g. photosynthesis, biological quenching of singlet oxygen) and photosensitized reactions. The principal aims of studies in this area are to develop a better understanding for phot-harvesting systems in organisms, and to provide a base for designing molecular photonic devices and synthetic solar energy conversion systems. The possibility to optimize optoelectronic devices, such as organic light-emitting diodes or solar cells, by exploiting the special characteristics of triplet electronic states and their migration ability is attracting increased attention. Fundamentals and some examples of the TTET processes were described in Chap. 2.

The following aspects of photophysical processes in donor-acceptor pairs (D-X-A) has long been a matter of interest of researchers: (1) an origin of creation of the donor triplet state, (2) quantum mechanical, structural and molecular dynamic factors affected on TT interaction, (3) effect of intermediate X bridge, (4) mechanism of long distance energy transfer, and (5) relationship between TTET and energy transfer (ET) in system under investigation.

8.4.2   Triplet-Triplet Energy Transfer in Chemical Systems

Several six-membered cyclic and [2.2.2] bicyclic organosilanes with varying proportions of silicon atoms in the bridges have been prepared and investigated [47]. Focus was placed on the bicyclic compounds which all have silicon atoms at the bridgehead positions. The possibility to enhance the coupling through a single cisoid tetrasilane cage segment by replacing one or two of the other $-SiMe_2SiMe_2-$ bridges with $-CH_2CH_2-$ bridges was theoretically predicted. To evaluate the properties of diabatic states of triplet–triplet energy transfer systems, in excited the C-1,4ee molecule the validity of locally diabatic states for the approximations of Marcus theory was employed [48]. The derivative couplings $d_{IJ}^{[Q]} \propto (E_J - E_I)^{-1}$, where E_J and E_I are the energy difference between the states that they couple, the adiabatic and diabatic representations where compared (Fig. 8.14) It was shown that the strictly diabatic states and the Condon approximation, assumed in the Marcus theory, are valid for systems under consideration (Fig. 8.15).

Highly efficient energy transfer was observed from the naphthalenimide (energy donor) to the perylenimide (energy acceptor) moiety (Fig. 8.16) [49]. It was found that, this process run predominantly through Coulombic coupling, Dexter-type interaction between the chromophoric units at carbon–carbon covalent distance (1.49 Å) (1.49 Å) with a barrier about 100 kJ/mol. An estimate of the energy-transfer rate from the naphthalenimide donor to the perylenimide acceptor of $k_{ET} = 2.2 \times 10^{10}$ s^{-1} was in agreement with observations.

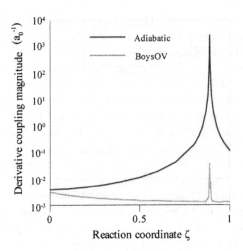

Fig. 8.14 Magnitudes of the DC vector along the linearly interpolated reaction pathway between A*D ($\zeta = 0$) and AD* ($\zeta = 1$) T_1 states of the C-1,4ee molecule. DC magnitudes are presented in both the adiabatic and diabatic (BoysOV) bases. While the DC magnitude is smaller in the BoysOV basis for every point sampled, the degree of reduction is greatest near the avoided crossing, where it peaks at $2.7 \times 10^3 a_0^{-1}$ in the adiabatic basis, and $3.6 \times 10^{-2} a_0^{-1}$ in the diabatic basis [48]

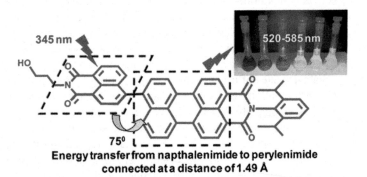

Fig. 8.15 Scheme illustrating energy transfer from the naphthalenimide (energy donor) to the perylenimide (energy acceptor) moiety [49]

8.4.3   Triplet-Triplet Energy Transfer in the Photosynthetic Light Harvesting Antennas

Photosynthetic light harvesting is achieved by a variety of pigments, primarily (bacterio)chlorophylls and carotenoids, bound to distinct types of antenna proteins depending on the organisms. Unlike bacterial and plant reaction centers, the carotenoids light-harvesting complexes differ substantially in terms of structure,

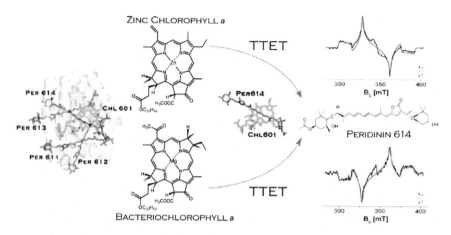

Fig. 8.16 Experimental TR-EPR spectrum of ^{T}Per for Chl a RFPCP (**a**) and ZnChl a RFPCP (**b**) at 150 K with the corresponding calculated TR-EPR spectra of TPer614 after a 12° rotation of the Per's X, Y ZFS axes around the Z axis as described in the text. $|Z| > |Y| > |X|$. A = absorption; E = emission [52]

pigment composition, spectral properties and mechanism of energy transfer/ [50–53].

The high resolution X-ray structure of the main form of PCP (MFPCP) from the dinoflagellate *Amphidinium carterae* (Fig. 8.16) [50] established the basis for detail investigation mechanism of triplet energy transfer in this area. To elucidate the details of the triplet–triplet energy transfer (TTET) mechanism, peridinin–chlorophyll–protein (PCP) complexes (Fig. 8.16), where the N-terminal domain of native PCP from *Amphidinium carterae* has been reconstituted with different chlorophyll (Chl) species, have been investigated [52]. The experimental TR-EPR spectra (9.7 GHz) were close to spectra stimulated considering only one triplet state species with average ZFS parameters $|D| = 47.6$ mT and $|E| = 4.6$ mT.

The pathways of excited-state energy transfer and relaxation in peridinin chlorophyll-*a* protein (PCP) were identified by means of femtosecond visible-pump, mid-infrared probe spectroscopy [53]. It was found that the singlet excited state of Chl-*a* undergoes intersystem crossing (ISC) to the triplet state on the nanosecond timescale, followed by rapid triplet excitation energy transfer (TEET) from Chl-*a* to peridinin, without detecting Chl-*a* triplet. It was suggested that the main channels of singlet and triplet energy transfer in PCP proceed through distinct peridinins. This suggestion is consistent with an energy transfer scheme where the ICT state mainly localizes on Per621/611 and Per623/613, the S_1 state on Per622/612 and the triplet state on Per624.

8.4.4   Triplet-Triplet Upconversion and Annihilation

Upconversion (Chap. 2) which creates population of excited states at higher energy upon photoexcitation at lower energy and generates of emission has attracted much attention due to its potential applications in photocatalysis, photovoltaics (dye-sensitized solar cells), nonlinear optics, and molecular probes [54–63]. Among these applications: wavelength shifting for spectroscopy, sensitized photoreaction by low-energy photons, photovoltaic devices, luminescent probes for bioimaging and improving the energy conversion efficiency of solar cells by harvesting photons below the energy threshold. The upconversion process can occur at an encounter of two chromophors, both in excited triplet state, and in a supramolecular structure in which the chromophors are covalently linked or squeezed in a solid matrix.

The up-conversion approach, first reported in the early 1960s by Parker and Hatchard [55] is based on sensitized triplet–triplet annihilation, which was performed by low power and incoherent excitation sources in organic chromophores absorbing and emitting in the UV region. Since then, a great variety of organic and inorganic chromophores have been investigated both in solution and in polymer matrix, spanning from UV to near infrared (NIR) spectral region [55–66].

In work [60], it was found that a substantial anti-Stokes energy gain up to 0.86 eV in the conjugated supermolecule ruthenium(II) [15-(4'-ethynyl-(2,2';6',2''-terpyridinyl))-bis[(5,5',-10,20-di(2',6'-bis(3,3-dimethylbutoxy)phenyl(porphinato) zinc(II)]ethyne][4'-pyrrolidin-1-yl-2,2';6',2''-terpyridine] bis(hexafluorophosphate) (Pyr(1)RuPZn(2)) in solutions containing N,N-bis(ethylpropyl)perylene-3,4,9,10-

Fig. 8.17 **a** The excitation dynamics in a hypothetical dimer consisting of a donor and an acceptor molecule: the prePump pulse creates an excitation on the donor; during the time between the prePump pulse and the Pump pulse, the excitation is transferred to the acceptor. The pump pulse produces another excitation on the donor which then annihilates with the existing excitation. The process is monitored by the probe pulse. **b** The pulse timing scheme for the prePump–Pump–Probe experiment [64]

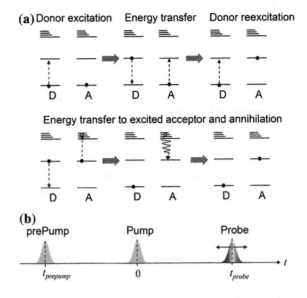

tetracarboxylicdiimide (PDI) or tetracene was reached by a selective near-IR excitation (780 nm). The Pyr(1)RuPZn(2)—causes producing this Upconverted fluorescence signal in Pyr(1)RuPZn(2) was produced by sensitized triplet-triplet annihilation photochemistry. For the PDI triplet sensitization process, the $T(1) \rightarrow T$ (n) excited state absorption decay of Pyr(1)RuPZn(2), monitored at 900 nm as a function of PDI concentration, revealed Stern-Volmer and bimolecular quenching constants of 10048 M^{-1} and 5.9×10^8 M^{-1} s^{-1}, respectively. ^{3}PDI* transient triplet absorption dynamics run with observed bimolecular (3)PDI*-(3)PDI* TTA rate constant $k_{TT} = 1.0 \times 10^9$ M^{-1} s^{-1}.

The mechanism of triplet–triplet annihilation (TTA)-induced up-converted (UC) delayed luminescence was studied by delayed luminescence spectroscopy in two binary organic systems consisting of platinum(II) octaethyl porphyrin (PtOEP) mixed with either poly(fluorene) (PF26) or ladder-type pentaphenylene (L5Ph) [61]. It was found that the energy of the lowest excited triplet state of L5Ph is 0.20 eV and energy of the triplet state of PtOEP is 1.90 eV. The different phosphorescence PtOEP lifetime indicated differences in PtOEP aggregation in the polymer matrices. A series of metal free organic triplet sensitizers derived from a single chromophore (BODIPY) for triplet-triplet annihilation upconversion were devised [62]. The systems absorption was optimized by chemical modification of the sensitizer molecular structures. Long-lived triplet excited state (τ_T up to 66.3 μs), populated upon excitation of the sensitizers, was detected by nanosecond time-resolved transient difference absorption spectra and DFT calculations. In this work, with perylene or 1-chloro-9,10-bis(phenylethynyl)anthracene as the triplet acceptors, upconversion (up to 6.1%) was observed for solution samples and polymer films. The anti-Stokes shift up to 0.56 eV was detected. The principle of TTA-based upconversion and the requirements of the sensitizer and its corresponding acceptor was discussed in review [63].

Specific features of energy transfer dynamics from an excited donor to an excited acceptor in an annihilation process was considered in [64]. on example of carotenoid-zinc-phthalocyanine dyad and a naturally occurring light-harvesting peridinin-chlorophyll protein complex from *Amphidinium carterae*. The energy transfer and annihilation dynamics were monitored by a novel three-pulse femtosecond prePump–Pump–Probe methods which can give 400 μJ, 60 fs, 790 nm pulses. This process is similar to bimolecular chemical reactions, in the case of the singlet–singlet annihilation $(A + A \rightarrow A)$ or for singlet–triplet annihilation $(A + B \rightarrow B)$ [65] and can occur also in single bichromophoric molecules [66]. The designed pump–dump–probe setup The excitation dynamics in a donor and an acceptor in a bichromophoric molecule and the pulse timing scheme for the prePump–Pump–Probe experiment are shown in Fig. 8.17.

Examples of the sensitizer molecules, including iridium complexes, palladium complexes, platinum complexes, ruthenium complexes as well as organic molecules, and their corresponding acceptor molecules were also summarized.

References

1. R.W. Fessenden, App. Magn. Reson. **45**, 483 (2014)
2. K.M. Salikhov, Yu.N. Molin, R.Z. Sagdeev, A.L. Buchachenko, in *Spin Polarization and Magnetic Effects in Radical Reactions*, ed. by Yu. N. Molin (Elsevier, Amsterdam, 19840
3. L.J. Berliner, E.G. Bagryanskaya, Chemically induced electron and nuclear polarization, in *Multifrequency Electron Paramagnetic Resonance*, ed. by S.K. Misra, pp. 947–999, 2011
4. G.L. Closs, M.D. Johnson, J.R. Miller, P.J. Piotrowiak, P. J. Am. Chem. Soc. **111**, 3751 (1989)
5. E. Bagryanskaya, M. Fedin, M.D.E. Forbes, J. Phys. Chem. A **109**, 5064 (2005)
6. R.C. White, M.D.E. Forbes, Org. Lett. **8**, 6027 (2006)
7. R. Das, Mol. Phys.: Int. J. Interface Between Chem. Phys. **104**, 1581 (2006)
8. G.L. Closs, M.D. Forbes, J.R. Norris, Spin-polarized electron paramagnetic resonance spectra of radical pairs in micelles: observation of electron spin-spin interactions. J. Phys. Chem. **91**, 3592–3599 (1987)
9. R.C. White, V.F. Tarasov, M.D.E. Forbes, Langmuir **21**, 2721 (2005)
10. P. Caregnato, L.E. Jarocha, H.S. Esinhart, N.V. Lebedeva, V.F. Tarasov, M.D.F. Forbes, Langmuir **27**, 5304 (2011)
11. M.S. Panov, A.N. Pravdivtsev, K.I. Ivanov, A.V. Yurkovskaya, H.-M. Vieth, App. Magn. Reson **45**(9), 893–900 (2014)
12. M.T. Colvin, R. Carmieli, T. Miura, S. Richert, D.M. Gardner, A.L. Smeigh, S.M. Dyar, S.M. Conron, M.A. Ratner, J. Phys. Chem. A **117**, 5314 (2013)
13. S. Bai, K. Song, Q. Shi, J. Phys. Chem. Lett. **6**, 1954 (2015)
14. M. Zarea, R. Carmieli, M.A. Ratner, M.R. Wasielewski, J. Phys. Chem. A **118**, 4249 (2014)
15. I.K. Kominis, New J. Phys. **15**, 075017 (2013)
16. V. Rane, R. Das, Phys. Chem. Chem. Phys. **16**, 5698 (2014)
17. P.W. Anderson, Phys. Rev. **115**, 2. (1959)
18. R.A. Marcus, J. Chem. Phys. **43**, 679 (1965)
19. K.I. Zamaraev, Yu.N. Molin, K.M. Salikhov, *Spin Exchange. Theory and Physicochemical application* (Springer-Verlag, Heidelberg, 1981)
20. S.M. Mickley Conron, L.E. Shoer, A.L Smeigh, A.B. Ricks, M.R. Wasielewski, J. Phys. Chem. B 01/2 J. Phys. Chem. B, **117**, 2195 (2013)
21. A.I. Kotel'nikov, V.R. Fogel, G.I. Likhtenshtein, G.B. Postnikova, E.A. Shlapnikova, Molekul. Biolo. (Moscow) **15**, 281 (1981)
22. G.I. Likhtenshtein, J. Photochem. Photobiol. A: Chem. **96**, 79 (1996)
23. G.I. Likhtenshtein, *New trends in enzyme catalysis and mimicking chemical reactions* (Kluwer Academic/ Plenum Publishers, Dordrecht, 2003)
24. G.I. Likhtenshtein, *Solar Energy Conversion. Chemical Aspects* (Weinhem, WILEY-VCH, 2012)
25. G.L. Closs, M.D. Johnson, J.R. Miller, P. Piotrowiak, J. Am. Chem. Soc. **111**, 3751 (1989)
26. G.L. Closs, P. Piotrowiak, J.M. MacInnis, G.R. Fleming, J. Am. Chem. Soc. **110**, 2652 (1988)
27. E.A. Weiss, M.J. Ahrens, L.E. Sinks, A.V. Gusev, M.A. Ratner, M.R. Wasielewski, J. Am. Chem. Soc. **126**, 5577(2004)
28. R. Calvo, E.C. Abresch, R. Bittl, G. Feher, W. Hofbauer, R.A. Isaacson, W. Lubitz, M.Y. Okamura, M.L. Paddock, Am. Chem. Soc. **122**, 7327 (2000)
29. M.S. Graige, M.L. Paddock, G. Feher, M.Y. Okamura, Biochemistry **38**, 11465 (1999)
30. M.T. Tomoyasu, D.M. Niedzwiedzki, S.A. Vinogradov, J. Phys. Chem. A **116**, 3598 (2012)
31. E. Göransson, J. Boixel, J. Fortage, D. Jacquemin, H.-C. Becke, E. Blart, L. Hammarström, F. Odobel, Inorg. Chem. **51**, 11500 (2012)
32. Y. Hiroaki, M.D.E. Forbes, Photochem. Photobiol. **91**, 672–677 (2015)
33. G.I. Likhtenshtein, D. Pines, E. Pines, V. Khutorsky, Appl. Magn. Reson. **35**, 459 (2009)
34. G.I. Likhtenshtein, Pure Appl. Chem. **80**, 2125 (2008)
35. G.I. Likhtenshtein, S. Nakatsuji, K. Ishii, Photochem. Photobiol. **83**, 871–881 (2007)

36. G.I. Likhtenshtein, Appl. Biochem. Biotechn. **152**, 135 (2009)
37. B.P. Paulson, J.R. Miller, W.-X. Gan, G. Closs, J. Am. Chem. Soc. **127**, 4860 (2005)
38. T. Colvin, R. Carmieli, T. Miura, S. Richert, D.M. Gardner, A.L. Smeigh, S.M. Dyar, S.M. Conron, M.A. Ratner, M.R. Wasielewski, Electron spin polarization transfer from photogenerated spin-correlated radical pairs to a stable radical observer spin. J. Phys. Chem. A **117**, 5314–5325 (2013)
39. D.M. Gardner, H.-F. Chen, M.D. Krzyaniak, M.A. Ratner, M.R. Wasielewski, J. Phys. Chem. A **119**, 8040 (2015)
40. M. Fathalla, J.C. Barnes, R.M. Young, K.J. Hartlieb, S.M. Dyar, S.W. Eaton, B.K. Rugg, A.A. Sarjeant, D.T. Co, M.R. Wasielewski, J.F. Stoddart, Chem. A Eur J. **20**, 14690 (2014)
41. M.A. Hagras, A.A. Stuchebrukhov, J. Phys. Chem. B (2015) [Ahead of Print]
42. T.C. Zhu, J.C. Finlay, Med. Phys. **135**, 3127 (2008)
43. W. Wu, J. Sun, X. Cui, J. Zhao, J. Mater. Chem. C: Mater. Opt. Electron. Devices **1**, 4577 (2013)
44. H. Sasabe, J. Kido, Chem. Mater. **23**, 621 (2011)
45. S.A. Vinogradov, D.F. Wilson, Porphyrin-dendrimers as biological oxygen sensors, in *Designing Dendrimers*, ed. by S. Capagna, P. Ceroni (Wiley, New York, 2010)
46. G.I. Likhtenshtein, *Biophysical Labeling Methods in Molecular Biology* (Cambridge, N. Y., Cambridge University Press, 1993)
47. A. Wallner, R. Emanuelsson, J. Baumgartner, C. Marschner, H. Ottosson, *Organometallics* **32**, 396 (2013)
48. E.C, Alguire, S. Fatehi, Y. Shao, J.E. Subotnik, J. Phys. Chem. A **118**, 11891 (2014)
49. R.T. Cheriya, J. Joy, A.P. Alex, A. Shaji, M. Hariharan, J. Phys. Chem. C **116**, 12489 (2012)
50. E. Hofmann, P.M. Wrench, F.P. Sharples, R.G. Hiller, W. Welte, K. Diederichs, Science **272**, 1788 (1996)
51. T. Niklas, S. Schulte, M. van Prakash, E. Gastel, E. Hofmann, W. Lubitz, J. Am. Chem. Soc. **129**, 15442 (2007)
52. M. Di Valentin, C. Tait, E. Salvadori, S. Ceola, H. Scheer, R.G. Hiller, D. Carbonera, J. Phys. Chem. B **115**, 13371 (2011)
53. C. Bonetti, M.T.A. Alexandre, I.H.M. van Stokkum, R.G. Hiller, M.L. Groot, R.J.T. van Grondelle, J.T.M. Kennis, Phys. Chem. Chem. Phys. **129**, 256 (2010)
54. P.E. Keivanidis, F. Laquai, J.W.F. Robertson, S. Baluschev, J. Jacob, K. Mullen, G. Wegner, J. Phys. Chem. Lett. **21**, 893 (2011)
55. C.A. Parker, C.G. Hatchard, Sensitized anti-Stokes delayed fluorescence. Proc. Chem. Soc. London 386–387 (1962)
56. F.N. Singh-Rachford, F.N. Castello, Coord. Chem. Rev. **254**, 2560 (2010)
57. W. Wu, J. Zhao, J. Sun, L. Huang, X. Yi, J. Mater. Chem. C **1**, 705 (2013)
58. S. Ji, W. Wu, W. Wu, H. Guo, J. Zhao, Angewan. Chem., Int. Ed. **50**, 1626 (2011)
59. H. Sun, H. Guo, W. Wu, X. Liu, J. Zhao, Dalton Trans. **40**, 7834 (2011)
60. T.N. Singh-Rachford, A. Nayak, M.L. Muro-Small, S. Goeb, M.J. Therien, F.N. Castellano, J. Am. Chem. Soc. **132**, 14203 (2010)
61. P.E. Keivanidis, F. Laquai, J.W.F. Robertson, S. Baluschev, J. Jacob, K. Müllen, G. Wegner, J. Phys. Chem. Lett. **2**, 2011 (1893)
62. W. Wu, H. Guo, W. Wu, S. Ji, J. Zhao, J. Org. Chem. **76**, 7056 (2011)
63. Z. Xinglin, Y. Huiran, S. Huibin, L. Shujuan, Z. Qiang, H. Wei, Prog. Chem. 1880–1889 (2012)
64. M. Vengris, D.S. Larsen, L. Valkunas, G. Kodis, C. Herrero, D. Gust, T. Moore, A. Moore, R. van Grondelle, J. Phys. Chem. B (Ahead of Print)
65. A.A. Ovchinikov, S.F. Timashev, A.A. Belyi, *Kinetics of Diffusion-Controlled Chemical Processes* (Nova, New York, 1989)
66. J. Hofkens, M. Cotlet, T. Vosch, P. Tinnefeld, K.D. Weston, C. Ego, A. Grimsdale, K. Mullen, D. Beljonne, J.L. Bredas, Proc. Natl. Acad. Sci. USA **100**, 13146 (2003)

Chapter 9
Magnetic and Electromagnetic Fields Effects on Chemical and Biological Processes

Abstract This chapter described important aspects of static magnetic and electromagnetic field influence on chemical and biological reactions. The relevant phenomena, physical principles, and main results in the area were the focus of attention. A wide variety of advanced methods including time-resolved ESR and NMR spectroscopy, luminescence and electrochemistry is based on the physical or chemical response of a system under interest by switching an appropriate magnetic field or fields, on or off. These methods allow one to shed light on mechanistic details of radical and ion-radical reactions, that is, the chemical nature and physical state of intermediates, its spin dynamic and kinetics. This knowledge paves the way for a focused understanding of spin correlated radical processes. The radical pair model for the molecular compass leads to a reasonable explanation of such phenomena as the capacity of animals to navigate in the Earth's magnetic field. Given the constantly increasing anthropogenic radio- and microwave background in our environment, the possibility of radiative impact on humans and animals is a concern that should be carefully.

9.1 Introduction

Chemical and biochemical reactions that involve radical or radical ion intermediates can be influenced by magnetic fields, which act to alter their rate, yield, or product distribution. These effects have been studied extensively in liquids, solids, micelles, nanoparticles and biological objects: naturally-occurring and photosynthetic model systems, in particular [1–10]. Magnetic field effects (MFEs) occur at field strengths ranging from upwards of 30 T, in superconducting magnets, to ~ 40 μT, which is comparable to the field strength of the Earth [7]. Considerable evidence now exists that exposure to strong static high magnetic fields (SMFs) causes marked changes in the properties of a number of biological systems, particularly those whose function is linked to the properties of membrane ion channels, transmembrane ion flux, for example [9].

© Springer International Publishing Switzerland 2016
G. Likhtenshtein, *Electron Spin Interactions in Chemistry and Biology*,
Biological and Medical Physics, Biomedical Engineering,
DOI 10.1007/978-3-319-33927-6_9

A great variety of theoretical and experimental work indicates the key importance of spin chemistry in both static magnetic field and electromagnetic field effects (Chaps. 4, 5 and 6). Molecular compass models of magnetoreception of birds and other living organisms is one of the existing challenging problems [7, 10]. Adequate data for proper human health risk assessment of static and electromagnetic magnetic fields are sparse and uncertain (see details in [11]) Discussion of this complicated area is beyond the scope of the present book.

9.2 Diamagnetic Effect of Strong Static Magnetic Fields

The effect of magnetic fields on lipid bilayers has been a subject of interest to researchers in recent years [9, 12, 13]. The mechanism suggested to explain these effects is based on the diamagnetic anisotropy features of membrane phospholipids. The exposure to static magnetic fields (SMF)s can result in a rotational displacement of the membrane's phospholipid molecule by virtue of their collective diamagnetic properties and, therefore, electron spin effects are not directly involved in the processes.

A fluctuation model of reactive bilayer membranes, taking into account the composition-curvature correlation in the presence of a homogeneous magnetic field, was proposed [12]. The model explained how the strength of the magnetic field in combination with the intensity of reaction induces instability in this system. The partial orientation of multilamellar vesicles (MLVs) in high magnetic fields has been studied and a method to prevent such effects was proposed [13]. The orientation effect was measured with ^{2}H-, ^{31}P-NMR and freeze-etch electron microscopy techniques in MLVs composed of dipalmitoyl phosphatidylcholine with 30 mol% cholesterol. The MLV samples that were frozen directly in the NMR magnet at a field strength of 9.4 T were investigated using freeze-etch electron microscopy techniques. These experiments showed that the MLVs in applied magnetic fields have an ellipsoidal shape.

An experiment to clarify the effects of multiple rapid temperature changes and magnetic fields (up to 8 T) on cell membrane fluidity by using red blood cell ghosts and a fluorescence dye, 1-aminonaphthalene-8-sulfonic acid (ANS) was performed [14]. The time course of ANS emission at 480 nm under the influence of a magnetic field at 5 T was observed. The results indicated that the fluidity of the molecules in the cell membrane was decelerated by exposure to magnetic fields at 5 T.

Though the aforementioned effects of the strong magnetic field are not directly related to spin phenomena, they need to be taken into account when analyzing the behavior of some specific spin systems.

9.3   Static Magnetic and Electromagnetic Fields Effects on Processes Involving Radicals and Radical Ions

9.3.1   General

The effect of an static magnetic field (SMF) on a system of singlet and triplet pairs was briefly described in Sect. 5.8.1. In references [1, 4, 5, 15–23], this phenomenon was considered in more detail. The observed static and electromagnetic field effects arose from fundamental properties of spin correlated radical pairs (SCRP) (Sections X and Y). In experiments on MFEs, the SCRP spins evolve according to a Hamiltonian [5]

$$\hat{H} = \sum_{N=\{A,B\}} \left\{ -\gamma_e B_0 \cdot \hat{S}^N - \gamma_e B_1(t) \cdot \hat{S}^N + \sum_i a_{Ni} \hat{S}^N \cdot \hat{I}^{Ni} \right\} - e^{-r/r_J} J_0 \hat{S}^A \cdot \hat{S}^B$$

(9.1)

where B_0 denotes the static (time-independent) magnetic field and $B_0(t)$ its time-dependent component, a^{Ni} is the isotropic hyperfine coupling constant for interaction between an electron spin S^N and nuclear spin I^{Ni}. S^A and S^B are the corresponding electron spins.

In a typical example, after photochemical excitation of the ground-state precursor species A–B, the transfer of an electron from A to B occurs, creating radical ions in a singlet state, a spin-correlated radical pair (SCRP) (Fig. 9.1) [5].

For about 500 ps after the excitation, the radicals, so called geminate pairs, are squeezed by a "cage" of surrounding solvent molecules forming spin correlated pair. The strong exchange interaction at short separation inhibits S ↔ T interconversion and this step is insensitive to magnetic fields Then, the singlet SCRPs undergo spin-selective reaction to produce the singlet initial product in the singlet state (SP) or they react to give a back-reaction product. The singlet radical pairs transforms back and forth into triplet radical pairs. Coherent evolution of the SCRP spin state provides S ↔T interconversion, which is driven by magnetic interactions in the SCRP and by the applied magnetic field.

The S ↔ T evolution can be characterized by the following three mechanisms [4–6]. In the **Δg mechanism**, in order for S–T_0 evolution to occur, the electron

Fig. 9.1 Radical pair mechanism. *AB* round-state precursor species, *SP* singlet product, *TP* the triplet product [5]

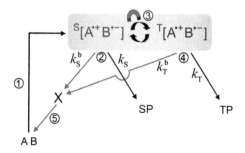

spins must differ in their g values. The frequency of the $S-T_0$ mixing is given by the difference in precession frequency of the two electron spins. The $S-T_0$ mixing frequency increases with increasing external magnetic field. To illustrate **the hyperfine mechanism**, a pair of radicals where only one of the electron spins, S_1, couples to a single magnetic nucleus of spin quantum number $I = 1/2$ was considered [4, 5, 16]. The magnetic moment of the nucleus induces an additional local magnetic field on S_1. In high magnetic fields, the magnetic moments of both nucleus and electron, S_1, are orientated along the field direction. While S_2 precesses with the unperturbed Larmor frequency, the precession frequency of S_1 differs from this value by $a/2$, where **a** is the hyperfine splitting constant. In low magnetic fields, when applied fields are weaker than or comparable with the hyperfine field, the hyperfine mechanism induces transitions between the singlet and all three triplet levels cause diminish of the singlet state population.

The **relaxation mechanism** (RM) of the S ↔ T evolution takes place due to the anisotropic **g**-tensor (δg), hyperfine interaction (HFC), and spin−spin dipolar interactions of radical pairs and includes the following steps [4, 5, 16]: (1) The spin−spin relaxation process leads to $S-T_+$ mixing, (2) the triplet radical pairs undergo their own spin-selective reaction to produce the triplet product TP or they react nonselectively to give the back-reaction product X or singlet pair SP or triplet pair TP, (3) internal convertion The system transfers back into the precursor A-B allowing for subsequent instances of the process to occur, and (4) fast spin relaxation leads to incoherent transitions between the different RP states with consequences for the RP kinetics in compounds.

9.3.2 Static Magnetic Field Effect

9.3.2.1 Theoretical Considerations

Understanding the aforementioned elementary processes allows one to predict changes in the singlet yield as a function of static magnetic field strength (Fig. 5.12, Sect. 5.8.1) The existence of a low-field effect (LFE) opposite in phase to the high field MFE was first predicted in 1976 by Brocklehurs [18, 19]. In low fields, $B_0 < 50$ mT, the singlet–triplet interconversion is governed by the isotropic electron Zeeman and the hyperfine interactions. The hyperfine mechanism drives the S → T transition and, therefore, diminishes the singlet state population (Fig. 1.9) [5]. For the low field effect, the following requirements should be fullfiled: (a) radical–radical interactions should be small compared with the average hyperfine coupling in the radical pair and (b) relaxation in the radical pair must be slow compared with both spin evolution and radical recombination.

Typical magnetic field effect in organic radical pairs is presented in Fig. 9.2 [5].

At higher fields where external static magnetic fields are much larger than the local one in the system ($B_0 > 50$ mT), the hyperfine interactions have practically no effect on the singlet–triplet interconversion and only the aforementioned

Fig. 9.2 Typical MFEs in
organic RPs from zero field
up to a few Tesla [5]

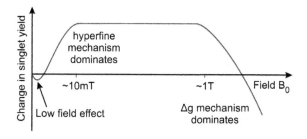

Δg-mechanism causes singlet–triplet interconversion [5, 20]. When the applied field
is further increased, the two triplet levels, T_+ and T_-, become progressively shifted
from one another. In such a condition, if the radical pair is originally created in a
singlet state, the singlet–triplet mixing at high fields decreases and the yield of the
singlet recombination product increases The effects of high magnetic fields on the
T–S spin conversion of radical pairs generated from a triplet precursor on the yield
of escaped radicals are illustrated by Fig. 9.3 [20].

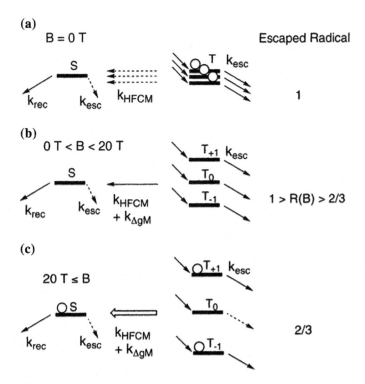

Fig. 9.3 T–S spin conversion of the present radical pair generated from a triplet precursor at
a $B = 0$ T, **b** 0 T $< B < 20$ T, and **c** 20 T $\leq B$ (see details in [20])

A theoretical simulation was used to investigate the magnetic field/microwave frequency dependence of the spin polarized EPR spectra of sequential spin correlated radical pair [21]. To describe the kinetics the following expressions were used

$$
\begin{aligned}
S(t, B_0) &= S_A + S_B, \\
S_A(t, B_0) &= \alpha(B_0)e^{-(k+w_A)t}, \\
S_B(t, B_0) &= \beta(B_0)\frac{k}{k+w_A-w_B}\left\{e^{-w_g t} - e^{-(k+w_A)t}\right\},
\end{aligned}
\tag{9.2}
$$

where S_A and S_B are the contributions from sequential radicalpairs; $\alpha(Bo)$ and $\beta(Bo)$ are their spin polarization patterns, w_A and w_B are the corresponding effective relaxation rates of the spin polarization and k is the electron transfer rate.

Figure 9.4 illustrates the magnetic field dependence of the net polarization p generated in sequential radical pairs derived from (2.9).

Transition metal compounds are involved in many processes such as biological and chemical redox reactions, corrosion, and catalysis [22]. In works [23–25], specifics of magnetic field effects in processes with participating transition metal complexes have been discussed In transition metal complexes, due to strong spin-orbit coupling, radical pairs formed in electron transfer have wave functions describing mixtures of the singlet or triplet multiplicities. In such systems spin relaxation times can be very short (down to picoseconds). In contrast, in organic radical pairs (RPs) the dominant interaction is the hyperfine interaction, with a typical magnitude of 10^7–10^8 rad s^{-1}, which can only affect kinetics of organic RPs with lifetimes longer than several nanoseconds [4].

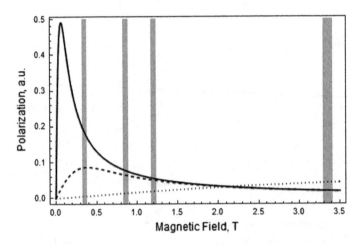

Fig. 9.4 Magnetic field dependence of the net polarization p generated in sequential radical pairs The field dependence is shown for three different life times of the precursor state. The values of the spin-spin coupling $b = 2 J + d$ are 1 mT for the short lived precursor case (lifetime = 0.611 s) and 0.1 mT for the intermediate (10 ns) and long lived (290 ns) precursor cases. The difference in the g factors of the precursor pair *is* $\Delta g = 0.003$. *Bo* is the value of magnetic field and in (9.2), and $q = \Delta g\ \beta Bo$ [21]

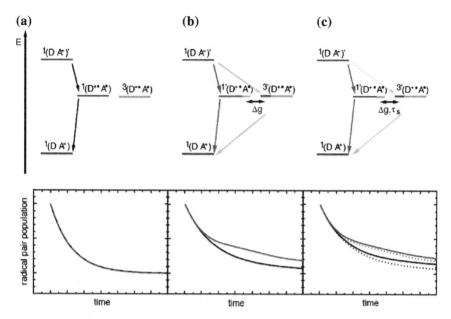

Fig. 9.5 Illustration of spin effects in fast RP recombination (*top*) and their influence on the RP population in zero (*central lines*) and high (*right lines*) magnetic field (*bottom*) [23]. **a** Organic RP: Starting from a singlet precursor (D A$^+$)*, only the singlet-phase RP state is populated and decays in single-exponential fashion. **b** RP bearing a transition metal moiety (static effect, no spin relaxation). **c** RP comprising a transition metal moiety (fast spin relaxation)

An external magnetic field changes the transition rates between the RP states and, therefore, has an effect on the RP magnetic properties and kinetics. The peculiarities of spin effects in fast RP recombination are illustrated in Fig. 9.5.

9.3.2.2 Experimental Data

Static magnetic field effects (SMFEs) on photochemical reactions through radical ion-radical pairs have been studied extensively in the last decades [23–29] and references therein. Some examples of investigations in this area are presented.

The radical pair recombination of an intramolecular electron-transfer system Fcp-Nb$^+$ bearing a ferrocenophane (Fcp) and a Nile blue (Nb$^+$) moiety has been investigated by femtosecond spectroscopy [23]. The radical pair was formed by ultrafast electron transfer (90 fs) from a ferrocene residue to a photoexcited Nb$^+$ moiety. Fast spin relaxation leads to incoherent transitions between the different RP states with consequences for the RP kinetics in compounds shown in Fig. 9.2. According to kinetics measurement, spin processes compete with electron transfer. Observed magnetic field effects on kinetics allowed to disentangle the contributions of spin processes competing with electron transfer (Fig. 9.6).

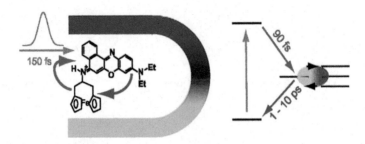

Fig. 9.6 Scematic illustration of fast kinetics in the intramolecular ET system Fcp-Nb$^+$ [23]

The MFE on the photoinduced electron transfer (PET) between phenazine (PZ) and the amines, N,N-dimethylaniline, N,N-diethylaniline, 4,4¢-bis(dimethylamino)diphenylmethane (DMDPM), and triethylamine was observed in micelles, reverse micelles, and small unilamellar vesicles [26]. The experiments showed that for different on the electron acceptors, the difference between optical density of intermediates (ΔOD), recorded at 0.8 μs after the laser flash, in the presence and absence of a magnetic field increases with an increase in field and then reaches saturation (ΔOD$_{sat} \approx 0.005$–0.007) at a magnetic field strength around 0.04–0.05 T. Experimental data were explained by the hopping electron transfer mechanism. For donor-acceptor pair PZ-DMA, the rate constants for the fast components of the absorption decay profile after 5 μs varied from 2.2×10^6 s^{-1} to 0.96×10^6 s^{-1} as the applied magnetic field was increased to 0.04 T.

The magnetic field effects on the photoinduced electron transfer (PET) reaction between the [Cu-(phen)$_2$]$^{2+}$ complex (I) and DNA in homogeneous buffer medium and in reverse micelles was also studied [25]. The sample was excited by 266 nm laser light with 8 ns FWHM followed by measurements of the nanosecond transient absorption spectra. The suggested mechanism of the reaction is as follows (Scheme 9.1).

Electron transfer in photosynthetic reaction centers is a subject of special interest [21, 27–30]. Magnetic field effects in photosynthetic reaction centers MFEs were first independently detected by Blankenship et al. [27] and by Hoff et al. [28, 29]. Under chemically reducing conditions, the primary acceptor Z of the photochemical reaction centers of *Rhodopseudomonas sphaeroides* was blocked [27]. An applied

$$[Cu(phen)_2]^{2+} \xrightarrow{h\nu} {}^1([Cu(phen)_2]^{2+})^* \xrightarrow{ISC} {}^3([Cu(phen)_2]^{2+})$$

$$^3([Cu(phen)_2]^{2+}) + DNA \xrightarrow{ET} {}^3\{[Cu(phen)$$

$$(phen^{\bullet-})]^{2+}\cdots DNA^{\bullet+}\} \xrightarrow[ISC]{MF} {}^1\{[Cu(phen)$$

$$(phen^{\bullet-})]^{2+}\cdots DNA^{\bullet+}\}$$

Scheme 9.1 Reaction mechanism between the [Cu-(phen)$_2$]$^{2+}$ complex (I) and DNA

magnetic field decreases the fraction of the transient state PF that decays by way of the bacteriochlorophyll triplet state radical pair. At room temperature, a 2-kG field decreases the quantum yield of a radical pair, which is generated in a singlet state and undergoes a rapid transformation into a mixture of singlet and triplet states, by about 40 %. In one of pioneering work in the area [28] it was shown that the yield of the triplet state in reaction centers of *Rhodopseudomonas sphaeroides*, in which the electron transfer was blocked by removing functional iron in its complex with the primary acceptor ubiquinone, is dependent on the strength of an applied magnetic field. The effect is decreased about twofold when the value of the magnetic field was about 50–100 G. The dependence of the triplet yield on magnetic field was discussed in terms of the Chemically Induced Dynamic Electron Polarization mechanism. The following factors were found to affect the dependence of the primary electron transfer in the reaction centers on magnetic field and on the yield of triplet products: (1) the rate constants of reversible electron transfer between the initially excited singlet state of the reaction center and an intermediate radical ion pair state; (2) the rate constants of irreversible electron transfer of the radical pair to the ground and excited triplet state of the reaction center; (3) the electron exchange interactions between the radical pair and the primary acceptor (Q_A).

Transient ESR and related theoretical simulations were used to investigate the magnetic field/microwave frequency dependence of spin polarized X-band (9.7 GHz) and K band (24 GHz) EPR spectra of the sequential spin correlated radical pairs primary donor—primary acceptor $(P^+A^-)_1$ and primary donor $(P^+F_x^-)$ in photosystem I (PSI) and reaction centres of heliobacteria [21]. In PSI, where the precursor $(9A^-)$ lifetime (290 ns) is much longer than the characteristic time of singlet-triplet mixing, the observable net polarization decreases with the field strength in this region. In heliobacteria, where the precursor lifetime (600 ps) is much shorter than the characteristic time of singlet-triplet mixing, state $P^+F_x^-$ is observed the net polarization increases in the same range of magnetic field. In another work, the spin exchange integral $J = 10^{-3}$ cm in the bacterial reaction center was estimated from magnetic field effects on recombination [30]. It was suggested that in the system under investigation there are two sites for the electron or for the hole, with hopping between sites.

In aforementioned works on photo-induced electron transfer in the reaction centers, radical pairs can be detected only if forward electron transport is blocked by some means.

9.4 Electromagnetic Microwave Irradiation Effects

Microwave technique widely used in organic and medical chemistry [31]. Commonly in the absence of applied static magnetic field trivial thermal and sometimes specific non-thermal microwave effects in chemical reaction were revealed [31]. The latter is still a controversial topic.

The dependence of resonance absorption of electromagnetic irradiation in the presence of static magnetic fields forms the basis for ESR spectroscopies and related methods (Chaps. 5 and 6). Moreover, this phenomenon offers the possibility not only to investigate in detail mechanisms of radical and ion-radical reactions but also to understand these processes which play a key role in many fields of chemistry and biology [4–9, 15, 32–35].

A brief outline of the behavior of a spin correlated pair in a static magnetic field, which is a prerequisite for understanding microwave frequency effects on spin selected radical processes and the MARY and RYDMR methods was presented in [15]. The Hamiltonian I of the pair including only isotropic electron Zeeman and hyperfine interactions in a magnetic field was written in the following form:

$$\mathscr{H} = g\beta H(\mathbf{S}_1 + \mathbf{S}_2)\sum a_i\mathbf{I}_{1i}\mathbf{S}_1 + \sum a_j\mathbf{I}_{2j}\mathbf{S}_2. \qquad (9.3)$$

In this equation dipole-dipole, exchange interactions and all anisotropic contributions were neglected and g-values for the two radicals were taken to be equal only in the region of weak magnetic field. The time dependent probability of finding the spin system in the singlet state $\rho_{ss}(t, H)$ [MARY spectrum G(H)], was calculated by solving the Hamiltonian I followed by convolution with the recombination function $f(t)$:

$$G(H) = \int \rho_{ss}(t, H)f(t)dt. \qquad (9.4)$$

As one can see from Fig. 9.7 microwave radiation effects on transition between triplet and singlet states and therefore on yield and kinetics of the spin selected radical processes.

Thus under conditions of normal ESR field strengths, applying microwave pumping to the described system effects on the reaction yield and makes it possible

Fig. 9.7 Schematic energy level layout of spin system of the pair showing the crossings of the eigenlevels (*top*), and a sketch of the magnetic field dependence of the observed signal that it would produce. See details in [15]

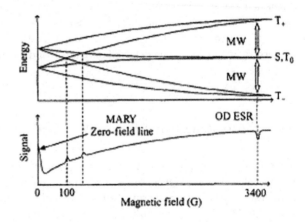

to detect the ESR signal by optical or other techniques. This also forms a theoretical basis for understanding effects of microwave radiation on leaving organisms.

The recently developed pulse version of Reaction Yield Detected Magnetic Resonance (RYDMR) based on refocusing the zero-quantum coherences in radical pairs by non-selective microwave pulses provide deeper insight in mechanism of microwave effects! [36]. In this approach, the population of a radical pair singlet spin state [$\rho_{SS}(t)$] depending on microwave radiation leads to the yield product. The suggested pulse sequences for pulse RYDMR, the primary echo, stimulated echo, Electron Spin Echo Envelope Modulation and pulse ENDOR are shown in Fig. 9.8. It was assumed that the radical pairs are formed instantaneously with a short laser flash and the refocusing of the zero-quantum coherences in radical pairs are produced by non-selective microwave pulses. Under such conditions, the spin evolution between pulses was described by means of the following Hamiltonian H^:

$$\hat{H} = \omega_1 S_{1z} + \omega_2 S_{2z} + J S_{1z} S_{2z} + A S_{1z} I_z + B S_{1z} I_x + \omega_1 I_z \tag{9.5}$$

Here, S_{1z} and S_{2z} are the operators for the z-projection of the electron spins; I_z and I_x are the operators for the z- and x-projections of the nuclear spin, respectively; w_1 and w_2 are the electron Zeeman interactions with the external field in frequency units; $J = J_0 + D(1 - 3 \cos^2 \theta)/2$ is the electron spin–spin interaction constant including contributions from exchange, J_0, and dipolar coupling, D; θ is the angle between the vector connecting the electron spins and the external field; A and B are the secular and pseudo-secular parts of the HFC tensor, respectively; w_I is the nuclear Zeeman interaction.

Fig. 9.8 Pulse sequences for RYDMR: primary echo (**a**), ESEEM (**b**), stimulated echo (**c**) and ENDOR (**d**). Laser pulses are shown in *black*, mw-pulses are shown in *dark grey*, rf-pulse is shown in *light grey*, echo is shown in *white*; delays and flip angles are indicated [36]

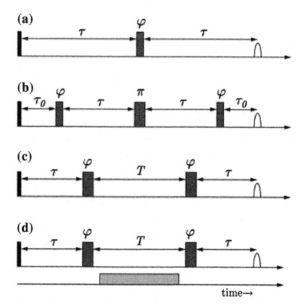

Effect of microwave radiation on human health have been attracting growing attention (see for example publication [33–35]. Discussion on this topic is out of scope of this book.

9.4.1 Radiofrequency Magnetic Field Effects

In the low magnetic field region, when the energy of hyperfine couplings is close to the radiofrequency (RF) irradiation frequency, a marked effect of the applied RF on the yield of radical reactions can be expected [37–41]. In the early book [38], Buchachenko and Frankevich considered the chemical generation and detection of radio- and microwave modified chemical kinetics, theoretical chemistry and metabolite chemistry. Experimental data on photoinduced electron transfer between a donor and acceptor in the presence of static and radiofrequency magnetic fields, can commonly be interpreted in the framework of the radical pair mechanism. This mechanism suggested a singlet $\leftrightarrow$ triplet interconversion in the cation- anion radical pair when the oscillating field is in resonance with hyperfine splittings in the anion radical.

For example, the influence of isotropic electron–nuclear hyperfine interactions and the isotropic Zeeman interactions of the electron spins with a static magnetic field strength of $B_0 = \omega_0/\gamma_e$, and a linearly polarized radio frequency field of peak strength $B_1 = \omega_1/\gamma_e$ and frequency $\omega_{RF}/2\pi$ on the evolution of a spin-correlated radical pair was analyzed [42, 43]. The following spin Hamiltonian (in angular frequency units) was suggested to govern the coherent evolution of the radical pair:

$$
\hat{H}(t;\gamma) = \sum_{j=1}^{2} \left\{ \sum_{k=1}^{N} a_{jk}\hat{\mathbf{S}}_j \cdot \hat{\mathbf{I}}_k + \omega_1\hat{\mathbf{S}}_{jx} \sin(\omega_{RF}t + \gamma) \\ + \omega_0\left[\hat{\mathbf{S}}_{jz} \sin\theta + \hat{\mathbf{S}}_{jx} \cos\theta\right] \right\}
\tag{9.6}
$$

Here j and k label the electron and nuclear spins respectively, $\hat{\mathbf{S}}_j$ and $\hat{\mathbf{I}}_k$ are the electron and nuclear spin angular momentum operators, a_{jk} are the hyperfine coupling constants, t is the time after the formation of the radical pair, θ is the angle between the two magnetic fields, and γ is the phase of the RF field at $t = 0$. $\hat{H}(t;\gamma)$ was represented by a $4M \times 4M$ Hermitian matrix of dimension M $(=2^N$ for N spin-$^1/_2$ nuclei). It was assumed that anisotropic magnetic interactions are averaged by molecular motion.

To demonstrate the sensitivity of a radical recombination reaction to the orientation and frequency (5−50 MHz) of a ~ 300 µT radio frequency magnetic field in the presence of a 0−4 mT static magnetic field, the photoinduced electron-transfer reaction of chrysene with isomers of dicyanobenzene was investigated [43]. Figure 9.9 schematically illustrated the photoinduced electron-transfer

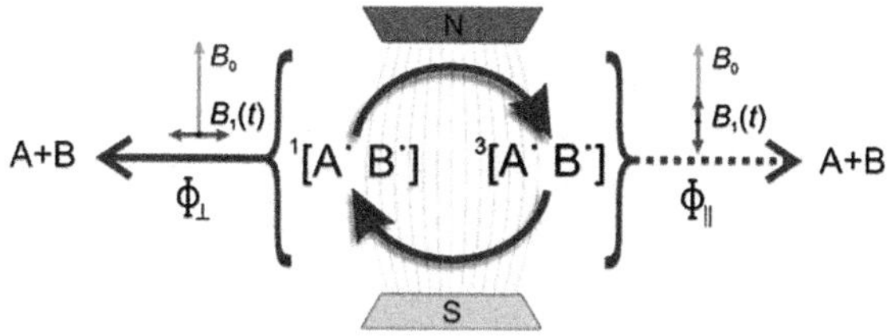

Fig. 9.9 Illustratiion of the photoinduced electron-transfer reaction of chrysene with isomers of dicyanobenzene and effect of radio frequency magnetic field in the presence of a static magnetic field on radical pairs recombination [42, 43]

reaction of chrysene with isomers of dicyanobenzene sequent processes. The recombination yield was detected via the fluorescence of the exciplex formed from the electronic singlet state of the radical ion pair (chrysene$^{\bullet+}$/dicyanobenzene$^{\bullet-}$).

The increase of the recombination probability of deuterated radical ion pairs Py-d$_{10}^{-}$/DMA-d$_{11}^{+}$ (DMA is N,N-dimethylaniline, Py is Pyren) exposed to radiofrequency magnetic fields (1–80 MHz) monitored via exciplex fluorescence was observed [42]. This effect was detected only when the radiofrequency matches energy-level hyperfine interactions in the radicals. The sensitivity of spin correlated radical pairs as intermediates in radical reactions to different applied static or oscillating magnetic fields was demonstrated on an example of photoinduced electron transfer between pyrene and 1,3-dicyanobenzene [42]. The experiments were performed using a linearly polarized radio frequency field of peak strength $B_1 = 0.3$ mT at three separate frequencies (5, 20, and 65 MHz) in the presence of a static magnetic field of 0–4 mT. Measurements on the photoinduced electron-transfer from perdeuterated pyrene (PY) to 1,3-dicyanobenzene (1,3-DCB) in static and the radio frequency of the time-dependent magnetic fields revealed factors affecting the singlet product yield detected via fluorescence of an exciplex which appears as a result of singlet–triplet interconversion altering the population of radical pairs. Scheme 9.2 of photochemical reactions of Pyren (Py) and 1,3-dicyanobenzene (DCB). S and T indicate singlet and triplet states of the radical pair; Py* is the excited singlet state of pyrene. The curved arrows represent the coherent interconversion of singlet and triplet radical pairs by hyperfine and electron Zeeman interactions.

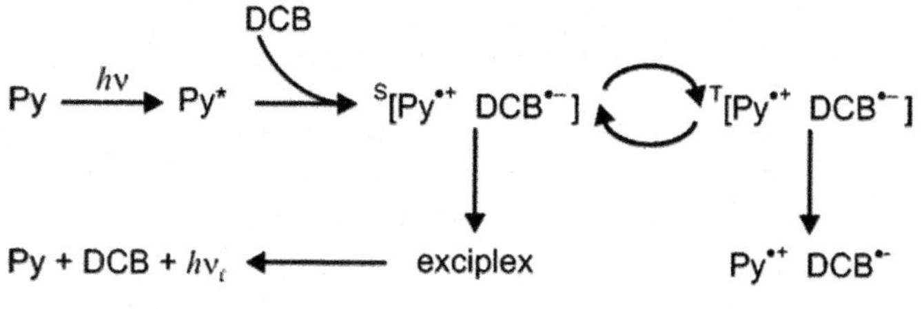

Scheme 9.2 The photo induced electron-transfer from perdeuterated pyrene (PY) to 1, 3-dicyanobenzene (1, 3-DCB)

The regeneration of Py and 1,3-dicyanobenzene (DCB) occurs when the free radicals that escape from the geminate radical pair eventually undergo back electron transfer. It was concluded that the reaction product yield depends on the polarized RF frequency, the strength of the applied static magnetic field, and the on the relative orientation of the two fields.

9.5   Magnetic Isotope Effect

9.5.1   Theoretical Grounds

The first mass-independent isotope effect which chemically discriminates isotopes by their nuclear spins and nuclear magnetic moments rather than by their masses was detected in 1976 by Buchachenko et al. [44] and, independently, in 1977 by the Sagdeev and Molin group [44].

In pioneered work [44], it was shown that photolysis of $(phCH_2)_2CO$ (I) to $PhCH_2CO^{\bullet}$ and $PhCH_2^{\bullet}$ radicals and recombination of the radicals leads to enrichment of I in ^{13}C, mainly at the carbonyl C atom. When the photolysis was performed in a magnetic field, the enrichment decreases with increasing field intensity. A magnetic ^{13}C isotopic effect, which considerably exceeds the known $^{12}C^{13}C$ kinetic isotopic effects, was also detected in the triplet sensitized photolysis of dibenzoyl peroxide [45]. This effect was explained in terms of the dependence of the recombination probabilities of radical pairs in the cage on the electron-nuclear hyperfine interaction.

This new phenomenon was named by Buchachenko as "the magnetic isotope effect", MIF and was independently confirmed within three years by the groups of Buchachenko, Turro and Pines [45–48]. The discovery opened the way for a new field of modern spin chemistry and led to an extensive literature [1, 49–51] and references therein. Since its discovery in 1976, MIE was detected in many chemical reactions, for many isotopic pairs and triads (H-D, ^{12}C-^{13}C, ^{16}O-^{17}O-^{18}O, ^{28}Si-^{29}Si-^{30}Si, ^{32}S-^{33}S-^{34}S, ^{72}Ge- ^{73}Ge-^{74}Ge, ^{235}U-^{238}U, $^{198,200}Hg$-$^{199,201}Hg$, ^{24}Mg-^{25}Mg-^{26-}Mg).

According to the fundamental spin conservation rule, the vast number of reactions with the participation of radicals, ion-radicals, paramagnetic metal ions, and molecules (such as NO, O_2) are spin selective and reactions choose spin-allowed channels. Nevertheless, for some processes, for example the recombination of the triplet radical pair (R· ·R) to produce the diamagnetic, zero-spin molecule R-R, a triplet–singlet spin conversion of the pair is required (Section Y). One of the magnetic interactions which transforms the nonreactive triplet radical pair $(R· ·R)^T$ into the reactive singlet pair $(R· ·R)^S$ is the hyperfine coupling and the reaction probability is a function of hyperfine coupling and nuclear magnetic moment. The key idea for the Buchachenko mechanism of MIE is that "*the radical pair functions as an electron and nuclear spin-selective chemical nanoreactor, which sorts magnetic and nonmagnetic nuclei and directs them into the different reaction products".*

For the simplest case of conversion between two states with zero-spin projections, S and T_0, the rate of triplet–singlet spin conversion of the pair may be estimated by solving the Schrodinger equation with spin Hamiltonian [52, 53]:

$$H = \beta H(g_1 S_1 + g_2 S_2) + a S_1 I \tag{9.7}$$

and wave functions

$$\begin{aligned} S &= 1/\sqrt{2(\alpha\beta - \beta\alpha)} \\ T_0 &= 1/\sqrt{2(\alpha\beta - \beta\alpha)} \end{aligned} \tag{9.8}$$

where the rate of triplet–singlet spin conversion ω is determined as a matrix element $< T_0|H|S >$; resulting in the equation

$$\omega = 1//2(\Delta g \beta H + am). \tag{9.9}$$

In (9.13–9.14), α and β are spin functions of unpaired electrons on the radicals of the pair; S_1 and S_2 are spins of these radicals; g_1 and g_2 are their g-factors; I and a are the nuclear spin and hyperfine coupling constant for one of the radicals carrying a magnetic nucleus; β is the Bohr magnetic moment; H is a magnetic field strength Δg is a difference of the g-factors of radicals in the pair, and m is a nuclear spin projection. Thus the singlet triplet mixing and therefor reaction yield can be depended on spin electron spin duclear interaction.

As an example, the MIE was shown to fractionate oxygen isotopes in the photo-oxidation of water by molecular oxygen [54]. According to experimental data, photolysis of ^{17,18}O-labeled water in the presence of molecular oxygen is accompanied by transfer of ^{17}O and ^{18}O isotopes from water to oxygen. The reaction exhibits a magnetic isotope effect: oxidation of $H_2^{17}O$ is faster by 3.3 % (in the Earth magnetic field) and by 3.7 % (in the field of 0.5 T) than that of $H_2^{18}O$. The effect was postulated to arise in the two spin-selective, isotope-sorting reactions— recombination and disproportionation—in the pairs of encountering $HO_2^{\cdot}$ radicals.

A multichannel kinetic description was used to study the magnetic isotope effect in zero magnetic field [54]. It was found that the maximal isotope effect can be realized via the hyperfine interaction and via the electron spin dipole-dipole interaction of the intermediate radical pair. Quantum mechanical calculations supported these conclusions and showed that a large MIE may be obtained even in the presence of a strong exchange interaction. Photo-reduction of mercury (Hg) is an important mechanism for removal of both Hg^{2+} species and monomethylmercury (MMHg) from surface waters [55, 56]. The effects of different regions of the solar spectrum on the expression of MIF caused by the MIE during Hg^{2+} and MMHg photo-reduction were investigated [55]. The experiments indicate that MIF produced during photo-reduction of Hg^{2+} is significantly influenced by both 290–320 nm and 320–400 nm radiation.

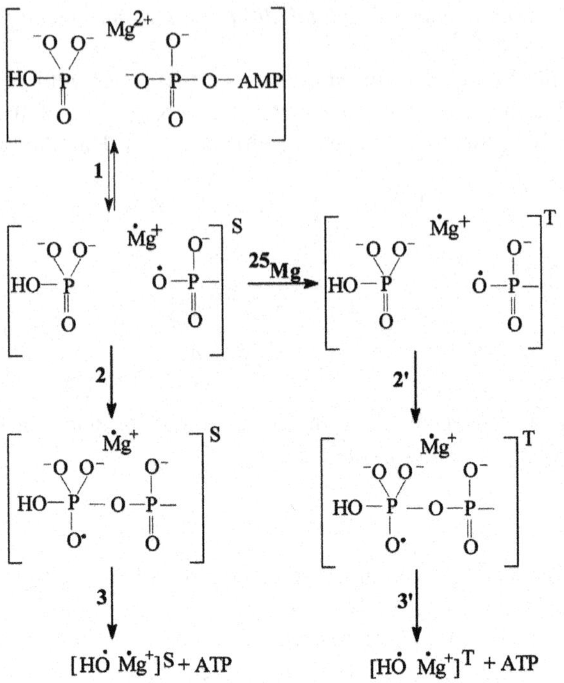

Scheme 9.3 Ion-radical mechanism of ATP synthesis

On the basis of observed magnetic isotope and magnetic field effects in enzymatic ATP synthesis catalyzed by magnesium, a new, ion-radical mechanism of ATP synthesis was formulated (Scheme 9.3) [1, 49].

The proposed mechanism was sustained by observed magnetic field effects on the enzymatic adenosine triphosphate (ATP) and desoxynucleic acid (DNA) synthesis by using catalyzing metal ions with magnetic nuclei (^{25}Mg, ^{43}Ca, ^{67}Zn) and was additionally supported by energy and kinetic arguments.

The challenging problems of the role of magnetic isotope effect in mechanism of the ATPase reaction and other enzymatic reactions and reproducibility of these processes in vivo and vitro have been discussed in detail the literature [49, 57–59]. In comprehensive review [49] three possible sources of irreproducible and contradictory of magnetic effects in enzymatic reactions were stressed: (1) the presence of paramagnetic metal ions as a component of enzymatic site or as an impurity in an uncontrollable amount, (2) the property of the radical pair mechanism to function when two or several ions (cluster) catalyzing metal ions enter into the catalytic site, and (3) the kinetic restrictions in chemical and spin dynamics in radical pair.

9.6 Molecular Compass

9.6.1 Biological Effects and Theoretical Considerations

The ability of living organism, including birds, mammals, reptiles, amphibians, fish, crustaceans and insects, to use the Earth's magnetic field for orientation and navigation is one of the most intriguing and challenging problem for biochemistry and chemical physics [60–76]. Two dominant working hypotheses on the molecular mechanism of the Earth's magnetic field for bionavigation generated much discussion: one based on deposits of iron oxide particle magnetite (Fe_3O_4) [67, 68] and the second based on a light dependent radical-pair mechanism [70]. In this section we will confine ourselves to a brief discussion of the radical-pair compass mechanism which is directly related to the subject of this book.

Based on the finding that electron transfer processes generate radical pairs in coherent electron spin states in weak magnetic fields, Schulten et al. [70] suggested a reaction mechanism for a chemical compass, which allows biological species to orient themselves in the geomagnetic field. According to this mechanism, a sensitivity to the orientation of the Earth's magnetic field originates from an anisotropy of the hyperfine or fine interactions in a weak magnetic field experienced by unpaired electron spins in a redox process. Taking into account the radical-pair mechanism, by which a magnetic field alters the product yields of radical-pair reactions,[4, 5, 10] (Sect. 5.8), a simple model providing evidence that the anisotropic hyperfine interaction is a key phenomenon in magnetosensory capability for an earth-strength magnetic field was considered.

In an attempt to explain the mechanism by which migratory birds are able to sense the direction of the Earth's magnetic field for the purpose of navigation, Ritz et al. [77] proposed cryptochromes (Fig. 9.10) as potential magnetoreceptors [78]. This idea has gained strong support from recent experimental and theoretical investigations [73, 79, 80] For example, signals generated by cryptochrome (CRY) blue-light photoreceptors are responsible for a variety of developmental and circadian responses. Structure of the domain of CRY-1 from *Arabidopsis thaliana* and its complex with AMP-PNP is presented in Fig. 9.10.

The cryptochrome/photolyase family contain the redox-active cofactor flavin adenine dinucleotide (FAD) triad of trytophans (Trp-triad) which are involved in electron transfer initiated by blue light giving a flavosemiquinone radical, FAD$^{\bullet-}$ or FADH$^\bullet$, and a radical derived from the Trp-triad [80–87]. Proposed photochemical reaction schemes for triad from Cry-1 of the plant *Arabidopsis thaliana*(AtCry) and of Escherichia coli photolyase (EcPL) is shown in Fig. 9.11.

The following mechanism of the compass process was suggested (1) photochemical electron transfer from the flavin cofactor to Trp 400, the electron "hoping" Trp 377 and then to Trp 324. At this point, there is a competition between back-electron transfer to the flavin or further reaction leading to generation of a biochemically signalling state of CRY [88, 89]. A quantum mechanical calculation of triplet yield decay rates in a radical pair with one electron and one nuclear spin

Fig. 9.10 AMP-PNP binding to CRY1-PHR. **a** Stereo representation of the electron density of the AMP-PNP bound to CRY1-PHR. The final refined coordinates for AMP-PNP are shown, colored as with the following changes: *yellow, carbon atoms*; *pink, phosphorus atoms*. **b** The AMP-PNP-binding site. Shown is a ball-and-stick representation of the final refined coordinates of AMP-PNP and nearby protein residues. *Dashed lines,* hydrogen bonds. Atoms are colored as in *A*, with the following exceptions: *green,* carbons belonging to the protein; *silver,* carbons from the FAD. Distances are given in Ångströms. Both hydrogen bonds from Arg-360 to the β-phosphate of the AMP-PNP measure 3.4 Å. **c** AMP-PNP binding in the FAD-access cavity. The AMP-PNP, shown as spheres, is colored as in *A*. A bound Mg^{2+} cation is not shown [78]

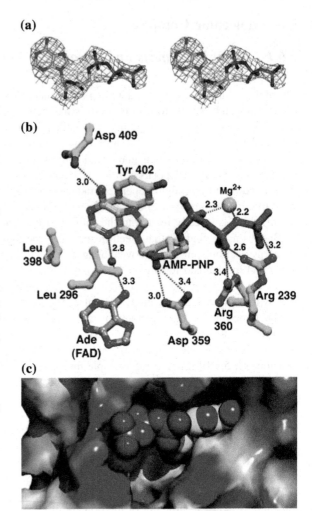

on each radical, assuming anisotropic hyperfine coupling strengths was performed [88]. According to the calculation, the time dependent triplet yield $\Phi^T(t)$ defined as the amount of product decaying via the triplet channel and the fraction of radical pairs in the triplet state at any given time t, was given as

$$\Phi^T(t) = \int_0^\infty k_T T(t) dt, \qquad (9.10)$$

where k_T is a first-order reaction rate constant, and

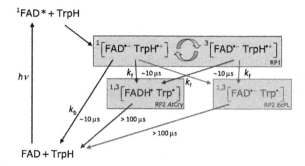

Fig. 9.11 Proposed photochemical reaction schemes for (Cry-1 from the plant Arabidopsis thaliana, AtCry) AtCry and EcPL. The *black arrows* and species are common to both proteins; the *blue* and *red* features refer to AtCry and EcPL, respectively. kb and kf are first-order rate constants for electron–hole recombination of RP1 and formation of RP2 from RP1, respectively. Although RP2 in AtCry is here drawn as [FADH·Trp·], the protonation state of the Trp radical is not certain. The curved *green arrows* indicate the coherent, magnetic field-dependent interconversion of the singlet and triplet states of RP1 [91] (colour online)

$$T(t) = \frac{1}{N} e^{-kt} \cdot \sum_{m=1}^{4N} \sum_{n=1}^{4N} Q_{mn}^{T} Q_{mn}^{S} \cos[(w_m - w_n)t] \qquad (9.11)$$

Here Q^T and Q^S are the triplet and singlet projection operators, respectively; N is the number of nuclear spin states; and w_i denotes the energy of the eigenstate of the Hamiltonian H. The cosine term in (9.11) explains the orientational effect of a weak magnetic field.

To elucidate the role of electron spin entanglement and coherence in determining the origins of the directional response, a minimal model consisting of two electron spins, one of which is coupled to a spin-1/2 nucleus, was considered [79]. It was shown that the proposed Scheme 9.4 can explain the sensitivity of a radical pair-based geomagnetic compass. In the case of an anisotropic mixed singlet-triplet initial state in which the triplet component is 100 % polarized along the molecular z axis and given an isotropic hf interaction, the fractional yield of the product formed via the singlet pathway, Φ^S, "the reaction yield," and $\Delta\Phi^S$, the magnitude of its anisotropy ($\Delta\Phi^S$, $= \Phi^S$ (max) $-$ Φ^S (min)) were given as

$$\Phi_S = \frac{3}{8} - \frac{1}{4}(1 - \eta)\sin^2\theta, \quad \Delta\Phi_S = \frac{1}{4}(1 - \eta), \qquad (9.12)$$

where θ is the angle between the triplet polarization axis (z) and the magnetic field vector. The anisotropy is maximized when $\eta = 0$ (pure triplet state, $\Delta\Phi^S = 1/4$) and

Scheme 9.4 Formation of mixed singlet-triplet initial state

$$AB \xrightarrow{h\nu} {}^{S}[AB]^* \xrightarrow{ISC} {}^{T}[AB]^* \xrightarrow{reaction} {}^{T}[A^{\bullet}B^{\bullet}]$$

is at a minimum when $\eta = 1$ (pure singletstate, $\Delta\Phi^S = 1/4$). Thus, the reaction yield is strongly depended on the orientation angle.

In work [82], it was assumed that the probability of radical pair formation (p) from a molecule with a given orientation in a compass system in the retina is proportional to the probability to absorb a photon. At the p calculation, the relative directions of the molecular transition dipole, μ, and the electric vector of the light was taken in account. In a suggested model of the magnetic field effect on an individual radical pair, two quantities—the p and the fractional yield of the reaction product formed from the singlet state of the radical pair (Φ_S) give the yield of the singlet reaction product:

$$S = \langle p \rangle \times \Phi_S.$$

A simplified version of Φ_S is

$$\tilde{\Phi}_S = 3\cos^2\xi - 1,$$

in which ξ defines the direction of the magnetic field vector with respect to the molecular axis system. Therefore, (9.10) represents the yield of the signalling state and that its dependence on the direction of the magnetic field supplies the directional information required for the compass sens to the retina surface

computational model of the FADH-tryptophan chain system was analyzed The suggested model was based on the available structure of Arabidopsis thaliana cryptochrome-1 [78]. For intermediate radical-pairs [FADH + Trp-400$^+$], [FADH + Trp-377$^+$], or [FADH + Trp-324$^+$], the Hamiltonian was written as is the sum of two Hamiltonians for each radical pair, a Hamiltonian $\hat{H}_{int}$ which takes into account the exchange and dipolar interactions within the radical pair [73]:

$$\hat{H} = \hat{H}_{FADH} + \hat{H}_{Trp} + \hat{H}_{int}. \tag{9.13}$$

The Zeeman interaction term and hyperfine coupling interaction terms of Hamiltonians $\hat{H}_{FADH}$ and $\hat{H}_{Trp}$, with dipolar and exchange interactions being neglected, are

$$\hat{H}_j = \mu_B(\vec{B} \cdot \hat{g} \cdot \vec{S}_j) + \mu_B \sum_i \vec{I}_i \cdot \hat{A}_i \cdot \vec{S}_j), \tag{9.14}$$

Calculations was performed with quantum chemical simulations of hyperfine coupling tensors for FADH and tryptophans, takin in 1on1ieration The stochastic Liouville equation and suggested values of reaction rate constants for electron forward and back transfers, and tryptophan deprotonation were taken in consideration in calculation of hyperfine coupling tensors for FADH and tryptophans, It was hown that the radical-pair mechanism in cryptochrome can explain an increase in the protein's signaling activity of $\sim 10\%$ for magnetic fields of the order of 5G.

Thus this work provides additional evidence that a radical-pair mechanism in cryptochrome is responsible for the magnetic field effect.

A feasible radical-pair mechanism in cryptochromes from *Drosophila melanogaster* and *Sylvia borin*, formed by excitation and electron transfer between a Trp-triad and FAD was analyzed on the base of modeling and molecular dynamics (MD) for structure refinement, high-level ab initio theory, and MD simulations using a polarizable force-field for prediction of pKa and the electron transfer rate [90].

9.6.2 Experimental Data

Escherichia coli photolyase contains two noncovalently bound cofactors, FAD and 5,10-methenyltetrahydrofolate (MTHF), which are responsible for redox chemistry and light harvesting, respectively [89, 91]. To identify transient intermediate species and to detect a magnetic field effect on the photochemical yield of a flavin–tryptophan radical pair in the *Ec*PL, optical transient absorption spectroscopy with picosecond time resolution was used. to record The light induced decay signals in a non-MTHF binding mutant of *E. coli* DNA photolyase (*Ec*PL) with and without an applied magnetic field (39 mT) were detected [89]. On the basis of the observed absorbance changes in *Ec*PL the following mechanism was suggested: (1) FAD previously oxygenated by ferricyanide (FADox) is excited by blue light to its excited state (FAD*); (2) an electron is transferred to the flavin from a nearby tryptophan forming a radical pair within a few picoseconds; (3) the generation of a radical pair comprising a flavosemiquinone anion radical and the W306 cation radical [FAD$^{\bullet-}$ – Trp(H)$^{\bullet+}$] by electron transfer between W306 and FADH$^{\bullet}$, probably via the neighboring two other tryptophans of the triad, W382 and W359 results in (4) in *Ec*PL releases a proton for 0.3 ms and Formation of the secondary radical pair, [FAD$^{\bullet-}$ – Trp$^{\bullet}$], which is stable for several hundred microseconds, after releasing proton from he surface-exposed Trp(H)$^{\bullet+}$, (5) [FAD$^{\bullet-}$ – Trp$^{\bullet}$], decays to the ground state.

As an illustration of implication of cryptochromes in multiple blue light-dependent signaling pathways, comparative photochemistry studies of magnetic-field effects on radical-pair kinetics in cryptochromes (Cry-1) from the plant *Arabidopsis thaliana*, AtCry, with that of *Escherichia coli* photolyase, EcPL, were performed [91] was found that photo-induced radical pairs in cryptochromes are sensitive in vitro to weak applied magnetic fields. The maximum magnetic field at the position of the sample was 29 mT. Experiments on transient absorption kinetic time profiles of AtCry and EcPL recorded at 510 nm with and without a 28 mT applied magnetic field up to 29 mT revealed effects at the level of 12–17 %. The electron transfer processes were described by Scheme 9.5, where WA, WB, and WC are the proximal, intermediate and distal tryptophans, respectively; k_{ET} and k'_{ET} are the rate constants for interconversion of radical pair (RP0 and RP1),

Scheme 9.5 Mechanism of
the electron transport process
in cryptochromes

$$^1FAD^{\bullet}W_AW_BW_C \rightarrow FAD^{\bullet-}W_A^{\bullet+}W_BW_C$$

$$\rightarrow \underbrace{FAD^{\bullet-}W_AW_B^{\bullet+}W_C}_{RP0} \underset{k'_{ET}}{\overset{k_{ET}}{\rightleftharpoons}} \underbrace{FAD^{\bullet-}W_AW_BW_C^{\bullet+}}_{RP1}$$

The interradical electron exchange interactions have been estimated as $J(RP0) \approx 10^{-3}$ mT and $J(RP1) \approx 10^{-1}$ mT. The authors stressed that magnetic sensitivity is a general feature of the cryptochrome/photolyase protein family, which can provide a mechanism for molecular magnetoreception.

Another challenging problem appeared to be findings that oscillating magnetic fields can disrupt the magnetic sense of migratory birds, and this effect strongly dependent on the angle between the geomagnetic and the oscillating fields [77]. For example, it was found that the bird migratory orientation can be disrupted by a radiofrequency 15 nT oscillating field matching the staticfield-induced energy-level splittings within the radical pair. The resonance feature of the rf effect was discussed in a recent publication [92].

A new principle step in the area was taken in recent works [93, 94] in which a carotenoid–porphyrin–fullerene triad (CPF) (Formula 9.1) was synthesized and investigated using newly developed time-resolved Low-Frequency Reaction Yield Detected Magnetic Resonance (tr-LF-RYDMR) technique adapted to application of radiofrequency (RF) fields. The CPF system was shown to function as a chemical compass. Figure 9.12 presents suggested photochemistry of the CPF triad which includes the following steps: (1) light generation a singlet state of exited the porphyrin, (2) rapid and irreversible electron transfer forming the primary radical pair and (3) a rapid electron transfer to produce secondary spin correlated pair in high yield. It is important that the final charge-separated state has a considerably lower electronic exchange coupling ($J = 0.17$ mT) allowing efficient singlet–triplet mixing. A radiofrequency field of strength B_1 and frequency $v_{rf} = 36$ MHz was applied perpendicular and parallel to a static magnetic field ($B_0 = 0$–4 mT). It was shown

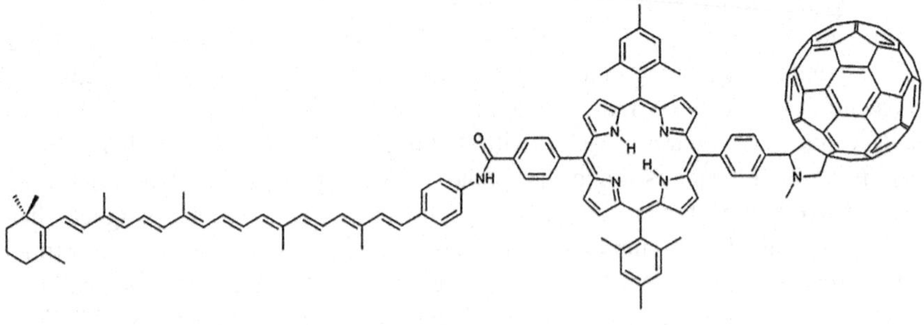

Formula 9.1 Carotenoid–porphyrin–fullerene triad

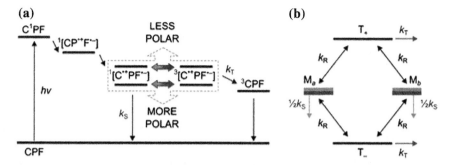

Fig. 9.12 **a** Photochemistry of the CPF triad. **b** Electron spin states of a radical pair under high field conditions. $k_S = 2 \times k_d = 1.8 \times 10^7$ s$^{-1} \gg k_T$, k_{di} is the decay rate of the signal [93]

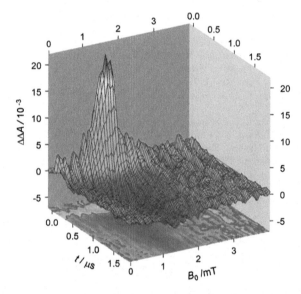

Fig. 9.13 Time-resolved LF-RYDMR spectrum of the carotenoid-porphyrin-fullerene triad in MeTHF with static field B_0 perpendicular to the oscillating field $B_1 = 0.1$ mT, rf = 36 MHz (see details in reference [93]

that in the case of perpendicular orientations of static and oscillating magnetic fields, sharp resonance of LF-RYDMR spectra centered at $B_0 = 1.3$ mT, along with a broader signal was observed (Fig. 9.13), while only the broad signal was detected for parallel orientations. The strength of the oscillating magnetic field was found to be the main factor affecting on the relative orientation of applied static and 36 MHz oscillating magnetic fields on the photogenerated carotenoid-fullerene radical pair.

References

1. A. Buchachenko, *Magneto-Biology and Medicine* (Nova Biomedical, New York, 2015)
2. F.S. Barnes, B. Greenebaum, *Bioengineering and Biophysical Aspects of Electromagnetic Fields* (CRC Press, 2007)
3. H. Hayashi, *Introduction to Dynamic Spin Chemistry: Magnetic Field Effects upon Chemical and Biochemical Reactions*. World Scientific Lecture and Course Notes in Chemistry, vol. 8, (World Scientific Pub Co Inc, 2004)
4. K.M. Salikhov, Y.N. Molin, R.Z. Sagdeev, A.L Buchachenko. *Spin Polarization and Magnetic Field Effects in Radical Reaction* (Elsevier, 1984)
5. C.T. Rodgers, Pure Appl. Chem. **81**, 19 (2009)
6. J.R. Woodward, C.R. Timmel, K.A. McLauchlan, P.J. Hore, Phys. Rev. Lett. **87**, 077602/1 (2001)
7. K. Maeda, K.B. Henbest, F. Cintolesi, I. Kuprov, C.T. Rodgers, P.A. Liddell, D. Gust, C.R. Timmel, P.J. Hore, Nature **453**, 387 (2008)
8. D.T. Edmonds, *Electricity and Magnetism in Biological Systems* (OUP, Oxford, 2001)
9. A.D. Rosen Studies on the effect of static magnetic fields on biological systems. PIERS ONLINE, **6**(2) (2010)
10. C.T. Rodgers, P.J. Hore, Proc. Natl. Acad. Sci. U.S.A. **106**, 353 (2009)
11. Possible effects of Electromagnetic Fields (EMF) on Human Health. Scientific Committee on Emerging and Newly Identified Health Risks. SCENIHR. March 2007
12. S. Dutta, D.S. Ray, Phys. Rev. E **75**, 016205 (2007)
13. T. Brumm, A. Möps, C. Dolainsky, S. Brückner, T.M. Bayerl, J. Phys. Chem. A **109**, 8113 (2005)
14. M. Iwasaka, M. Yaoita, T. Iwasawa, S. Ueno, J. App. Phys. **99**, 08S104 (2006)
15. E.V. Kalneus, D.V. Stass, YuN Molin, Appl. Magn. Reson. **28**, 213 (2005)
16. C.R. Timmel, K.B. Henbest, Phi. Trans. R Soc. Lond A **362**, 2573 (2004)
17. D.I. Stass, S.V. Anishchik, V.N. Verkhovlyuk, *Coherent Spin Control of Radiation-generated Radical Ion Pairs in Liquid Alkanes*, ed. by D.I. Stass, V.I. Feldman, Selectivity, Control, and Fine Tuning in High-Energy Chemistry (2011), p. 143
18. B. Brocklehurst, J. Chem. Soc., Faraday Trans. **72**, 1869 (1976)
19. B. Brocklehurst, K.A. McLauchlan, Int. J. Radiat. Biol. **69**, 3 (1996)
20. M. Wakasa, K. Nishizawa, H. Abe, G. Kido, H. Hayashi, J. Am. Chem. Soc. **121**, 9191 (1999)
21. E. Yu, W. Kandrashkin, D. Vollmann, K.Salikhov Stehlik, A. Van Der Est, Molec. Phys. **100**, 1431 (2002)
22. F.E. Mabbs, D.J. Machin, *Magnetism and Transition Metal Complexes* (Dover Publications Inc., 2008)
23. T.P. von Feilitzsch, P. Harter, O. Schiemann, M.E. Michel-Beyerle, U.E. Steiner, P. Gilch, J. Am. Chem. Soc. **127**, 15218 (2005)
24. Hironobu Tahara, Hiroaki Yonemura, Satoko Harada, Akio Nakashima, Sunao Yamada, Chem. Phys. Lett. **524**, 42 (2012)
25. D. Dey, A. Bose, N. Pramanik, S. Basu, J. Phys. Chem. A **112**, 3943 (2008)
26. S. Dutta Choudhury, S. Basu, J. Phys. Chem. A **109**, 8113 (2005)
27. R.E. Blankenship, T.J. Schaafsma, W.W. Parson, Biochim. Biophys. Acta **461**, 297 (1977)
28. A.J. Hoff, J.H. Rademaker, R. van Grondelle, L.N.M. Duysens, Biochim. Biophys. Acta **460**, 547 (1977)
29. A.J. Hoff, J. Deisenhofe, Phys. Rep. **287**, 2 (1997)
30. R. Haberkorn, M.F. Michel-Beyerle, R.A. Marcus, Proc. Natl. Acad. Sci. U.S.A. **76**, 4185 (1979)
31. O. Kappe, A. Stadler, D. Dallinger, R Mannhold, H. Kubinyi, G. Folkers, *Microwaves in Organic and Medicinal Chemistry*, 2nd, Completely Revised and Enlarged Edition (Wiley-VCH, Weinheim, 2012)

32. V.I. Borovkov, I.S. Ivanishko, V.A. Bagryansky, Y.N. Molin, J. Phys. Chem. A **117**, 1692 (2013)
33. A.M. Domijan, D. Flajs, M. Peraica, Int. J. Hyg. Environ. Health **214**, 59 (2011)
34. P.S. Deshmukh, N. Nasare, K. Megha, B.D. Banerjee, R.S. Ahmed, D. Singh, M. P. Abegaonkar, A.K. Tripathi, P.K. Mediratta, Cognitive impairment and neurogenotoxic effects in rats exposed to low-intensity microwave radiation. *Int. J. Toxicol* **34**, 284 (2015)
35. M.K. Razavi, A.R. Raji, M. Maleki, H. Dehghani, A. Haghpeima, Comp. Clin. Pathol. **24**, 1271 (2015)
36. E.A. Nasibulov, L.V. Kulik, R. Kaptein, K.L. Ivanov, Phys. Chem. Chem. Phys. **14**, 13325 (2012)
37. J.M. Canfield, R.L. Belford, P.G. Debrunner, K.J. Schulten, Chem. Phys. **182**, 1–18 (1994)
38. A.L. Buchachenko, E.L. Frankevich, *Chemical Generation and Reception of Radio- and Microwaves* (VCH Publishers, N.Y., 1994)
39. B. Brocklehurst, Magnetic fields and radical reactions: recent developments and their role in nature. Chem. Soc. Rev. **31**, 301–311 (2002)
40. C.J. Wedge, C.T. Rodgers, S.A. Norman, N. Baker, K. Maeda, K.B. Henbest, C.R. Timmel, P. J. Hore, Phys. Chem. Chem. Phys. **11**, 6573 (2009)
41. K.B. Henbest, P. Kukura, C.T. Rodgers, P.J. Hore, C.R. Timmel, J. Am. Chem. Soc. **126**, 8102 (2004)
42. C.T. Rodgers, K.B. Henbest, P. Kukura, C.R. Timmel, P.J. Hore, J. Phys. Chem. A **109**, 5035 (2005)
43. J.R. Woodward, C.R. Timmel, P.J. Hore, P.J. McLauchlan, Mol. Phys. **100**, 1181 (2002)
44. A.L. Buchachenko, E.M. Galimov, V.V. Ershov, G.A. Nikiforov, A.D. Pershin, Dokl. Akad. Nauk SSSR **228**, 379 (1976)
45. R.Z. Sagdeev, T.V. Leshina, M. Kamkha, O. Belchenko, Y.N. Molin, A.A. Rezvukhin, Chem. Phys. Lett. **48**, 89 (1977)
46. V.A Belyakov, V.I. Mal'tsev, E.M. Galimov, A.L. Buchachenko, Dokl. Akad. Nauk SSSR **243**, 924 (1978)
47. L. Sterna, D. Ronis, S. Wolfe, A. Pines J. Chem. Phys. **72**, 5493 (1980)
48. N.J. Turro, B. Kraeutler, Acc. Chem. Res. **13**, 369 (1980)
49. A.L. Buchachenko, Bioelectromagnetics **37**, 1 (2016)
50. A.L. Buchachenko, J. Phys, Chem. B **117**, 2231 (2013)
51. V.K. Koltover, Russian Chemical Bulletin **63**, 1029 (2014)
52. V.L Berdinskii, L.L. Yasina, A.L. Buchachenko, Khim. Fiz **24**, 35 (2005)
53. A.S. Letuta, V.L. Berdinskii, Dokl. Phys. Chem. **457**, 120 (2014)
54. A.L. Buchachenko E.O. Dubinina, J. Phys, Chem. A **115**, 196 (2011)
55. C.H. Rose, S. Ghosh, J.D. Blum, B.A. Bergquist, Chemical Geology **405** (2015)
56. H. Hintelmann, W. Zheng, in *Environmental Chemistry and Toxicology of Mercury,* ed. by G. Liu, Y. Cai, N. O'Driscoll, (2015) p. 293
57. D. Crottya, G. Silkstonec, S. Poddara, R. Ransonc, P.M. Adriele, M.T. Wilsonc, J.M.D. Coeya, Proc. Natl. Acad. Sci. **109**, 1437 (2012)
58. P.J. Hore, Are biochemical reactions affected by weak magnetic fields? *Proc. Natl. Acad. Sci.* **109**, 1357–1358 (2012)
59. A.R. Jones, N.S. Scrutton, J.R. Woodward, J. Am. Chem. Soc. **128**, 8408 (2006)
60. A. von Middendorff, Mem. Acad. Sci. St. Petersbourg VI, Ser. Tome. **8**, 1 (1859)
61. W. Wiltschko, R. Wiltschko, Science **176**, 62 (1972)
62. M. Iwasaka, Y. Mizukawa, J. Appl. Phys. **115**, 17B501/1 (2014)
63. I. Chaves, R. Pokorny, M. Byrdin, N. Hoang, T. Ritz, K. Brettel, L.O. Essen, G.T. van der Horst, A. Batschauer, M. Ahmad, Annu. Rev. Plant Biol. **62**, 335–364 (2011)
64. G. Hong, R. Pachter, J. Phys. Chem. B **119**, 3883 (2015)
65. S. Johnsen, K.J. Lohmann, The physics and neurobiology of magnetoreception. Nature. Rev. Neurosci. **6**, 703 (2005)
66. S. Johnsen, K.J. Lohmann, Phys. Today **61**, 29 (2008)

67. J.M. Kirschvink, M.M. Walker, S.-B. Chang, A.E. Dizon, K.A. Peterson, J. Comp. Physiol. A. **157**, 375 (1985)

68. M. Iwasaka, Y. Mizukawa, J. Appl. Phys. **115**(17), 17B501/1 (2014)

69. J.L. Kirschvink, M. Winklhofer, M.M. Walker, Biophysics of magnetic orientation: strengthening the interface between theory and experimental design. J. Roy. Soc. Interface Roy. Soc. 7(Suppl 2), S179 (2010)

70. K. Schulten, C.E. Swenberg, A. Weller, Z. Phys, Chem. Neue. Fol. **111**, 1 (1978)

71. O.E. Efimova, P.J. Hore, Biophys. J. **94**, 1565 (2008)

72. E.M. Gauger, E. Rieper, J.J.L. Morton, S.C. Benjamin, V. Vedral, Phys. Rev. Lett, **106**, 040503 (2011)

73. I.A. Solov'yov, D.E Chandler, K. Schulten, Biophys. J. **92**, 2711 (2007)

74. H. Mouritsen, U. Janssen-Bienhold, M. Liedvogel, G. Feenders, J. Stalleicken, P. Dirks, R. Weiler, Proc. Natl. Acad. Sci. USA **101**, 14294 (2004)

75. J.A.S. Lau, C.T. Rodgers, P.J. Hore, J.R. Soc, Interface **9**, 3329 (2012)

76. T. Ritz, M. Ahmad, H. Mouritsen, R. Wiltschko, W. Wiltschko, J.R. Soc, Interface 7(Suppl. 2), S135 (2010)

77. T. Ritz, P. Thalau, J.B. Phillips, R. Wiltschko, W. Wiltschk, Nature **429**, 177 (2004)

78. C.A. Brautigam, B.S. Smith, Z. Ma, M. Palnitkar, D.R. Tomchick, M. Machius, M. Deisenhofer, Proc. Natl. Acad. Sci. USA **101**, 12142 (2004)

79. H.J. Hogben, T. Biskup, P.J. Hore, Phys. Rev. Lett. **109**, 220501 (2012)

80. M. Ahmad, P. Galland, T. Ritz, R. Wiltschko, W. Wiltschko, Magnetic intensity aVects cryptochrome-dependent responses in Arabidopsis thaliana. *Planta* **225**, 615–624 (2007)

81. S. Weber, T. Biskup, A. Okafuji, A.R. Marino, T. Berthold, G. Link, K. Hitomi, E.D. Getzoff, E. Schleicher, J.R., Jr. Norris, J. Phys. Chem. B **114**, 14745 (2010)

82. R. David, Nat. Rev. Mol. Cell Biol. **14**, 547 (2013)

83. Z. Dominguez, H. Dang, M.J. Strouse, M.A. Garcia-Garibay, J. Am. Chem. Soc. **124**, 2398 (2002)

84. H. Senoo, M. Iijima, Commun. Integr. Biol. **6**, e27681 (2013)

85. Y.M. Gindt, E. Vollenbroek, K. Westphal, H. Sackett, G.T. Babcock, Biochemistry **38**, 3857 (1999)

86. T. Biskup, E. Schleicher, A. Okafuji, G. Link, K. Hitomi, E.D. Getzoff, S. Weber, Angew. Chem. Int. Ed. **48**, 404 (2009)

87. M. Liedvogel, H. Mouritsen, Cryptochromes—a potential magnetoreceptor: what do we know and what do we want to know? J Roy. Soc. Interface **7**, S147 (2010)

88. T. Ritz, S. Adem, S, and K. Schulten. Biophys. J. **78**, 707 (2000)

89. K.B. Henbest, K. Maeda, P.J. Hore, M. Joshi, A. Bacher, R. Bittl, S. Weber, C.R, Timmel Schleicher E. Magnetic-field effect on the photoactivation reactionof Escherichia coli DNA photolyase. *Proc. Natl. Acad. Sci. USA* 105, 14395–14399 (2008)

90. Y. Zhang, G. Berman, S. Kais, Int. J. Quant. Chem. **115**, 1327 (2015)

91. K. Maeda, A.J. Robinsona, K.B. Henbesta, H.J. Hogbenb, T. Biskup, M. Ahmad, E. Schleichere, S. Webere, C.R. Timmela, P.J. Hore, Magnetically sensitive light-induced reactions in cryptochrome are consistent with its proposed role as a magnetoreceptor. Proc. Natl. Acad. Sci. **109**, 4774–4779 (2012)

92. S. Engels, N.-L. Schneider, N. Lefeldt, C.M. Hein, M. Zapka, A. Michalik, D. Elbers, A. Kittel, P.J. Hore, H. Mouritsen, Nature **509**, 353 (2014)

93. K. Maeda, J. Storey, P.A. Liddell, D. Gust, P.J. Hore, C.J. Wedge, C.R. Timmel, Phys. Chem. Chem. Phys. **17**, 3550 (2015)

94. G. Kodis, P.A. Liddell, A.L. Moore, T.A. Moore, D. Gust, Synthesis and photochemistry of a carotene-porphyrin-fullerene model photosynthetic reaction center. J. Phys. Org. Chem. **17**, 724 (2004)

Chapter 10
Electron Spin Interactions in Investigations of the Structure and Spin State of Organic and Metalloorganic Compounds

Abstract This chapter has briefly reviewed recent progress in the investigation of electronic structure and spin effects in various compounds such as bi-and polyradicals, paramagnetic metal complexes, molecular magnets, doped and undoped conjugated polymers and super-paramagnetic compounds. To study such systems, a whole arsenal of advanced physical methods including magnetic susceptibility, CW and pulse ESR, NMR, neutron scattering, Mossbauer spectroscopy, X-ray and photoelectron spectroscopy, resonance absorption of phonons, etc. has been used. It is a measure of the complexity of the magnetic structure of these systems that such a variety of techniques must be brought to bear. Applications and new research fields will surely continue to expand and develop from this active area of research.

10.1 Introduction

Organic and, in particular, metalloorganic compounds bearing spins appear to be promising materials for optical data storage, spintronics applications (spintronic-logic devices, semiconductor-based devices, magnetic-tunnel transistor, antiferro-magnetic storage media), optical sensing, spin valves, magnetic switching systems, optical information processing devices, conducting and semiconducting magnetic materials, catalyzers, drugs, etc. [1–3]. X-ray crystallography, a structural method of great importance, is extensively used in physics and chemistry. The achievements of this technique are especially striking particularly for establishing the structure of such huge molecular machines as nitrogenase and ribosomes. Nevertheless, X-ray crystallography suffers a number of significant limitations. The X-ray analysis of a species can only be performed using stable monocrystals labeled with heavy atoms to address, e.g., phase ambiguities, the preparation of which is often a time consuming and complicated limiting step. Compounds in solution, amorphous objects and biological systems are beyond the capabilities of the technique. Molecular dynamics, properties of spin states, and spin dynamics problems are also out of side the framework of X-ray crystallography.

© Springer International Publishing Switzerland 2016

G. Likhtenshtein, *Electron Spin Interactions in Chemistry and Biology*,
Biological and Medical Physics, Biomedical Engineering,
DOI 10.1007/978-3-319-33927-6_10

On the other hand, approaches based on the measurement of electron spin interactions, especially advanced pulse techniques, described in previous chapters, are able to cover the aforementioned gap in investigations of the structure and spin states of a large number of objects including radicals, transition metal complexes, polymers, surfaces, and biological systems.

10.2   Spin State of Excited Radicals. Spin Switching Systems

Photo-magnetochromism can be defined as the light-induced reversible transformation of chemical species between two isomers having different magnetic properties. Magnetization reversal in organic compounds, as a basis for magnetic data storage, has attracted substantial attention [4–10]. The references in these works are also a good resource for further study.

With a goal to control the intramolecular magnetic interaction by irradiation of photochromic diarylethenes having two nitronyl nitroxides, 1,2-bis[6-(1-oxyl-3-oxide-4,4,5,5-tetramethylimidazolin-2-yl)-2-methyl-1-benzothiophen-3-yl]hexafluorocyclopentene (2a), were synthesized and investigated (Fig. 10.1) [8]. An increase of the antiferromagnetic interaction between two nitronyl nitroxides from $2 \; J/k_B = -2.2$ K to $2 \; J/k_B = -11.6$ K, when the diarylethene spin coupler was switched from the open-ring isomer 2a to the closed-ring isomer 2b, was revealed by magnetic measurements and ESR.

The results indicated that the intramolecular interaction was switched by the photochromic spin couple.

The compound 2,5-bis(arylethynyl)-4-methyl-3-thienyl with two tethered nitroxide radicals showed an efficient photochromic reactivity as monitored by ESR spectroscopy [4]. Three kinds of aryl groups, 2,5-thienylene, p-phenylene, and m-phenylene groups, were used in the arylethynyl moiety. It was found that the

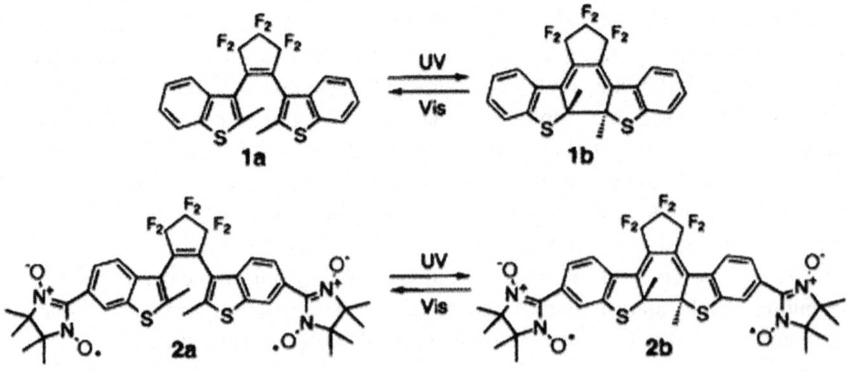

Fig. 10.1 Photochromic diarylethenes used in work [8]

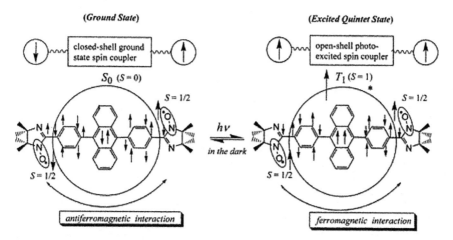

Fig. 10.2 Schematic diagram of the spin alignment utilizing the excited triplet molecular field in in the nitroxide biradical [9]

photochromic reaction of the diarylethenes is accompanied by a change in the magnetic interaction between two unpaired electrons due to the change of the hybrid orbital at the 2-position of the thiophene ring from sp(2) to sp(3). At room temperature, the following values of exchange integrals were calculated: $J = 540.6$ cm^{-1} for the NN-TTF diradical with 15,16-dinitrile pyrene as the coupler in dihydropyrenes form; $2 J/k_B T \sim -2.6$ for the substituted and unsubstituted cyclophanediene form; and $2 J/k_B T \sim 5.4$ for the DHP species.

Time-resolved electron spin resonance (TRESR) spectra of the first excited states with resolved fine-structure splittings of a series of nitronyl derivatives, the purely organic π-conjugated spin system were detected [9]. Figure 10.2 shows the schematic diagram of the spin alignment utilizing the excited molecular field in the nitroxide biradical. Two mechanisms for the intramolecular spin alignment, a spin polarization, and spin delocalization mechanism, as well as the relationship between the spin alignment in the excited states and the topology of the π-electron networks were discussed.

In the ground state, when the diphenylanthracene spin coupler is a closed shell, antiferromagnetic coupling between the two dangling radical spins takes place. In the photoexcited states the spin coupler becomes an open-shell triplet excited state and the spin delocalization mechanism leads to ferromagnetic coupling between the two dangling radical spins in the photoexcited state. In work [11]. CW ESR, echo detected pulse ESR, time-resolved ESR (TRESR), the molar magnetic susceptibility χ_{mol}, and laser-excitation pulsed ESR techniques were employed to establish electronic structures in the ground state and photoexcited states, and spin alignment in biradical **1** (Fig. 10.3).

On photoirradiation, a lowest photoexcited state with intermediate spin $(S = 1)$ arising from four unpaired electrons with low-lying quintet $(S = 2)$ photoexcited

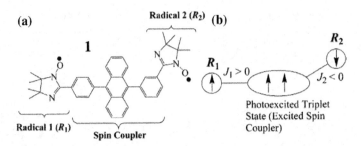

Fig. 10.3 Molecular structure of **1** and schematic picture of the expected spin alignment in the lowest photoexcited state [11]

state was detected. For compound 1, values of g = 2.0061, D = 0.036 cm^{-1}, E = 0.0 and hyperfine coupling constants $A_{N(1)}$ = 0.465, $A_{N(2)}$ = 0.210; $A_{N(3)}$ = 0.414, $A_{N(4)}$ = 0.250 were found in the ground state. In the triplet, the excited state was characterized by the following parameters: g = 2.0045, D = 0.036 cm^{-1}, E = 0.0.

Spin crossover, which refers to the transitions between high to low, or low to high, spin states, can be belonged to spin switching phenomena [12–15]. An arsenal of physical methods such as electron microscopy, powder X-ray diffraction, optical reflectivity, Raman, FTIR, ^{57}Fe Mössbauer, and broadband (10^{-2}–10^6 Hz) dielectric spectroscopies were utilized for investigation of structure charge transport, and spin transition properties of the $[Fe_{1-x}Zn_x(Htrz)_2(trz)](BF_4)$ (trz = triazole, x = 0, 0.26, or 0.43) compound [12]. The following metal substitution effects were revealed for increasing values of x: (1) the thermal hysteresis width decreases from 45 to 8 K, (2) a strong overall decrease in conductivity. Additional important result of this work is a detection of the electrical conductivity droping when the iron(II) electronic configuration is switched from the low-spin to the high-spin state. The strong spin-state dependence of the electrical properties in this compound was explained by the direct participation of ferrous ions participate directly in the charge transport mechanism. Another illustration of the spin inversion process is the reaction of nonheme Fe^{IV}O species $[(N4Py)FeIV(O)]^{2+}$ and $[(Bn-TPEN)FeIV(O)]^{2+}$, used in study [13] (N4Py = N,N-bis(2-pyridylmethyl)-N-bis(2-pyridyl)methylamine, Bn-TPEN = N-benzyl-N,N′,N′-tris(2-pyridylmethyl)-1,2-diaminoethane.]) (Fig. 10.4). It was shown that selective hydrogen atom transfer reactions with C≡C epoxidation occurs in reaction of these complexes with cyclohexene. Experiments and DFT calculations revealed that an interplay of tunneling and spin-inversion probability prefers exclusive S = 1 hydrogen atom transfer, over mixed selective hydrogen atom transfer and epoxidation on S = 2.

General trends in the magnetic behavior, spin alternation, effect of planarity and absorption wavelength in the system of interest were discussed in [10].

Fig. 10.4 Schematic illustration of reactions reaction of C≡C epoxidation by nonheme $Fe^{IV}O$ species. HAT—hydrogen atom transfer, OAT—oxygen atom transfer [13]

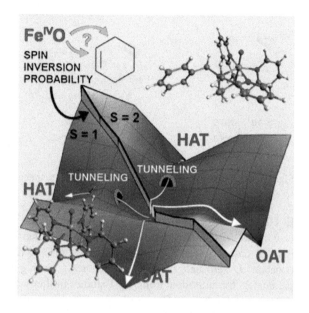

10.3 Structure and Spin State of Transition Metal Complexes

10.3.1 Mono- and Binuclear Complexes

Data on investigations of structure and the electronic state of mono binuclear transition metal complexes derived using ESR, and Mossbauer techniques in combination with other methods can be found in recent publications and references therein [15–20].

The electronic structure of cationic electron-rich ethynylpyridyl Fe(III) of the formula $[(\eta^2\text{-dppe})(\eta^5\text{-}C_5Me_5)FeC \equiv C(x\text{-}C_5H_4N)][PF_6]$ (x = 4, 3, 2; **1a − c[PF$_6$]**) and $[(\eta^2\text{-dppe})(\eta^5\text{-}C_5Me_5)FeC \equiv C(2,5\text{-}C_5H_3NX)][PF_6]$ (X = Cl, Br; **2a,b[PF$_6$]**) (Fig. 10.5) was characterized by Mössbauer, NMR, and ESR techniques and verified by DFT calculations [16].

The positive spin density delocalized on the pyridylethynyl ligand of **1a−c[PF$_6$]**, which depends on the position of the nitrogen in the heterocycle, was established by ESR or Mössbauer spectroscopy. The contact hyperfine coupling constants and the spin densities were derived for selected protons of the pyridylalkynyl linker from the ^1H NMR contact shifts. The large asymmetry of the rhombic ESR spectra observed for **1a−c[PF$_6$]** ($\Delta g \approx 0.53$) indicated the metallic character of the unpaired electron. In work [17], a combine approach including X-ray crystallography, EPR, and Mössbauer spectroscopy was used for characterization of the diiron(II,III) centers in the active sites of diiron enzymes, the (μ-alkoxo)(μ-carboxylato)diiron(II,III) complexes $[Fe^{II}Fe^{III}(N\text{-Et-HPTB})(O_2CPh)(NCCH_3)_2](ClO_4)_3$ **(1)** and $[Fe^{II}Fe^{III}(N\text{-}$

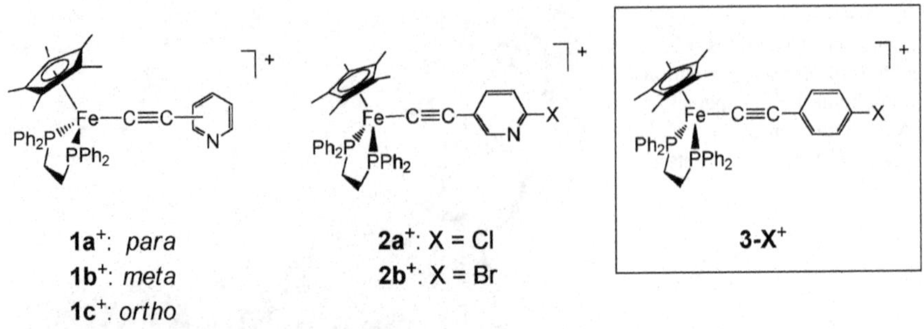

1a⁺: *para*
1b⁺: *meta*
1c⁺: *ortho*

2a⁺: X = Cl
2b⁺: X = Br

3-X⁺

Fig. 10.5 Selected Pyridyl- and Aryl-Based Fe(III) Compounds investigated in [16]

Et-HPTB)(O₂CPh) (Cl)(HOCH₃)](ClO₄)₂ (**2**) (*N*-Et-HPTB = *N,N,N′,N′*-tetrakis(2-(1-ethyl-benzimidazolylmethyl))-2-hydroxy-1,3-diamino propane. From the experimental data 16 spin Hamiltonian parameters, including the exchange coupling constant *J*, zero-field splitting parameters, local *g*-values, (hyper)fine structure parameters for individual irons. DFT studies further were revealed. The anisotropy of the g-values for the iron(II) site, an admixture of excited $S > 1/2$ states into the ground $S = 1/2$ state through D/*J* mixing as the origin of the anisotropies of the g values were discussed.

The structure and spin properties of a series of coordination binuclear compounds comprising manganese, iron, nickel, and zinc bound by a hexaanionic cryptand (Fig. 10.6). where carboxamides are anionic N-donor, were investigated by a combined approach including IR, NMR, ESR Mössbauer spectroscopy, X-ray crystallography and SQuID magnetometry [18]. It was shown that (1) The antiferromagnetic coupling between the diiron(II) and the dicobalt(II) centers when bridged by cyanide does not increase significantly relative to the unbridged congeners, (2) a one-site model satisfactorily fits Mössbauer spectra of unbridged

Fig. 10.6 Solid state structure of the core of compound 10. Thermal ellipsoids at 50 % probability level. H atoms, dipropoxyphenoxyl substituents, crown ethers, and solvents of crystallization omitted for clarity. One phenylene spacer has been *grayed* for ease of viewing [18]

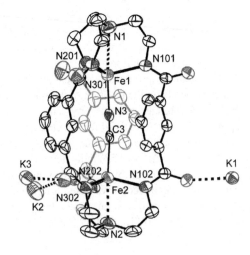

diiron(II) and diiron(III) complexes, and a two site fit was needed to model the iron (II) centers that are bridged by cyanide.

Mössbauer spectroscopy and magnetic susceptibility were utilized for investigating the structure and exchange interactions in the dinuclear complex [Fe $(III)_2(\mu$-OH$)_2$ (bik)$_4$](NO$_3$)$_4$ (1) (bik, bis(1-methylimidazol-2-yl)ketone) [19]. experimental data were consistent with the presence of two identical high-spin iron(III) sites and revealed antiferromagnetic exchange (J = 35.9 cm^{-1}, H = JS$_1$ · S$_2$) of the metal ions.

In work [20], the spin multiplicities of inverse sandwich-type complexes (ISTCs) of ethylene and dinitrogen molecules with 3d transition metal elements (Sc to Ni), $(\mu$-C$_2$H$_4)$[M(AIP)]$_2$ and $(\mu$-N$_2)$[M(AIP)]$_2$ (AIPH = (Z)-1-amino-3-iminoprop-1-ene; M = Sc to Ni were established. It was found that in both ethylene and dinitrogen ISTCs of the early 3d transition metals (Sc to Cr), sandwiched ethylene and dinitrogen ligands coordinate with two metal atoms in an η^2-side-on form and their ground states have an open-shell singlet spin multiplicity. For the late 3d transition metals (Mn to Ni), ethylene and dinitrogen ISTCs: in ethylene ISTCs of Mn to Ni, the ground state has an open-shell singlet spin multiplicity like those of the ISTCs of early transition metals.

10.4 Molecular Magnets

10.4.1 General

Molecular magnets are systems showing long-range magnetic order and exhibit a spontaneous magnetization [21–25]. Molecular magnets can be used as magnetic sensors and in magneto-optic applications. They are also useful as core materials in transformers, for guiding magnetic fields, and magnetic shielding of low-frequency magnetic fields, and in magnetic data storage applications by aligning the magnetic moments perpendicular to the plane of the substrate of magnetic disks, and in optical disks relying on the magneto-optic effect [25]. Four major classes of magnetism, paramagnetism, ferromagnetism, antiferromagnetism, and ferrimagnetism, describe the macroscopic magnetic behavior of adjacent magnetic moments interacting with each other at absolute zero. Scheme 1.10 presents tow-dimensional representations of possible spin alignments: ferromagnetic ordered spins), antiferromagnetic ordered (opposed) spins and ferrimagnetic ordered (opposed) spins of different magnetic moments. The alignments govern the macroscopic magnetic behavior of systems at absolute zero.

In ferromagnetic systems with bulk magnetism, all spins align in the direction of the internal magnetic field. A weak ferromagnetism appears in an antiferromagnetic structure in which there exists a canting angle between the magnetic moments of different sub-lattices with opposite signs of different magnetic moments. Ferrimagnets are composed of spins of different magnetic moments (Scheme 10.1).

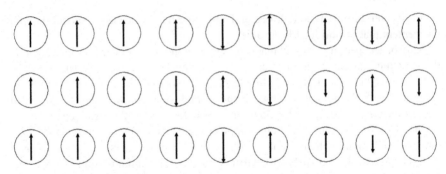

Scheme 10.1 *Left* Ferromagnetism, *Center* Antiferromagnetism, *Right* Ferrimagnetism [26].

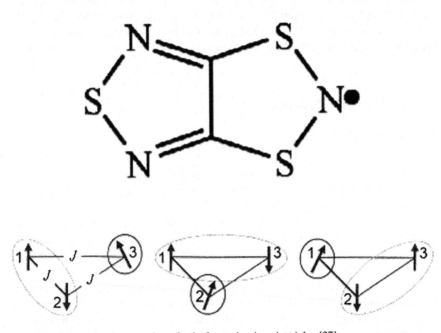

Fig. 10.7 Schematic representation of spin frustration in spin triples [27]

In the case of three nonlinear spins in which two of the spins align antiferro-magnetically, the third spin cannot be aligned antiferromagnetically with both of the spins. This situation is termed antiferromagnetic spin frustration (Fig. 10.7) [27].

A great diversity of known molecular magnets fall into the following classes: single molecular magnets, single chain magnets and bulk magnets. Molecular magnets are classified on the basis of their chemical structure as pure organics and metallorganics. The former are divided into three groups: undoped conjugation polymers, doped conjugation polymers, and high-spin superparamagnetic polymers.

In purely organic molecular systems, Heisenberg exchange can be responsible for magnetic anisotropy of the such systems because in these molecules the magnetic dipolar interactions and spin-orbit coupling are relatively weak compared to spin exchange interactions. In paramagnetic metal containing systems the magnetic dipolar interactions and spin-orbit coupling may need to be taken into consideration.

In McConnell molecular magnetic model 1 [28, 29], the spin polarization phenomenon can arise from the different exchange interactions between the unpaired electron in the singly occupied molecular orbital (SOMO) with the 'spin up' (α) and spin down' (β) electrons in the fully occupied molecular orbitals (FOMO) and between the α SOMO spin and the α FOMO spin. This leads to different spatial distributions of the α and β spins. A ferromagnetic coupling is expected, when a region of negative spin density in one molecule overlaps with a region of positive spin density in another molecule. Exchange interaction charge-transfer salts composed of chains of alternating electron-donor and electron-acceptor molecules in which the SOMO orbitals of the donor in the neutral state are degenerate is the basis of the McConnel model 2 [30].

According to Miller and Epstein [31], the following molecule-based magnet structure types have been pointed out: magnets with (a) only p orbitals, (b) isolated p and d orbitals, (c) isolated d orbitals (d) p and d orbitals connected via covalent bonds and (e) only d orbitals connected via covalent bonds. In molecular magnets, the following spin-coupling mechanisms were considered: (1) Ferromagnetic ordering caused by ferromagnetic coupling and is governed by exchange interactions in a system of spins residing in orthogonal orbitals. This state can be stabilized by through-space, dipole–dipole spin interactions and (2) Antiferromagnetic coupling of adjacent spin sites a and b, each with a differing number of spins per site can stabilize magnetic ordering and lead to ferrimagnetic behavior (abab).

10.4.2 Single Molecular Magnets. Transition Metal Clusters

Due to the unique physical properties of transition metal clusters, single molecular magnets on the length scale of approximately 1–100 nm have driven a large number of experimental and theoretical investigations with promising practical applications in high-density storage, catalysis and molecular electronics [32–34]. Typical transition or rare metal clusters consist of magnetic ions surrounded by ligands. Commonly the exchange interaction between magnetic ions inside of a cluster ($J \sim 10^2$ K) is stronger then that between molecules in a crystal ($\sim 10^{-3}$ K). Because in many cases the exchange interaction is found to be antiferromagnetic, on the molecular level nanoclusters can be considered as molecular ferrimagnets.

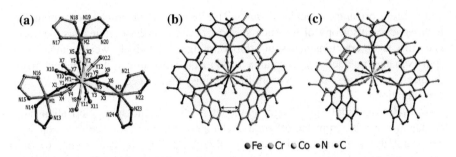

●Fe ⊙Cr ◖Co ●N ●C

Fig. 10.8 a Trigonal bipyramidal cluster [M(tmphen)$_2$]$_3$[M'(CN)$_6$]$_2$ viewed approximately along the axis of the bipyramid and showing the general atom labeling scheme for compounds **1–5**. For the sake of clarity, only four atoms of each tmphen ligand are shown; X, Y = C, N. **b, c** The TBP clusters in the structures of **5** and **4** in panels **b** and **c**, respectively. Intramolecular $\pi - \pi$ contacts are shown with *black arrows*. H atoms are omitted for the sake of clarity [45]

A whole arsenal of physical methods including magnetic susceptibility [35] ESR [33], Electron Spin Transient Nutation spectroscopy [33], NMR [34], neutron scattering [35], Mossbauer spectroscopy [36], X-ray and photoelectron spectroscopy [37], resonance absorption of phonons [38], muon scattering [39], and electro chemistry [40], in conjunction with advanced theoretical and experimental approaches has been utilized for the detailed investigation of electron spin interactions in molecular magnets [41–44].

A combined approach using Mössbauer spectroscopy, magnetic susceptibility measurements, infrared spectroscopy, thermogravimetric analyses and single-crystal X-ray analysis was used for investigating pentanuclear, cyanide-bridged clusters [M(tmphen)$_2$]$_3$[M'(CN)$_6$]$_2$ (M/M' = Zn/Cr (**1**), Zn/Fe (**2**), Fe/Fe (**3**), Fe/Co (**4**), and Fe/Cr (**5**); tmphen = 3,4,7,8-tetramethyl-1,10-phenanthroline) (Fig. 10.8) [45]. Magnetic susceptibility measurements were carried out in an applied field of 0.1 T in the 2–300 K range an reiel The contribution of the low spin Fe^{III} sites in the $\left[Fe_3^{II}Fe_2^{III}\right]$ core was revealed by the χT measurement and byMössbauer spectroscopy. These results indicate that a low spin state for three Fe^{II} sites of the cluster **3** up to 100 K. The exchange integral value J, derived from the temperature dependence of the internal field at the Fe sites measured at 8 T by Mössbauer spectroscopy and the magnetic susceptibility measured at 0.1 T, was found to be $J = 0.65$ cm^{-1}.

In work [46], high resolution ESR experiments showed that a single molecule magnet, Mn$_{12}$O$_{12}$(CH$_3$COO)$_{16}$(H$_2$O)$_4$] · 4H$_2$O · 2CH$_3$COOH, (Mn12Ac), a cluster of $S4$ symmetry consisting of eight manganese(III) ions, bearing $S = 2$, is antiferromagnetically coupled to the remaining four manganese(IV) with $S = 3/2$, to give a ground $S = 10$ state. A very large magnetic anisotropy of the cluster was EPR of molecular nanomagnets was revealed a large negative zero field splitting of $D/k = -0.65$ K. Another molecular mesoscopic magnet [(C$_6$H$_{15}$N$_3$)$_6$Fe$_8$O$_2$(OH)$_{12}$] Br$_7$(H$_2$O)Br · 8H$_2$O were investigated in [47]. By measuring the temperature, magnetic field, and angular dependences of the magnetization of a single crystal of a molecular mesoscopic magnet, it was found that the cluster consists of eight Fe^{3+} ions with spins of $S = 5/2$ with total spin $S = 10$. The experimental results of the magnetization at low temperatures showed a large anisotropy which is characterized by $D = -0.276$ K and $E = -0.035$ K. In other work [48], for $V_{15}4$, the existence of a zero-field gap of 30 mK and strong anisotropy in the angular dependence of the resonance field was established by low frequency ESR (0.6–3 GHz) at temperature (~ 0.5 K). The value of the symmetric exchange coupling was estimated as $J = 1.22$ K.

The polyoxovanadate $K_6[V_{15}As_6O_{42}(H_2O)] \cdot 8H_2O$ (V_{15}) (Fig. 10.9) cluster synthesized in [49] was demonstrated to be a convenient model system for the investigation of the magnetic properties of clusters using modern experimental physical methods and quantum mechanical approaches [41]. The stable cluster comprises a lattice of molecules with 15 V^{IV} ions of spin $S = 1/2$ placed in a quasi-spherical layered structure forming a triangle, sandwiched by two hexagons possessing overall symmetry D_3 and possesses antiferromagnetic couplings of 15 spins ½. Each hexagon contains three pairs of strongly coupled spins ($J \cong -800$ K). In low fields, susceptibility measurements give the following effective paramagnetic moments $\mu_{eff} = 1.75 \pm 0.02$ μ_B: below 0.5 K, and $\mu_{eff} = 3 \pm 0.02$ μ_{eff} corresponding to three independent spins $S = 1/2$ below 100 K. The experimental magnetization data agreed with theoretical results calculated with an exchange integral corresponding to isotropic exchange $J_0 = -2.445$ K. When the applied field reaches $-3J_0/(2\ g\mu_B)$ the system ground state switches from $S = /1/2$ to $S = 3.2$.

Figure 10.10 illustrates characteristic features of the energy levels of the V_{15} cluster [41].

Effects of isotropic and asymmetric spin exchange (AS) and Jahn-Teller distorsion (JT) on the instability and magnetic anisotropy of spin-frustrated systems exhibiting non-collinear spin structure have been considered within the context of

Fig. 10.9 The cluster anion $[K_6[V_{15}As_6O_{42}(H_2O)] \cdot 8H_2O$ (V_{15}): ball-and stick representation without the water central molecules emphasizing the V3 triangle [49]

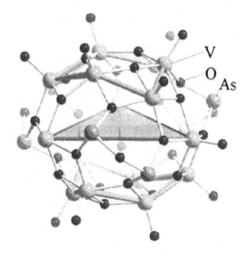

Fig. 10.10 Energy levels of the V_{15} cluster calculated in [47]. The levels are grouped according to the total spin. *Inset* the two lowest levels = 1/2 and S = 3/2 [41]

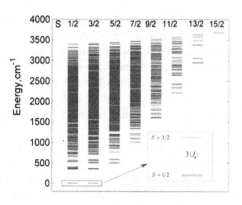

the three-spin model of the V_{15} cluster [50]. It was shown that in the V_{15} single crystals, the vibronic interaction is small and the orientation of the AS exchange vector affects the magnetic behavior of spin-frustrated systems. In this system, AS exchange plays a crucial role in understanding the field and temperature dependence of the adiabatic magnetization The best fit parameters are as follows: J = 0.855 cm^{-1}, g = 1.94, D⊥ = 0.238 cm^{-1}, Dn = 0.054 cm^{-1}.

10.4.3   Purely Organic Magnets

In organic molecular systems, the magnetic dipolar interactions and spin-orbit coupling are relatively weak compared with spin exchange interactions. Therefore, only Heisenberg exchange can be responsible for magnetic anisotropy of purely organic systems. A number of studies have been done on single-chain magnets (SCMs), which are superparamagnet-like, isolated, one-dimensional (1D) materials. Stability of these magnet is important for: the development of new devices, for use in data storage, light and heat sensing applications, and switching applications [51–60]. To design SCMs, three requirements should be fulfilled [31] (1) The spin carriers must exhibit a strong uniaxial anisotropy, (2) the material needs to exhibit a spontaneous magnetization, and (3) the chains must be isolated magnetically to avoid tridimensional ordering.

In an investigation of a purely organic molecule-based magnet, a crystal of the thiazyl radical 1,3,5-trithia-2,4,6-triazapentalenyl, TTTA, [51],

the main result of the calculation is that the strongest inter-chain interaction is the exchange J_{AB} interaction (J_{CN} = 183.8 cm^{-1}) for the contact S-N (p-p). All other interactions have a value smaller than 11 cm^{-1}. It was suggested that a 1D magnetic alignment gives the origin of the stability of this material. In this analysis, the inter-stack interactions as well as the difference in packing within the stacks **was takin in consideration.**

Neutral organic radicals based on nitronylnitroxide(p-NPNN radical) (2-substituted 4,4,5,5-tetramethyl-4,5-dihydro-3-oxido-1H-imi-dazol-3-ium-1-yloxyl)   synthesized in [52] were proved be a suitable reporter group in many organic magnets.. Kinoshita and coworkers were the first who demonstrated ferromagnetic properties of p-NPNNAn organic radical molecule p-NPNN1–3) [2-(40-nitrophenyl)-4,4,5,5-tetramethyl-4,5-dihydro-1H-imidazol-1-oxyl 3-N-oxide [53]. A transition to a ferromagnetic long-range ordered state was found at 0.60 K in the orthorhombic β-phase crystal and in the triclinic γ-phase crystal of p-NPNN. In work [54], the crystal structures and bulk ferromagnet properties of the α- and δ-phases of an organic radical p-NPNN were compared with those of the β- and γ-phases. These phases show distinct magnetic ordering due to intermolecular ferromagnetic couplings of J/k_B = 0:3 K in the AC plane of -p-NPNN.

In another investigation [55], it was found that the quasi-one-dimensional organic ferromagnet 2-benzimidazolyl nitronyl nitroxide (2-BIMNN) with the intrachange exchange interaction undergoes a phase transition to long-range magnetic order below T_C = 1.0 K Electron-spin-resonance and muon-spin rotation measurements provided evidence for the one-dimensional character of the magnetic fluctuations in 2-BIMNN and for the broad phase transition in the material. These results, discussed in the context of spin-1/2 quasi-one-dimensional Heisenberg ferromagnets, demonstrated that 2-BIMNN behaved as a typical spin-1/2 one-dimensional Heisenberg ferromagnet. Magnetic behavior of binary compositions of 2-(4,5,6,7-tetrafluorobenzimidazol-2-yl)-4,4,5,5-tetramethyl- 4,5-dihydro-1H-imidazole-3-oxide-1-oxyl   (F4BImNN)   and 2-(benzimidazol-2-yl)-4,4,5,5-tetramethyl-4,5-dihydro-1H-imidazole-3-oxide-1-oxyl (BImNN) (Fig. 9.11) crystallized as solid solutions was studied in [56]. Measurements carried out by the Magnetic dc-susceptibility techniques over 1.8–300 K allowed to measure magnetization versus field temperature of over 1.8–300 K. The data were fitted to spin quantum numbers ranging over S = 6–11 because the spins in the chain are strongly oriented by the intrachain. Magnetic analysis over 0.4–300 K indicated ordering with strong one-dimensional 1D positive ferromagnetic exchange along the chains with J/k = 12–22 K). It was suggested that, the effective spin orbital overlap at the $C\beta \cdots O\alpha$ contact provided ferromagnetic exchange coupling (Fig. 10.11).

The use of liquid crystals (LC), as an ordered environment for chiral nitroxides, provides possibilities for the formation of magnetic domains in applied magnetic and electric fields. and for the realization of [57–60]. and references therein. The main goal of this research is to enhance the effect of magnetic fields on the electric and optical properties of liquid crystals. Realization of paramagnetic susceptibility anisotropy ($\Delta\chi_{para}$)-controlled molecular orientation by weak magnetic fields opened the way for the design of soft materials with such properties. An anisotropic and inhomogeneous magnetic interaction, with the average spin–spin interaction constant $\bar{J} > 0$, was observed in the various liquid crystalline (LC) phases of nitroxides 1a and 1b (Fig. 10.12) [58].

Fig. 10.11 Spin orbital overlap at the Cβ⋯Oα contact that leads to ferromagnetic chain exchange coupling [56]

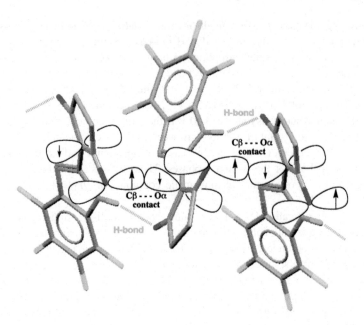

Fig. 10.12 Molecular structures and phase transition temperatures determined by DSC analysis of racemic and (2S,5S)-enriched **1**. Cr, SmC, SmC*, N, N*, and Iso denote the crystalline, smectic C, chiral smectic C, nematic, chiral nematic, and isotropic phases, respectively racemic and nonracemic all-organic radical LC compounds **1a** and **1b** [58]

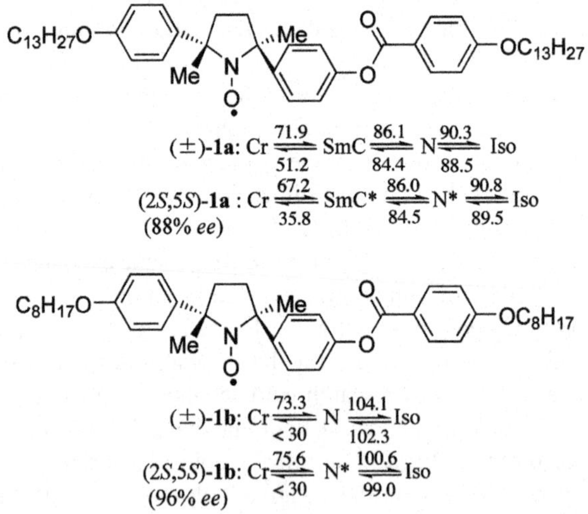

For **1a** and **1b,** the and the temperature dependence of the magnetic suscepti-bility χ_{para} between 25 and 115 derived from g-value, and peak-to-peak line width (ΔH_{pp}) of X-band ESR spectra indicated An anisotropic and inhomogeneous magnetic interaction in **1a** and **1b** was proved by a series of physical investigations including the measurement of temperature dependence of the magnetic

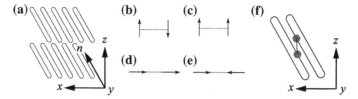

Fig. 10.13 Schematic representation of the spin–spin dipole interaction in a paramagnetic SmC phase. **a** Rod-like molecules form layered (x–y plane) structures in the SmC phase, and their director (n) is not parallel to the layer normal (z-axis). **b–e** Possible spin–spin dipole interactions between two spins. A magnetic field is applied perpendicular to the direction of the side-by-side interaction between two spins in (**b**) and (**c**), while it is applied parallel to the direction of the head-to-tail interaction between two spins in (**d**) and (**e**). **f** Spin–spin interaction between spins localized in radical moieties in the SmC phase [58]

Fig. 10.14 Motion of LC droplets on water under the action of a permanent magnet (maximum 0.5 T). **a** Schematic representation of the experimental setup for observing the attraction by a permanent magnet of a paramagnetic LC droplet on water in a shallow laboratory dish. **b** Photographs showing the attraction of the *yellow* paramagnetic N droplet of (±)-**1b** on water at 73 °C to the magnet. **c** Photographs showing the repulsion of the white diamagnetic N droplet of ZLI-1132 on water at 25 °C from the magnet. See details in [58]

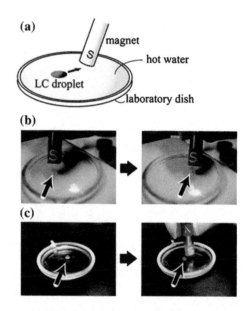

susceptibility χ_M using superconducting quantum interference device (SQUID) magnetometry at a magnetic field of 0.05 or 0.5 T A and the temperature dependence of the magnetic susceptibility χ_{para} derived from g-value, and peak-to-peak line width (ΔH_{pp}) of X-band ESR spectra. Suggested schematic representation of the spin–spin dipole interaction in a paramagnetic SmC phase is presented in Fig. 10.13.

Figure 10.14 demonstrates that a rod-like rare-earth magnet (0.5 T) attracted all of the four LC droplets enriched with **1a** and **1b** are floating on water, whereas the non-modified crystallized particles on water never moved under the influence of the same magnet.

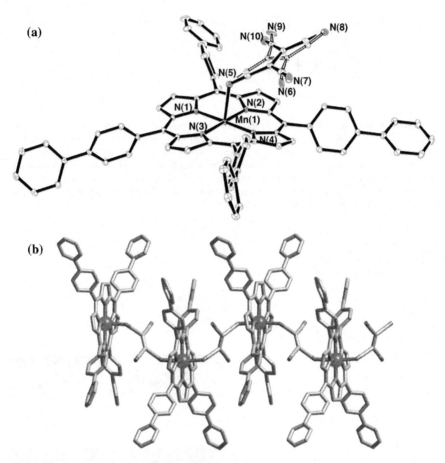

Fig. 10.15 **a** ORTEP drawing with 30 % probability thermal ellipsoids of the single-crystal X-ray crystal structure and **b** chain motif of **1**. Hydrogen atoms and interstitial solvent molecule are omitted for the sake of clarity [61]

10.4.4   Magnetic Metallorganics

Metallorganic chemistry provides a practically unlimited diversity of structures with synthesized and conjectured magnetic materials [21–25]. This section is restricted to several recent illustrative examples.

The AC magnetic susceptibility method has revealed a magneto-structural correlation in a single-chain magnet (SCM) of complex [Mn(TBPP)- (TCNE)] · 4 m-PhCl$_2$ (1) (Fig. 10.15) where TBPP$_2$ − = meso-tetra(4- biphenyl)porphyrinate; TCNE• − = tetracyanoethenide radical anion; m-PhCl$_2$ = meta-dichlorobenzene with a one-dimensional alternating MnIII(porphrin)—TCNE [61]. On the basis of the DC susceptibility data, the following results were obtained for complex **1**: exchange integral (J/k_B = −136 K and Cuirie-Weiss constant θ′ = +91 K), magnetic anisotropy (D_{Mn}/k_B = -3.7 K), magnetic relaxation with energety barrier Δ_τ/k_B = 146.3 K, and τ_0 = 4.1 × 10^{-10} s.

A comparison of the static magnetic properties of **1** with other Mn^{III}(porphyrin)–TCNE$^{\cdot-}$ chains allowed to suggest a model of a magneto-structural correlation between the intrachain magnetic exchange and the dihedral angle between the mean plane on [Mn(TBPP)(TCNE)] and Mn–N $\equiv$ C.

Magnetic susceptibility and ESR parameters of two six-coordinated high-spin Co(II) complexes, [Co(dpamH)$_2$(5-NO$_2$-salo)]NO$_3$ (1) and [Co(dpamH)$_2$(3-OCH$_3$-salo)]NO$_3$ · 1.3 EtOH · 0.4H$_2$O (2) were measured in the temperature range 2–300 K [62]. Obtain results are consistent with a suggestion about existing a spin quadruplet with largely unquenched angular momentum. This suggestion was supported by values $\chi_M T$ is 2.75 emu mol^{-1} K for complex 1 and $\chi_M T$ is 3.68 emu mol^{-1} K for complex 2 measured at room temperature. In zero-field, dynamic magnetic susceptibility measurements show slow magnetic relaxation below 5.5 K for compound 2 characterized by an Arrhenius law with a single energy barrier Δ_r/k$_B$ = 55 K and τ_o = 1.15 × 10^{-11} s. An analysis of the powder EPR spectrum of compound 2 indicated a ferromagnetic exchange interaction between the Co(II) ions with |J| = 4.5 cm^{-1}.

A technique based on frequency-dependent out-of-phase signals of alternating current magnetic susceptibilities were applied to the investigation of magnetic properties of a mononuclear lanthanide–nitronyl radical complex [Tb(hfac)$_3$-(NITPhSCH$_3$)$_2$] (**1**) [63]. In this compound, two NITPhSCH$_3$ radicals are coordinated as monodentate ligands to the Tb(III) ion through their NO groups to form a tri-spin complex, and a one-dimensional chain complex [Tb(hfac)$_3$(NITPhSCH$_3$)] n (**2**) {hfac = hexafluoroacetylacetonate, NITPhSCH$_3$ = 2-[4-(methylthio)phenyl]-4,4,5,5-tetramethylimidazolin- 1-oxyl 3-oxide. The temperature dependence of the magnetic susceptibilities of complexes **1** and **2**, measured in the 2–300 K range under an applied magnetic field of 2000 G at room temperature showed the $\chi_M T$ = 12.87 cm^{3}Kmol^{-1} for **1**. This value is consistent with an uncoupled Tb(III) ion (S = 3/2) and two nitroxide radicals (S = 1/2). The observed relaxation time τ_0 value supports the super-paramagnetic-like character of the relaxation dynamics.

Magnetic properties of the metal-free radicals and the metal complexes of Rare-earth-metal (M = **Nd, Gd** and Dy) complexes of 15-crown-5-Ph nitronyl nitroxide 1 and 15-crown-5- or 18-crown-6-Ph iminonitroxide 2, studied using SQUID magnetometry, have also been studied [64]. Antiferromagnetic (AFM) intrasublattice and ferromagnetic (FM) intersublattice interactions with Curie constant of Γ 10–100 K over the temperature range between 1.8 and 300 K were inferred. The metal-free radicals exhibit only weak AFM intermolecular interactions with the Weiss constant $\theta < -1$ K. Magneto-structural anomalies observed in "breathing crystals" composed of exchange-coupled copper two nitroxide [Cu(hfac)$_2$LR] (hfac is hexafluoroacetyl acetonate) were discussed in [65]. The experimental $\mu_{eff}(T)$ and $\chi T(T)$ dependences showed that the polymer exhibits reversible thermally induced structural rearrangements. A strong exchange

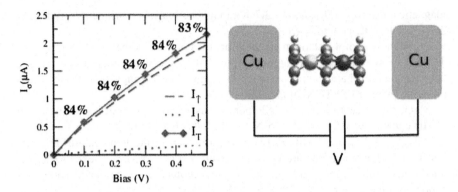

Fig. 10.16 Schematic illustration of filtering properties of FeCoCp$_3$ [23]

interaction ($J \approx 10$–100 cm^{-1}) takes place in the "head_to_tail" coordination one-spin systems >N–Cu^{2+}–N < and three-spin clusters > N–•O–Cu^{2+}–O–N < coupled in different complexes.

Metallorganic molecules can serve as effective spin filters (devices with capacity to put specifically polarized electron current) -in molecular spintronics because of the large spin polarization of their electronic structure. For example, for molecule FeCoCp$_3$ containing a Co and an Fe atom stacked between three cyclopentadienyl rings, (Fig. 10.16), which presents a large magnetic anisotropy and a $S = 1$, was theoretically predicted excellent spin-filtering properties both in tunnel and contact transport regimes [23].

10.4.5   Magnetic Polymers

10.4.5.1   Undoped Conjugation Polymer

In magnetic light sensitive conjugated conducting polymers, such as polyaniline, polyacetylene, polypyrrole, polythiophene, polyphenylene, and many others, strong pairwise exchange couplings between unpaired electron spin are mediated through the conjugated system [21, 25, 55–77]. Their physical properties, magnetic susceptibility, luminescence and conductivity offer possibilities for diverse practical applications in microelectronic devices, photodiodes, sensors, batteries, technological membranes, etc. Electron spin interactions in conjugated polymers are closely related to the nature of light induced charge carriers, positive and negative bearing spin polaron (electron with the accompanying deformation of atoms), and to fine details of polymer structure. In these systems the singlet triplet splitting controls optoelectronic properties through spin-dependent dissociation and recombination of charge carriers. The light sensitive conjugated polymers and doped conjugation polymers form the basis for spintronics, a technology which

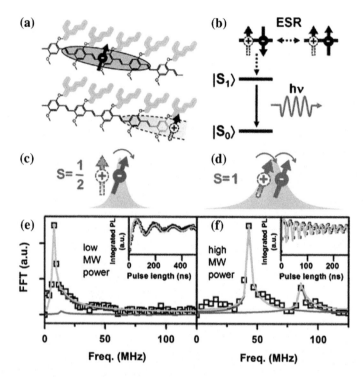

Fig. 10.17 Effect of deuteration of the polymer side groups on the ODMR resonance spectrum and on spin beating. **a, b** Structures of the polymers studied (C_8X_{17} = 2-ethylhexyl, where X = H or D). **c, d** The differential resonance spectrum is accurately described by a superposition of two Gaussians, representing electron and hole resonances. **e, f** Fourier analysis of the beating transients allows the extraction of the spin-1 (*red*) and spin-$^1/_2$ (*blue*) contributions to the resonance [69]

exploits spin-dependent electron transport phenomena in solid-state devices. The following examples illustrate some specific properties of undoped conjugation polymer.

The dynamics of electron and hole spins in a film of MEH-PPV (poly [2-methoxy-5-(2'-ethylhexyloxy)-1,4-phenylene-vinylene]) (Fig. 10.17) in which optical excitation of the polymer leads to the formation of tightly bound excitons with subsequent emission of a photon was investigated by optically detected magnetic resonance (pE/ODMR) [69]. Spin beating in the polymer was detected directly by its photoluminescence. polymer. Two principle findings in a conjugated polymer were pointed out: the equivalence of spin-dependent observables under optical and electrical excitation, proved by pulsed electrically and pE/ODMR techniques and effect of charge carriers and deuteration on the hyperfine field strength.

A method for obtaining the polaron spin-lattice relaxation time in π-conjugated polymers by measuring the ODMR dynamics as a function of microwave power and laser intensity was applied to 1/2 ODMR in films of pristine 2-methoxy-5-(2{'}-ethylhexyloxy) phenylene vinylene [MEH-PPV] polymer, as well as MEH-PPV doped with various concentrations of radical impurities [70]. The experimental results were fit to a spin dependent recombination model where both recombination and spin relaxation rates determine the response dynamics. In work [71], the photoconductivy of poly(p-phenylenevinylene) and 2 derivative, poly[1,4-phenylene-1,2-bis (4-methoxyphenyl)vinylene] and poly(2-phenyl-1,4-phenylenevinylene) at temperaturs 130–350 K were attributed to the formation of interchain pairs involving a negative polaron and a positive polaron [71]. Thermal dissociation of a polaron pair with a lifetime of 10^{-8}–10^{-9} s produced free charge carriers, and recombination of the pair regenerated a singlet or triplet exciton on a single conjugated segment of a chain.

Light-induced ESR action spectra of polarons in two dialkoxy derivatives of poly(p-phenylenevinylene) (PPV), MEH-PPV and CN-PPV, with cyano groups detected in [72] showed a threshold energy around 3 eV [Evidence that the spin density of the polaron predominantly resides on the vinylic carbons was also provided. In another work, the frequency response of the photoinduced absorption (PA) and the spin-1/2 photoluminescence-detected magnetic resonance (PLDMR) in the archetypal π-conjugated polymer poly[2-methoxy-5-(2-ethylhexyloxy)-1,4-phenylenevinylene] was consistent with a quenching model mediated by spin-dependent interactions between triplet excitons and polarons [73]. Short-lived paired polarons that may recombine after spin-dependent collisions with triplet excitons, and long-lived unpaired polarons that are unaffected by microwave resonance were identified. The recombination and escape yields of the carriers caused magnetic field effects (MFEs) on the charge-transfer fluorescence and transient photocurrent of a 1,2,4,5-tetracyanobenzene-doped poly(N-vinylcarbazole) film were observed in [74]. Two types of magnetic field effects, growth with increasing B due to the hyperfine mechanism (HFM) and a negative dip due to the level-crossing mechanism (LCM) were underlined. Values for the recombination of separated charges and hole transfer rate were obtainedas 7.0×10^7 and 4.5×10^8 s^{-1}, respectively,

10.4.5.2 Doped Conjugation Polymers

Since the e discovery of highly conductive doped polyacetylene, magnetic properties of conducting doped polymers such as polyacetylene, polyaniline (PANI), polypyrrole (PPy), and polythiophene, etc. have been extensively studied [78–91]. In pioneering works of the MacDiarmid, Heeger and Shirakawa groups [78] an essential increase in the magnetic susceptibility and metallic behavior of doped polyacetylene compared with undoped samples was detected. This discovery led to a program of work on the synthesis of organic polymers being p- or n-doped chemically, electrochemically, or with photo-doping and charge-injection doping acid-base chemistry to states possessing magnetic, optical and electronic properties of metals including significant conductivity. It was found that the doping induced electrical conductivity of conducting polymers results from mobile charge carriers introduced into the π-electron system [87–89].

In conducting polymers, the experimental susceptibility χT can be divided into two components: a temperature independent part $\chi_P T$ and a part obeying the Curie law $\chi = C/T$ (data for polyaniline [79]. Polythiophene [80], and polypyrrole [81].) The origin of these two components was explained within the context of the "metallic" model, which treats doped conducting polymers as highly ordered metallic domains immersed into amorphous domains. It was suggested that the temperature-independent component (the Pauli susceptibility) is attributed to metallic domains while the Curie susceptibility arises from defects in the amorphous domains. In the framework of the "metallic" model, the field dependence of the magnetization of conducting polymers at helium temperatures is described by the Brillouin function with S = 1/2. This function describes the dependency of the magnetization M on the applied magnetic field B and the total angular momentum quantum number J of the microscopic magnetic moments of the material under interest.

An alternative "triplet" mode [82] suggested that conducting doped polymers polyaniline, polypyrrole, polyacetylene and polythiophene, consist of fragments solely in the singlet or triplet state with a wide distribution of singlet-triplet splitting. In this model, magnetic properties of conducting polymers are described by an integral over the fragment magnetizations. The model explains features of magnetic properties of conducting polymers obtained by ESR and SQUID The analysis shows that the field dependence of the magnetization of conducting polymers at helium temperatures was described by the Brillouin function with S = 1, whereas the "metallic" model predicts S = 1/2. The "triplet" model also explains the non-linearity of the temperature dependences of χT, and the effect of heating and gases on these dependences.

Magnetic measurements at different temperatures and magnetic force microscopy of polymer, PANiCNQ produced from polyaniline (PANi) and an acceptor molecule, tetracyanoquinodimethane (TCNQ) provided evidence for ferri- or ferro-magnetic domain wall formation and motion [83]. A Curie temperature of over 350 K, and a maximum saturation magnetization of 0.1 $JT^{-1} kg^{-1}$ were determined by magnetic susceptibility methods. In addition, the spin density distribution was estimated from integrated electron spin resonance lines. In work [84], a poly[(N-vinylimidazole)-co-(1-pyrenylmethyl methacrylate)] (VI-co-PyMMA) ferric complex prepared by complexing a VI-co-PyMMA copolymer with 1-chlorobutane and $FeCl_3$ was synthesized and its super-paramagnetic magnetic properties were characterized. The authors suggested that the developed single-component fluorescent–magnetic polymer material they may find biotech applications.

Effects of temperature and magnetic field **on** magnetic susceptibility in conducting polyaniline and polypyrrole nanostructures were investigated [85]. This parameter was measured from 300 K down to 3 K at a dc magnetic field of 10 or 0.5 kOe with different dopant type and doping level. The magnetic field dependence of the magnetization was also observed from −90 to 90 kOe at different temperatures. The following important results were established: (1) The dc susceptibility $\chi_{dc} = M_{dc}/B$ is dependent on the applied magnetic field for samples in which the magnetic field dependence of the magnetization was also detected, (2) The temperature dependence of the susceptibility showed unusual transitions which can be explained by the coexistence of paramagnetic polarons and spinless bipolarons and the possible formation of bipolarons (or polarons) with changes in doping level and temperature.

In work [82], the samples of polyaniline films with camphorsulfonic acid (CSA), exhibiting a conductivity of 100 S cm^{-1}, were characterized by ESR spectroscopy at 133–423 K and by optical spectroscopy in the range $\lambda = 350$–1100 nm. It was found that the interchain distances correlate with the conductivity of the films and with the broadening of their ESR lines. Detail studies of charge transfer in H_2SO_4-doped polyaniline employing 3-cm (9.7 GHz) and 2-mm (150 GHz) EPR spectroscopy, and AC, SQUID, and DC conductometries were performed in [86]. A steady-state saturation method at 2-mm waveband EPR at 90–330 K showed the existence in non-doped polyaniline (PA) polarons with different relaxation and mobility and values of spin-lattice and spin-spin relaxation times. Intrachain diffusion and interchain hopping rates of these centers was also observed. When the polymer was doped H_2SO_4, the effective spin relaxation decreases due to the possible formation of metal-like domains of well-coupled chains with three-dimensional (3D) delocalized charge carriers. To explain the detected macroconductivity and microconductivity of the medium-doped PA, models 3D interdomain and intradomain hopping of charge carriers were suggested. A strong interaction of the charge with lattice phonons and a bipolaron as the preferred charge carrier were proposed to explain Observed specificity of a heavily doped polymer.

A number of organic solid-state spin transport electronics (spintronics) experiments, exploring the intrinsic spin of the electron, its magnetic moment and its electronic charge have been designed using doped polymers [87–91].

10.4.5.3 Super Paramagnetic Compounds and Polymers

In the last two decades, π-conjugated superparamagnetic organic compounds including polymer magnets with stability at ambient temperature and/or higher magnetic ordering temperatures have been attracting attention as models of multispin systems and potential magnetic devices [92–97]. Figure 10.18 demonstrates the progress in the area [93]. Bellow a few typical examples are described.

In pioneering work by Rajca et al. [92], the synthesis and investigation of high-spin polyradicals (Fig. 10.19), in which multiple radical sites coupled by exchange produced large net values of S in the ground state were presented. In these compounds, the multiple radical sites are coupled by exchange to produce large net values of S in the ground state. Studies of high-spin polyradicals, single molecular

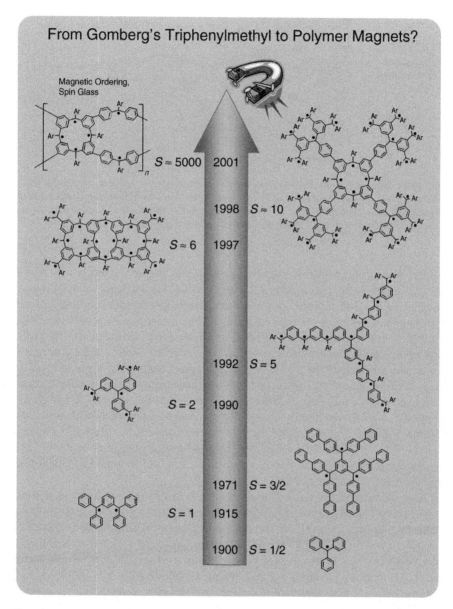

Fig. 10.18   16 Structure polyradicals and corresponding spin number S in the chronological order [94]

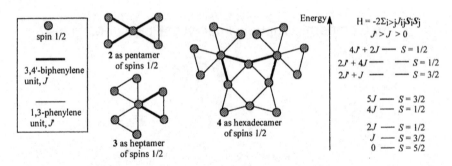

Fig. 10.19 Pentamer, heptamer, and hexadecamer of spins $^1/_2$, corresponding to pentaradical **2**, heptaradical **3**, and hexadecaradical **4**. Energy eigenvalues from the Heisenberg Hamiltonian for pentaradical **2** are shown for J, $J' > 0$ (both couplings ferromagnetic). J and J' are spin coupling through 3,4'-biphenylene and 1,3-phenylene units, respectively [93]

Spin clusters, polyarylmethyl tri-, penta-, hepta-, and hexadeca radicals showed that the strong ferromagnetic coupling through via 3,4'-biphenylene (or 3,5,4'-biphenylyne) units 1,3-phenylene units and weak ferromagnetic spin coupling between the component spins for tri- and pentaradical. The magnetization versus temperature data were fit to the value of J/k ≈ 90 K for tri- and pentaradicals. Data for hepta- and hexadecaradical were characterized as a dimer of the $S' = {}^5/_2$ and 1 component spins and a trimer of the $S' = {}^5/_2$, 3 and $^5/_2$ component spins, respectively with respective values of J/k of 13 and 4 K. Measurement of the hexadecaradical. magnetization vs magnetic field data to Brillouin functions at low temperatures, showed $S = 7.2$ for the high-spin ground states in a good agreement with theoretical value $S = 8$. Star-branched polyradicals were characterized by values of large total spin S = 5 − 10. The spin values and arrangement of pentamer, heptamer, and hexadecamer as well as the energy eigenvalues from the Heisenberg Hamiltonian for pentaradical **2** are presented in Fig. 10.19 [93].

A series of polyradicals and two high-spin polymers **12** and **13**, in which bis (biphenylene)- methyl groups link calix [4] arene macrocycles (Fig. 10.20) were prepared and investigated [94]. The polymer 12 quasi-linear chain is composed of unequal spins of S = 3 and S = ½ possessing an average value of S = 18 per macromolecule, while cross-linked polymer 13 is a network of unequal spins, with individual modules having S = 2 and linking modules having S = 1/2. For a long polymerization time, values of the total spin on a macromolecule can reach S = 7000!

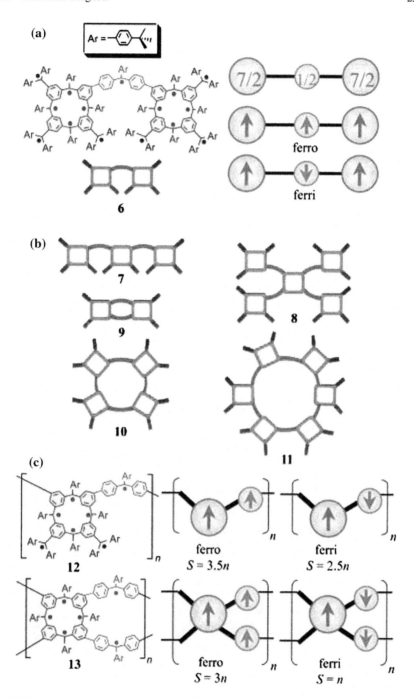

Fig. 10.20 Polyradicals and polymers with the ferromagnetic-ferromagnetic coupling scheme: **a** pentadecaradical 6 as a trimer of spins 7/2, 1/2, 7/2; **b** linear, branched, and annelated polyradicals 7–11; **c** polymers 12 and 13 with the ferromagnetic—ferrimagnetic coupling scheme with the macrocyclic (S = 3 and S = 2) and cross-linking (S = 1/2) modules [94]

References

1. C. Elschenbroich, *Organometallics* (Wiley-VCH, Weinheim, 2006)
2. M. Shiraishi, T. Ikoma. *Molecular Spintronics* (FreeScience.info., 2011)
3. L.R. MacGillivray, C.M. Lukehart, *Metal-Organic Framework* (Materials Wiley, Chichester, 2014)
4. N. Tanifuji, M. Irie, K. Matsuda, J. Am. Chem. Soc. **12**, 13344 (2005)
5. A. Saha, I.A. Latif, S.N. Datta, J. Phys. Chem. A **115**, 1371 (2011)
6. S. Nakatsuji, Preparation, reaction and properties of functional nitroxides, in *Nitroxides: Application in Chemistry, Biomedicine, and Materials Science*, ed. by G.I. Likhtenshtein, J. Yamauchi, S. Nakatsuji, A. Smirnov, R. Tamura (WILEY-VCH, Weinhem, 2008) p. 161
7. C. Li, J. Liu, S. Zhang, G. Lefkidis, W. Hubner, Carbon **87**, 153–162 (2015)
8. K. Matsuda, M. Irie, A diarylethene with two nitronyl nitroxides: photoswitching of intramolecular magnetic interaction. J. Am. Chem. Soc. **122**, 7195 (2000)
9. Y. Teki, S. Miyamoto, M. Nakatsuji, Y. Miura, J. Am. Chem. Soc. **123**, 294–(2001)
10. C. Huebner, B. Baxevanis, A.A Khajetoorians, D. Pfannkuche, Phys. Rev. B: Condens. Matter Mater. Phys. **90**, 155134/1- (2014)
11. Y. Teki, T. Toichi, S. Nakajima, Chemistry **12**, 2329 (2006)
12. C. Lefter, S. Tricard, H. Peng, G. Molnar, L. Salmon, P. Demont, A. Rotaru, A. Bousseksou, J. Phys. Chem. C **119**, 8522 (2015)
13. Y.H. Kwon, B.K. Mai, Y.-M. Lee, S.N. Dhuri, D. Mandal, K.-B. Cho, Y. Kim, S. Shaik, W. Nam, J. Phys. Chem. Lett. **6**, 1472 (2015)
14. C.R. Tichnell, D.A. Shultz, C.V. Popescu, I. Sokirniy, P.D. Boyle, Inorg. Chem. **54**, 4466 (2015)
15. H. Phan, S.M. Benjamin, E. Steven, J.S. Brooks, M. Shatruk, Angewan. Chem. Intern. Ed. **54**, 823 (2015)
16. F. Paul, F. Malvolti, G. da Costa, S. Le Stang, F. Justaud, G. Argouarch, A. Bondon, S. Sinbandhit, K. Costuas, L. Toupet, L. Laude, Organometallics **29**, 2491 (2010)
17. F. Li, M. Chakrabarti, Y. Dong, R. Kauffmann, E.L. Bominaar, E. Munck, L. Que Jr., Inorg. Chem. **51**, 2917 (2012)
18. G.E. Alliger, P. Müller, L.H. Do, C.C. Cummins, D.G. Nocera, Inorg. Chem. **54**, 107 (2011)
19. P.C. Bruijnincx, I.L. Buurmans, Y. Huang, G. Juhász, M. Viciano-Chumillas, M. Quesada, J. Reedijk, M. Lutz, A.L. Spek, E. Münck, E.L. Bominaar, R.J. Klein Gebbink, Inorg. Chem. **50**, 9243 (2011)
20. M. Nakagaki, S. Sakaki, Physic. Chem. Chem. Phys. (2015). Ahead of Print
21. C. Benelli, D. Gatteschi, in *Introduction to Molecular Magnetism: From Transition Metals to Lanthanides* (Wiley, 2015)
22. E.M. Pineda, N.F. Chilton, F. Tuna, R.E.P. Winpenny, E.J.L. McInnes, Inorg. Chem.(2015), Ahead of Print
23. P.N. Abufager, R. Robles, N. Lorente, J. Phys. Chem. C **119**, 12119 (2015)
24. C.-I. Yang, Z.-Z. Zhang, S.-B. Lin, Coord. Chem. Rev. **289–290**, 289 (2015)
25. V.V. Lukov, V.A. Kogan, S.I. Levchenkov, I.N. Shcherbakov, L.D. Popov, Russ. J. Coord. Chem. **41**, 1 (2015)
26. https://en.wikipedia.org/wiki/Magnetism
27. J.L. Manson, E. Ressouche, J.S. Miller, Inorg. Chem. **39**, 1135 (2000)
28. H.M. McConnell, J. Chem. Phys. **39**, 1910 (1963)
29. M. Deumal, J.J. Novoa, M.J. Bearpark, Celani Olivucci M. Robb MA. J. Phys. Chem. A **102**, 8404 (1998)
30. D. Gatteschi, R. Sessoli, J. Villain, *Molecular Nanomagnets* (Oxford University Press, Oxford, 2006)
31. J.S. Miller, A.J. Epstein, Molecule-based magnets An overview. www.mrs.org/publications/bulletin. MRS BULLETIN/NOVEMBER 2000, 21–28

32. A.L. Barra, D. Gatteschi, L. Pardi, A. Mueller, J. Doering, J. Am. Chem. Soc. **114**, 8509 (1992)
33. E.J.L. McInnes, Recent development in ESR spectroscopy of molecular nonomagnets. In: *Molecular Cluster Magnet*, ed. by R. Winnpenny (World Scientific, New Jersey, 2012) p. 215
34. Y. Furukawaa, Y. Nishisaka, K. Kumagai, P. Kögerler, J. Magn. Magn. Mater. **310**, 1429 (2007)
35. T. Guidi, Neutron spectroscopy of molecular nanomagnets, in *Molecular Cluster Magnet*, ed. by R. Winnpenny (World Scientific, New Jersey, 2012), p 109
36. M. Shatruk, A. Dragulescu-Andrasi, K.E.Chambers, S.A. Stoian, E.L. Bominaar, A.C. Catalina, K.R. Dunbar, J. Am. Chem. Soc. **129**, 6104 (2007)
37. D.W. Boukhvalov, E.Z. Kurmaev, A. Moewes, D.A. Zatsepin, V.M. Cherkashenko, S.N. Nemnonov, L.D. Finkelstein, Yu.M. Yarmoshenko, M. Neumann, V. Dobrovitski, M.I. Katsnelson, A.I. Lichtenstein, B.N. Harmon, P. Kogerler, Phys. Rev. B: Condens. Matter **67**, 134408 (2003)
38. W. Wernsdorfer, A. Müller, D. Mailly, B. Barbara, Europhys. Lett. **66**, 861 (2004)
39. Z. Salman, R.F. Kiefl, K.H. Chow, W.A. Mac Farlane, T. Keeler, T. Parolin, S. Tabbara, Phys. Rev. B **72**, 214415 (2008)
40. R. della Pergola, A. Fumagalli, F. Fabrizi de Biani, L. Garlaschelli, F. Laschi, Malatesta,. M.C, M. Manassero, E. Roda, M. Sansoni, P. Zanello, Eur. J. Inorg. Chem. **119**, 3901 (2004)
41. B. Tsukerblat, A. Tarantul, The nanoscopic V_{15} cluster: A unique magnetic polyoxometalate, in *Molecular Cluster Magnets*, ed. by R. Winnpenny (World Scientific, New Jersey, 2012), p. 109
42. E.M. Pineda, N.F. Chilton, F. Tuna, R.E.P. Winpenny, E.J.L. McInnes, Inorg. Chem. (2015). Ahead of Print
43. P.N. Abufager, R. Robles, M. Lorente, J. Phys. Chem. C **119**, 12119 (2015)
44. E.J.L. McInnes, R.E.P. Winpenny, Comprehensive Inorg. Chem. II **4**, 371 (2013). C.-I. Yang; Z.-Z. Zhang, S.-B. Lin, A review of manganese-based molecular magnets and supramolecular architectures from phenolic oximes. Coord. Chem. Rev. **289–290**, 289–314 (2015)
45. M. Shatruk, A. Dragulescu-Andrasi, K.E. Chambers, S.A. Stoian, L. Emile, E.L. Bominaar, C. Achim, K.R. Dunbar, J. Am. Chem. Soc. **129**, 6104 (2007)
46. D. Gatteschi, A.L. Barra, A. Caneschi, A. Cornia, R. Sessoli, L. Sorace, L. Coord, Chem. Rev. **250**, 1514 (2006)
47. M. Ueda, S. Maegawa, H. Miyasaka, S. Kitagawa, J. Phys. Soc. Jpn. **70**, 3084 (2001)
48. K. Kajiyoshi, T. Kambe, M. Mino, H. Nojiri, P. Koegerler, M. Luban, J. Magn. Magn. Mater. **310**, 1203 (2007)
49. A. Mueller, J. Doring, Angew. Chem. Int. Ed. Eng. **27**, 1719 (1988)
50. B.S. Tsukerblat, A. Tarantul, A. Mueller, J. Mol. Struct. **838**, 124 (2007)
51. C.S. Clarke, J. Jornet, N.J.J. Deumal, Polyhedron **28**, 1614 (2009)
52. E.F. Ullman, H. Osiecki, D.G.B. Boocock, R. J. Am. Chem. Soc. **94**, 7049 (1972)
53. M. Tamura, Y. Nakaz awa, S. Di, K. Nozawa, Y. Hosokoshi, M. Ishikawa, M. Takahashi, M. Kinoshita, Chem. Phys. Lett. **186**, 401 (1991)
54. M. Tamura, Y Hosokoshi, D. Shiomi, M. Kinoshita, Y. Nakasawa, M. Ishikawa, H. Sawa, T. Kitazawa, A. Eguchi, Y. Nishio, K. Kajita, J. Phys. Soc. Jpn. **72**, 1735 (2003)
55. T. Sugano, S.J. Blundell, T. Lancaster, F.L. Pratt, H. Mori, Phys. Rev. B: Condens. Matter Mater. Phys. **82**,180401/1 (2010)
56. G. Seber, R.S. Freitas, J.T. Mague, A. Paduan-Filho, X. Gratens, V. Bindilatti, F.O. Nei Jr., N. Yoshioka, P.M. Lahti, Magnetic tuning of all-organic binary alloys between two stable radicals. J. Am. Chem. Soc. **134**, 3825–3833 (2012)
57. Y. Uchida, K. Suzuki, R. Tamura, Y. Aoki, H. Nohira, J. Phys. Chem. B, **117,** 3054 (2013)
58. Y. Uchida, K. Suzuki, R. Tamura, N. Ikuma, S. Shimono, Y. Noda, J. Yamauchi, Anisotropic and inhomogeneous magnetic interactions observed in all-organic nitroxide radical liquid crystals. J. Am. Chem. Soc. **132**, 9746–9752 (2010)

59. Y. Takemoto, T. Yamamoto, N. Ikuma, Y. Uchida, K. Suzuki, S. Shimono, H. Takahashi, N. Sato, O. Nobuhiro; Y. Oba, R. Inoue, Rintaro et al. Soft Matter (2015), Ahead of Print
60. R. Tamura, Y. Uchida, K. Suzuki, in *Handbook of Liquid Crystals*, ed. by J.W. Goodby, 2nd edn. **8**, 837 (2014)
61. R. Ishikawa, K. Katoh, B.K. Breedlove, M. Yamashita, Inorg. Chem. **51**, 9123 (2012)
62. V. Tangoulis, M. Lalia-Kantouri, M. Gdaniec, Ch. Papadopoulos, V. Miletic, A. Czapik, Inorg. Chem. **52**, 6559, (2013)
63. X. Wang, X. Bao, X. Panpan, Li Li, From discrete molecule to one-dimension chain: Two new nitronyl nitroxide–lanthanide complexes exhibiting slow magnetic relaxation. Eur. J. Inorg. Chem. **2011**, 3586 (2011)
64. T. Sugano, S.F. Blundell, W. Hayes, H. Tajima, H. Mori, Physica status solidi. **9**, 1205 (2012)
65. M.V. Fedin, KYu. Maryunina, R.Z. Sagdeev, V.I. Ovcharenko, E.G. Bagryanskaya, Inorg. Chem. **51**, 709 (2012)
66. A. Rajca, Chemistry – Eur. J. **8**, 4834 (2002)
67. Nishide, Hiroyuki; Suga, Takeo. Magnetic polemers Edited by Mark, Herman F Encyclopedia of Polymer Science and Technology (4th Edition) **7**, 692 (2014)
68. R.T. Araujo, G. Ferreira, T. Segura, F.G. Souza, F. Machado, Eur. Polym. J. **68**, 441 (2015)
69. S.-Y. Lee, S.Y. Paik, D.R. McCamey, J. Yu, P.L. Burn, J.M. Lupton, C. Boehme, J. Am. Chem. Soc. **133**, 2019 (2011)
70. C.G. Yang, E. Ehrenfreund, Z.V. Vardeny, Phys. Rev. Lett. **99**, 157401 (2007)
71. E.L. Frankevich, A.A. Lymarev, I.I. Sokolik, F.E. Karasz, S. Blumstengel, R.H. Baughman, H.H. Hörhold, Phys. Rev. B: Condens. Matter **46**, 9320 (1992)
72. S. Kuroda, K. Marumoto, N.C. Greenham, R.H. Friend, Y. Shimoi, S. Abe, Synth. Met. **119**, 655 (2001)
73. M. Segal, M. A. Baldo, M. K. Lee, J. Shinar, and Z. G. Soos Phys. Rev. B **71**, 245201. (2005)
74. F. Ito, T. Ikoma, K. Akiyama, A. Watanabe, S. Tero-Kubota, J. Phys. Chem. B **109**, 8707 (2005)
75. V. Callegari, S. Demoustier-Champagne, A.C.S. Appl, Mater. Interfaces **2**, 1369 (2010)
76. T.D. Nguyen, G. Hukic-Markosian, F. Wang, L. Wojcik, X.G. Li, E. Ehrenfreund, Z.V. Vardeny, Nat. Mat. **9**, 345 (2010)
77. S.-Y. Lee, S.Y. Paik, D.R. McCamey, J. Yu, P.L. Burn, J.M. Lupton, C. Boehme, J. Am. Chem. Soc. **133**, 2019 (2011)
78. C.K. Chiang, C.R. Fincher Jr., Y.W. Park, A.J. Heeger, H. Shirakawa, E.J. Louis, S.C. Gau, A. G. MacDiarmid, Phys. Rev. Lett. **39**, 1098 (1977)
79. Z.H. Wang, E.M. Scherr, A.G. MacDiarmid, A.J. Epstein, Transport and *EPR* studies of polyaniline: A quasi-one-dimensional conductor with three-dimensional "Metallic" states. Phys. Rev. B **45**, 4190 (1992)
80. P.K. Kahol, J.C. Ho, Y.Y. Chen, C.R. Wang, S. Neeleshwar, C.B. Tsai, B. Wessling Synth. Met. **151**, 65 (2005)
81. J. Joo, J.K. Lee, J.S. Baeck, K.H. Kim, E.J. Oh, J. Epstein, Synth. Met. **117**, 45 (2001)
82. A.V. Kulikov, A.S. Komissarova, A.G. Ryabenko, L.S. Fokeeva, I.G. Shunina, O.V. Belonogova, Russ. Chem. Bull. **54**, 2794 (2005)
83. N.A. Zaidi, S.R. Giblin, T. Terry, A.P. Monkman, Polymer **45**, 5683 (2004)
84. J. Cui, S. Yang, J. Zhang, S. Zhao, Y. Yan, RSC Adv. **2**, 12224 (2012)
85. Y. Long, Z. Chen, J. Shen, Z. Zhang, L. Zhang, H. Xiao, M. Wan, Jean Luc Duvail. J. Phys. Chem. B **110**, 23228 (2006)
86. V.I. Krinichnyi, H.-K. Roth, G. Hinrichsen, F. Lux, K. Lüders, Phys. Rev. B **65**, 155205 ((2002)
87. A.G. MacDiarmid, Angew. Chem. Int. Ed. **40**, 2580. (2001)
88. A.J. Heeger, Angew. Chem. Int. Ed. **40**, 2590 (2001)
89. H. Shirakawa, Angew. Chem. Int. Ed. **40**, 2574 (2001)
90. I. Žutić, S. Das, Sarma. Rev. Modern Phys. **76**, 323 (2004)
91. S.D. Bader, S.S.P. Parkin, Ann. Rev. Condens. Matter Phys. **1**, 71 (2010)
92. S.A. Wolf, A.Y. Chtchelkanova, D.M. Treger, IBM J. Res. Develop. **50**, 101(2006)

93. J.S. Rajca, J. Wongsriratanakul, S. Rajca, J. Am. Chem. Soc. **119**, 11674 (1997)
94. A.J. Rajca, Chem.—Eur. J. **8**, 4834–4841 (2002)
95. W. Adam, M. Baumgarten, W. Maas, J. Am, Chem. Soc. **122**, 6735 (2000)
96. C. Khemtong, C.W. Kessinger, O. Togao, J. Ren, M. Takahashi, A.D Sherry, J. Gao, Conf. Proc. IEEE Eng. Med. Biol. Soc. 4095 (2009)
97. X. Zhou, L. Ding, Y. Long, Xu LiGuang, Libing Wang, Xu Chuailai, Anal. Methods **3**, 1737–1744 (2011)

Chapter 11
Electron Transfer in Biological Systems. Light Energy Conversion

Abstract In this chapter, two aspects of fundamental importance not only for biology but for modern chemistry and for artificial photosynthesis related to spin phenomena are briefly considered. The first aspect is the structure and action mechanism of the system of conversion of light energy into chemical energy in the primary charge photoseparation in bacterial and plant photosynthesis. This system is an unique device which almost ideally organized for collection and goal-direction of light upon combination of active groups into an ordered structure. Light energy conversion in the photosynthetic reaction centers is characterized by the high energetic efficiency and the quantum yield close to 100 %. The second aspect is structure and the possible mechanisms of the participation of polynuclear manganese systems in the photooxidation of water. This system accomplishes one of the most surprising reactions that occur in nature, the production of a strong reducing agent from water on account of the quanta of low energy. Here we survey the recent progress in understanding electron spin effects in the processes of light biological energy conversion.

11.1 Introduction

Nature captures and stores solar energy through photosynthesis—a biological process of converting light into chemical energy—that is ultimately responsible for all the biomass on earth. The photosynthetic organisms are represented by diverse classes of bacteria, algae, and plants that are very well adapted to a wide range of environmental conditions [1–3].

While different species store chemical energy differently, the primary process of capturing photons by reaction centers (RC), which are membrane proteins, involves the same molecular mechanism. Specifically, absorption of light by pigment molecules like chlorophyll a and b and carotenoids produces an electronic excited state that quickly (<0.1 ps) decays via vibrational relaxation to the first excited singlet state. This energy is then transferred by resonant dipolar mechanism to the neighboring RC protein that serves as a trap for the exciton. Chemical energy is

© Springer International Publishing Switzerland 2016

G. Likhtenshtein, *Electron Spin Interactions in Chemistry and Biology*,
Biological and Medical Physics, Biomedical Engineering,
DOI 10.1007/978-3-319-33927-6_11

generated by RC through formation of charge separation in the donor-acceptor (D-A) pair. Such (D^+A^-) pair is formed by D^+ cation-radical, which is a strong oxidant, and A^- anion-radical, which is a strong reducing agent. The (D^+A^-) pair should be relatively long-lived in order to enable subsequent chemical reactions eventually producing such stable chemical compounds as ATP and NADPH. The most important problems essential for understanding photosynthesis as well as developing biomimetics and synthetic light-harvesting systems are the exact molecular structure and specific physical mechanisms that prevent immediate recombination of D^+ and A^- centers. Over the years, essential details on dynamics and molecular structures involved in electron transfer reactions continue to emerge owing to the progress in time-resolved and multi-frequency EPR [4–16].

In this text, we start with reviewing electron transfer in proteins and model systems and then proceed with more detailed consideration of three major natural light-harvesting and light conversion systems: bacterial reaction center (BRC), photosystem I (PS I), and photosystem II (PS II) and conclude with the oxygen evolving complex [17–38].

11.2   Electron Transfer in Model Systems and Proteins

Numerous biological processes including photosynthesis involve electron transfer (ET)—a process that usually occurs via a quantum mechanical tunneling of an electron from one spatial location (at an atom or a chemical moiety) to another. For a simple barrier of a constant height the rate of quantum tunneling decays exponentially with the length of the barrier but is also affected by the height of the barrier in the energies of the initial and the final states (see Sect. 1.3 for fundamentals of modern theories of ET). In proteins, a majority of electron transfer occurs over relatively long distances and, therefore, is affected by molecular structure of the tunneling bridge. The latter aspects have been comprehensively and elegantly studied by Gray and coworkers using experimental and theoretical approaches [17–20] and references there in. Those authors employed an electronic excitation of ruthenium complexes and redox-active sites of metalloproteins to demonstrate effects of a number of factors on the electron tunneling rate, k_{ET}, that could be described by semiclassical Marcus-Levich equation (1.39). In this equation chemical composition and the structure of redox centers is given by $(-\Delta G^o)$, the contribution to the Gibbs free energy describing the driving force of the electron transfer, and by parameter λ accounting for the extent of nuclear orientation in the D-A pair. The distance between the redox centers contributes to the parameter H_{AB}, which is the electronic coupling between the reactants [D, A] and products [D^+, A^-] at the transition state (see also 1.33, 1.39). Data for the electron tunneling demonstrate nearly perfect exponential distance dependence for through proteins (Fig. 11.1).

Redox cycles are typically completed by proteins within a few milliseconds or even microseconds. This scale of electron tunneling time limits the maximum

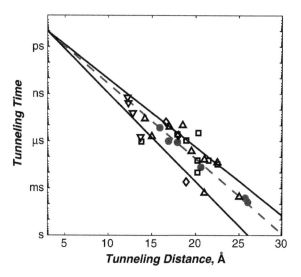

Fig. 11.1 Tunneling timetable for intraprotein ET in Ru-modified azurin (•), cytochrome c (□), myoglobin (◇), cytochrome b_{562} (△), HiPIP (▽). The *solid black lines* illustrate distance decay factors of 1.3 (*lower*) and 1.0 $Å^{-1}$ (*upper*); the *dashed blue line* illustrates a 1.1 $Å^{-1}$ decay [19]

distance of a single tunneling event to about 20 Å as evidenced by the experimental data of Fig. 11.1. However, the distance range could be expanded and the time of transfer decreased substantially by hopping mechanism in which ET is achieved by a chain of short tunneling events. For example, ET over the same 20 Å distance is about two orders of magnitude faster if it proceeds via a hole hopping through an intervening tryptophan residue. Multiple electron hopping events would result in even faster ET. Indeed, nanosecond ET was measured between a Ru-diimine wire, [(4,4′,5,5′-tetramethylbipyridine)$_2$Ru(F9 bp)]$_2^+$ (tmRu-F9 bp, where F9 bp is 4-methyl-4′-methylperfluorobiphenylbipyridine), bound tightly to the oxidase domain of inducible nitric oxide synthase (iNOSoxy) [20]. Photoreduction of an imidazole-bound active-site heme iron in the enzyme-wire conjugate ($k_{ET} = 2 \times 10^7 \text{ s}^{-1}$) was seven orders of magnitude faster than that in the non-modified enzyme (Figs. 11.2 and 11.3).

However, ET in proteins does not always follow the exponential dependence upon the donor-acceptor distance as has been shown in recent studies of cytochrome c absorbed on a self-assembled monolayer (SAM) formed on silver electrodes [20]. ET kinetics was measured by time-resolved surface-enhanced resonance Raman and resonance Raman spectroscopy. Mutation the positively charged K87 of WT cytochrome c to a neutral amino acid allowed for elucidating the role of the strength of the interfacial electric field. The observed anomalous nonexponential distance dependence (Fig. 11.4) was explained by suggesting a coexistence of at least two protein conformations with distinctly different average electronic couplings. Indeed, conformations of cytrochrome c could affect orientation of the heme with respect to the electrode plane (defined by two angles α and φ) and this would change the ET rate.

The influence of the mutual orientation of the donor and acceptor orbitals in the Ru(bpy)$_2$ in HisX-cytochrome c on the ET rate transfer was also analyzed by the

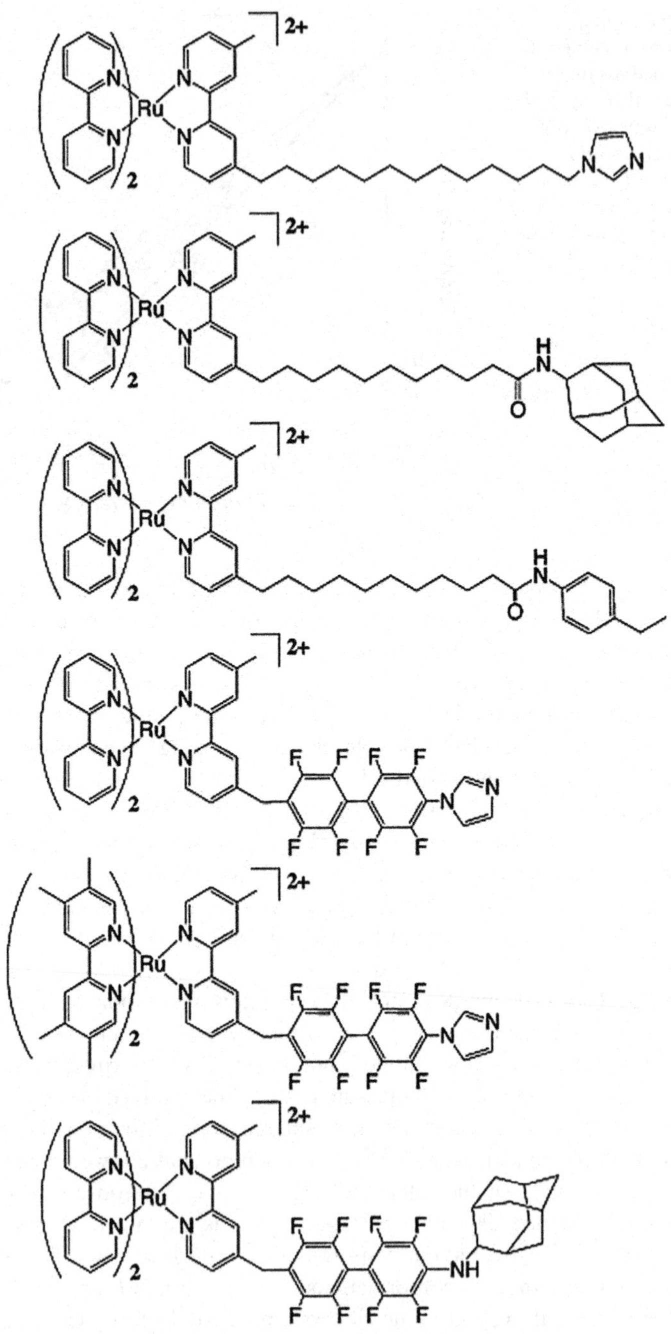

Fig. 11.2 Sensitizer wires (*top* to *bottom* bpyRu-C$_{13}$-Im, bpyRu-C$_{11}$-Ad, bpyRu-C$_{11}$-EB, bpyRu-F$_8$bp-Im, tmbpyRu-F$_8$bp-Im, and bpyRu-F$_8$bp-Ad) [17, 18]

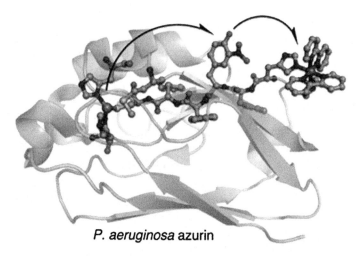

P. aeruginosa azurin

Fig. 11.3 Electron transfer (ET) pathways in the Ruthenium-modified Pseudomonas aeruginosa azurins that incorporate 3-nitrotyrosine (NO$_2$YOH) between Ru(2,2′-bipyridine)2(imidazole) (histidine) and Cu redox centers [14]

Fig. 11.4 Nonexponential distance dependence of k$_{ET}^{app}$ for both the oxidation (*red*) and reduction (*blue*) processes of WT cytochrome c (*circles*) and K87C (*triangles*). *Full and half-filled triangles* represent the fast and slow component of k$_{ET}^{app}$ for K87C, respectively [20]

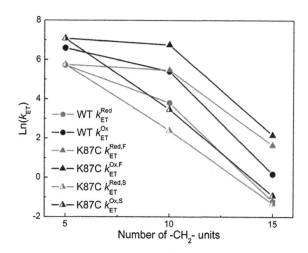

artificial intelligence-superexchange method that takes into account details of the protein electronic structure [21]. Calculations of the superexchange matrix element between the donor and acceptor states in seven electron-transfer ferroproteins: Fe(heme)-cyt-Ru(HisX)(bpy)zim, where X = 79, 72, 66, 62, 58, 39, 33 that included all the atoms in the systems were carried out. The results were analyzed to establish the most effective electronic coupling pathways in these proteins. The distances between donor (Fe^{2+}) and acceptor (Ru^{3+}) varied in the range 13–20 Å. An approximate quantitative agreement within a factor of two was found between all-atom calculations and a reduced model that included a small number of amino acids. Effect of the mutual orientation of the donor and acceptor orbitals in the $Ru(bpy)_2$ in HisX-cytochrome c on the rate of electron transfer was also analyzed.

Overall, these studies illuminate effects of several factors that control electron transfer in proteins that in addition to the distance in the donor–acceptor pair include [19–24]: (i) ET mechanism (single tunneling or hopping), (ii) contributions of multiple pathways, (iii) effect of immediate chemical environment of D and A as well as the properties of the tunneling channel, and last but not least, (iv) the mutual orientation of the D and A that provide control of ET through electric field effects. These finding should guide researchers to uncover further details of molecular mechanisms that allow for effectively controlling and possible switching electron transfer in proteins.

11.3   Light Energy Conversion

The process of photosynthesis embrace practically all the aspects of modern biochemistry, biophysics and molecular biology [1–3]. In all living organisms the photosynthesis is carried out by photosynthetic reaction centers (RCs) that represent a remarkable examples of how assembly of different functional groups evolved into an ordered structure with fundamentally new properties.

As has been noted by several authors [25–31], for a system containing a cascade of tightly packed of donor and acceptor pairs, the reactions of charge separation and recombination are expected to be very fast and the efficiency of the light energy conversion to be very low. Further, for suggested cascade structures with optimal distance between the centers, the forward electron transfer between adjacent D-A_1 and A_i-A_{i+1} pairs was predicted to slow down with the distance compared with the electron transfer in a system having close contacts between the centers. Nevertheless, for the optimum length of the "nonconducting" zones not exceeding 0.5–0.7 nm, the direct electron transfer can be sufficiently fast. Essential feature enabling the photosynthetic cascade to function is that recombination of the each D^+-A_i^- pair becomes progressively slower as A_i moves away from the donor. The aforementioned cascade mechanism is fulfilled in all the biological RCs and, thus, should be taken into consideration when designing artificial system of light energy harvesting

This brief analysis yields following two conclusions. (1) A fast high-quantum-yield conversion of light into energy of chemical bonds is provided by cascade photochemical systems in which photo-and chemically active centers (aromatic photo chromophores, transition metal clusters) are separated by "insulating" 6–10 Å long zones consisting of nonsaturated molecules and bonds. (2) The electron transfer between the donor and acceptor centers has to occur by a long-range, most likely, nonadiabatic mechanism.

This principal scheme of the mechanism realized in photosynthetic RCs for the effective light energy conversion has been further supported in subsequent experiments with a bacterial system *R rubrum*, in which electron transfer can be described by the following sequence of events [1]:

$$(BChl)_2 \rightarrow BCl \rightarrow BPh \rightarrow Q_A \rightarrow Q_B$$
$$\quad I \qquad\quad II \qquad III \qquad IV \qquad V$$

where Q_A and Q_B are the primary and secondary quinine acceptors, respectively. The distances between donor–acceptor components of this cascade were estimated from EPR measurements. Specifically, by evaluating effect of Q_A^--Fe on the spin-lattice relaxation time of $(Bchl)_2^+$, the distances between $(Bchl)_2^+$ and Q_A^- was shown to be 32–35 Å [26, 27]. Other distances, r, were estimated using an (8.3) that relates r with experimentally measured spin exchange coupling (J_{SE}) [27]. The estimated distances (r [BPh^--$Q1^-$] = 1.45 nm, r[$BChl)_2^+$ -BPh^-] = 1.37 nm, r[$BChl)_2^+$ Bcl^- = 0.3 nm) were found to be in good agreement with structural data obtained several years later by x-ray of RC monocrystals [32].

It should be noted here that some of the most important chemical reactions for sustaining the Earth ecosystem (photosynthesis, water splitting, nitrogen fixation) require multiple ET events. The concept of four electron mechanism was first suggested back in seventies [33, 34] and then applied to such energy-demanding enzymatic and chemical reactions as reduction of molecular nitrogen and water splitting under mild conditions [28–31].

For example, the biological oxygen evolution complex (OEC), $CaMn_4O_x$, achieves water splitting into O_2 and H_2 by withdrawing four electrons from two water molecules and realizing four protons [35–38]. The oxidation of two water molecules into dioxygen proceeds in a stepwise manner as described in the four S-state cycle with the following intermediate states:

$$S_0 \rightarrow S_1 \rightarrow S_2 \rightarrow S_3 \rightarrow S_4$$

The radical cation P_{680}[sup+] serves as a primary acceptor and has a very high oxidizing potential, estimated to be 1.13 V [39]. Potentials of oxidation of water by one, two and four electron mechanisms are equal to 2.7 V (hydroxyl radical), 1.36 (hydrogen peroxide) and 0.81 (dioxygen). Thus, the only thermodynamically favorable mechanism of the photosynthesis has to involve a four-electron oxidation of deprotonated water molecules in a four-nuclear cluster, which was previously positively charged in course of the photocycle.

11.3.1   Bacterial Reaction Center (RC)

The primary photochemical processes of photosynthesis take place within membrane-bound complexes of pigments and protein, reaction centers [1–3]. Currently, a wealth of structural data on bacterial RCs is readily available with over 50 x-ray crystal structures of RC from *Rhodopseudomonas (Blastochloris) viridis*,

Rhodobacter sphaeroides, and *Thermochromatium tepidum* published. Further, these structures have been compared on the basis of data quality and quantity, resolution, and structural features revealed [40, 41]. The structure of the principal components of RC from *Rb. sphaeroides* is shown in Fig. 8.3.

The process of photosynthesis starts with an accumulation of light quanta by the light-harvesting complex (LHC) [3]. The singlet electronic excitation migrates along the antenna and enters the primary acceptor P_{870}, the dimer of bacteriochlorophyl (D_A) that also passes over into the singlet state. This is followed by the following chain of events: (1) an electron from the excited P_{870} is transferred to bacteriochlorophyl (Bcl) within a femtosecond and then (2) to bacteriopheophytin (Bph) in picoseconds, (3) to the primary acceptor ubiquinone (Q_A) in *ca.* 200 ps, and (4) to the secondary acceptor Q_B in milliseconds. During this time the electron from the secondary donor, which is the type c chytochrome, transforms from the reduced cytochrome c to P_{870}^+. As a result, the energy of a solar photon is transformed into chemical energy of the reduced secondary (Fig. 11.5).

Method of time-resolved strongly enhanced nuclear polarization with laser-flash ^{13}C photo-CIDNP magic-angle spinning NMR was used to study of spin effects in the carotenoid-less mutant R26 composed of the two bacteriochlorophyll cofactors, PL and PM, in bacterial reaction centers of purple bacteria of *Rhodobacter sphaeroides* [42]. A photocycle of processes under investigation and values of time constant for of corresponding spin dynamic processes is presented in Fig. 11.6. From the photochemically excited primary donor P*, an electron is transferred to

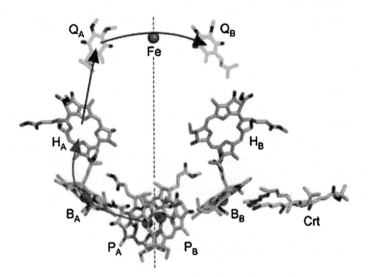

Fig. 11.5 Structure of principal components and sequential electron transfer mechanism of the RC from *Rb. sphaeroides.* The cofactors are arranged around an axis of two-fold symmetry (*dotted line*) running perpendicular to the plane of the membrane and comprise of four BChls (B_A, B_B, P_A, P_B), two BPhes (H_A, H_B), two quinones (Q_A, Q_B), a carotenoid (Crt), and an iron atom (Fe). *Arrows* show the route of electron transfer [41]

Fig. 11.6 Kinetics and spin
dynamics of electron transport
in quinone-depleted RCs of
Rb. sphaeroides R26 (See
details in [42])

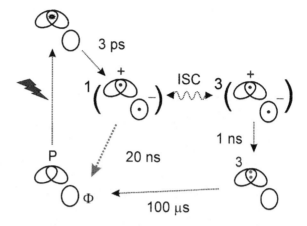

the primary acceptor. This light-induced spin-correlated radical pair is created in a
nonstationary and highly electron polarized singlet state. The radical pair undergoes
two processes: (1) Intersystem crossing (ISC) driven by the isotropic iteration *a*iso
and the difference in *g* values of the electrons leads to a coherent oscillation
between the singlet (S) and the triplet (T_0) states of the radical pair and (2) Electron
back-transfer to the special pair can occur in the S and T_0 states having different
lifetimes of the two decay channels. It was found that in the radical cation state, the
ratio of total electron spin densities between PL and PM is 2:1, 2.5:1 for the pyrrole
carbons, 2.2:1 for all porphyrin carbons, and 4:1 for the pyrrole nitrogen.

In the reaction centers of *Rb. sphaeroides*, the excited singlet state of the primary
donor P, *P, donates an electron to the initial electron acceptor Q_A to form the
initial radical-pair state $P^{+\cdot} Q_A^-$ [43]. When $P^{+\cdot} Q_A^-$ is blocked for the latter reaction,
either by removal or prior reduction of the quinine, this pair lives for 10–20 ns
before the charge recombination occurs. Over this time the singlet electron spin
states of the radical pair can evolve into a triplet. Such an evolution would occur
through singlet-triplet mixing in the initial radical-pair state by the hyperfine
mechanism at low magnetic fields and by both the hyperfine and Δg mechanisms at
high magnetic fields (>1 kg). Since the hyperfine field felt by the electron spins, $P^{+\cdot}$
Q_A^-, is dependent upon the nuclear spin state in each radical, the nuclear spin
configuration would affect the relative probabilities of charge recombination to the
triplet state of the primary electron donor. As a result, these recombination products
will have non-equilibrium distributions of nuclear spin states (nuclear spin
polarization).

Fast time-resolved EPR spectroscopy was used to study electron spin polarization
(ESP) in perdeuterated native, Fe^{2+}-containing reaction centers (RCs) of photosynthetic purple bacteria [44]. The proposed spin-correlated radical pair model
included the large anisotropy arising from the magnetic interactions between Fe^{2+}
and the reduced primary electron–acceptor quinone ($Q_A^{-\cdot}$), which results in different
quantization axes for the $P^{+\cdot}$ and the ($Q_A^{-\cdot} Fe^{2+}$) spins. The ESR spectra were calculated with the point–dipole approximation for $P^{+\cdot}$ and $Q_A^{-\cdot}$ taken yields

$D_{PQ} = -0.12$ mT and $J_{PQ} = 0$. It was shown that the stimulated ESP spectrum is solely due to the $P^{\bullet+}$ part of the spin-correlated radical pair $[P^{\bullet+}(Q_A^{\bullet-}Fe^{2+})]$, whereas the rapid decay of the spin-polarized signal is due to spin-lattice relaxation of the $(Q_A^{\bullet-}Fe^{2+})$ complex. Simulations obtained for the lowest two quartets of states of $[P^{\bullet+}(Q_A^{\bullet-}Fe^{2+})]$, associated with the $m_{S,Fe} = -2$ and $m_{S,Fe} = -1$ levels,

Photochemically induced dynamic nuclear polarization (photo-CIDNP) was observed in photosynthetic RCs of the carotenoid-less strain R26 of the purple bacterium *Rb. sphaeroides* by ^{13}C solid-state magic-angle spinning (ssMAS) NMR at three different magnetic fields (4.7, 9.4, and 17.6 T) [45]. The spectra were simulated assuming lifetimes of triplet radical and singlet radical pairs to be 1 and 20 ns respectively, exchange coupling of $J = 7$ G, and dipole-dipole coupling $d = 5$ G. The principal advantage of time-resolved photo-CIDNP MAS NMR as an experimental method to study electron transfer in photosynthesis, is that enables to elucidate spin dynamics of spin-correlated radical pairs and map the corresponding spin densities at the atomic resolution and at microsecond time scales. Specifically, NMR signals of the donor showed an absorptive (positive) enhancement while the enhancement for the acceptor was emissive (negative) (Fig. 11.7). Time-resolved photo-CIDNP ssMAS NMR data of reaction centers of quinine-depleted *Rb. sphaeroides* were also reported [45]. It was demonstrated that the build-up of nuclear polarization on the primary donor and the bacteriopheophytin acceptor depends on the presence and lifetimes of the molecular triplet states of the donor and carotenoid. The electron-nuclear spin dynamics in the solid state is governed by a combination of electron-electron nuclear three-spin mixing (TSM), electron-nuclear differential decay (DD), and differential relaxation (DR) mechanisms. The mechanisms of photochemical reactivity of the special pair in the bacterial RC have been further studied by ssNMR in conjunction with density functional theory (DFT) modeling.

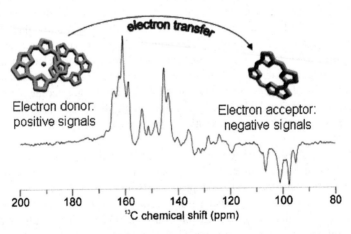

Fig. 11.7 ^{13}C photo-CIDNP ssMAS NMR spectra of RCs of *Rb. Sphaeroides* and a scheme of electron transfer [45]

The first pulsed EPR detection of quantum oscillations in photosynthetic RCs, initially predicted in [46], was achieved using an electron spin echo (ESE) sequence [47]. The latter study was focused on spin-correlated radical pair $[P^+ Q_A^-]$ in protonated zinc-replaced reaction centers of Rb. sphaeroides R26. Upon applying a (flash-t-pulse 1-τ-pulse 2-T) pulse sequence, it was found that the shape of the echo at T = τ is influenced markedly by the delay t, for 0 < t < 40 ns. In addition a second echo signal was found at T = 2τ for t = τ. The change in shape of the first echo (T = τ) is ascribed to quantum beats; the second echo (T = 2τ) is due to double-quantum coherence. D = −175 mT J = 0.0016 mT at 20 K from ESEEM (electron spin echo envelope modulation).

Marked advances in understanding charge separation in photosynthetic RCs was achieved by high frequency (HF) TR-ENDOR and high time resolution pulsed EPR [6]. These authors studied completely deuterated or protonated Rb. Sphaeroides RCs, in which Fe was removed and substituted by Zn and the primary quinone acceptor, Q_A, ubiquinone-10, was substituted by deuterated ubiquinone-10. Pulsed TR EPR spectra of the spin-correlated radical pair (SCRP) were recorded as a function of magnetic field by monitoring two pulse ESE which followed a 5 ns laser pulse at a fixed delay after laser flash (DAF) time. Pulsed ENDOR spectra were recorded using a Mims-type sequence of microwave (MW) and radiofrequency (RF) pulses (laser- DAF- rt/2Mw-r-n/2Mw-nRv-n/2Mw) by monitoring the ESE intensity as a function of the frequency of the RF pulse. A new phenomenon manifesting in HF TR-ENDOR spectra was observed and described.

Spin-correlated radical pairs formed in course of photosynthesis are known to produce quantum oscillations that is a pure quantum mechanical phenomenon [48]. This phenomenon was rationalized by an analytical EPR model developed by the density matrix approach for these short-lived intermediates. The suggested mechanism of the nuclear spin oscillations is based on electron–electron–nuclear three-spin mixing and implies that the nuclear modulation amplitude depends on the static magnetic field of the EPR experiment [49]. The formation of the quantum oscillations was explained in terms of the non-adiabatic change of the spin Hamiltonian at the instant of the laser pulse. At time zero, the radical pair is created in singlet state as a consequence of spin conservation in an ultra-fast photochemical reaction. The radical pair emerges as a coherent superposition of the eigenstates that manifest themselves as quantum oscillations in an EPR experiment with adequate time resolution. At higher microwave frequencies, the modulation amplitude rapidly decreases with increasing the magnetic field and, at Q- (34 GHz/1.2 T) and W-band (95 GHz/3.4 T) resonant frequencies/field, only the electron spin oscillations are expected.

The main advantages of EPR methods lay in ability to provide structural data (distances and mutual orientations) on SCRPs from non-crystalline samples so RCs could be studied in native lipid environment. For example, a combination of pulsed HF EPR W band (95 GHz, 3.4 T), pulsed electron-electron double resonance (PELDOR), and relaxation-induced dipolar modulation enhancement (RIDME) led to three-dimensional structure of laser-flash-induced transient radical

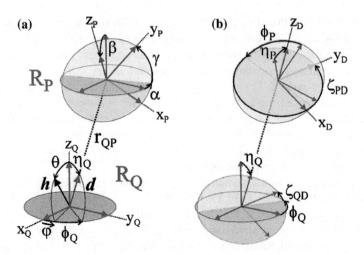

Fig. 11.8 Relative positioning of coordinate frames of radicals R_Q (x_Q, y_Q, z_Q) and R_P (x_P, y_P, z_P). **a** Conventional representation in terms of distance r_{QP}, two polar angles η_Q and φ_Q, and three Euler angles α, β, γ. The polar angles θ and φ define the orientation of the external magnetic field director h in the Q frame. **b** Representation involving the intermediate dipolar axis system (x_D, y_D, z_D) in terms of the angles η_Q, φ_Q, η_P, φ_P, and the angle $\zeta_{QP} = \zeta_{QD} - \zeta_{PD}$, which defines the relative turn of the Q and P frames around the dipolar axis [50]

Table 11.1 Comparison of the geometry parameters characterizing the paired radicals $P_{865}^{\cdot+}Q_A^{\cdot-}$ of RCs from *Rb. Sphaeroides* as determined by dipolar HF EPR and x-ray methods [50]

Structure	R_{QP} (nm)	φ_P (°)	η_P (°)	φ_Q (°)	η_P (°)	ζ_{QP} (°)
Dipolar EPR	2.89(2)	82	59	242	110	50
x-ray	2.84(2)	79	66	246	97	49
Turned quinine model	2.84(3)	79	66	245	110	44

pairs $P_{865}^{\cdot+}Q_A^{\cdot-}$ in frozen-solution RCs derived from *Rb. Sphaeroides* (Fig. 11.8) [50]. The structural parameters derived from EPR were found to be in a good agreement with those obtained from x-ray crystal structures (Table 11.1) (ibid).

11.3.2   Reaction Centers of Photosystem I

In oxygenic photosynthetic organisms, plant and green bacteria, the reaction centers of two systems Photosystem I (PS I) and Photosystem II (PS II) convert the absorbed four light quanta into energy of stable products, that is ferredoxin in reduced state and dioxygen [3, 51–59] PS I from plants and cyanobacteria mediates light-induced electron transfer from plastocyanin to ferredoxin (flavodoxin) at the stromal membrane side, while PS II is a photoenzyme that catalyzes oxidation of

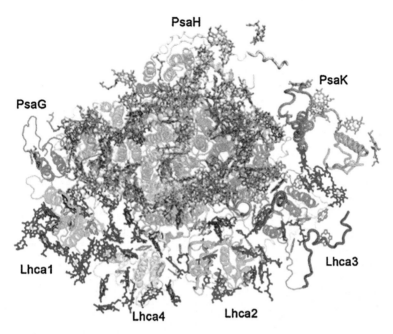

Fig. 11.9 The arrangement of 167 chlorophyll molecules of plant PSI as seen from the stromal side (See detail in [52])

water in a water-splitting Mn-containing system. PS I from cyanobacteria consist of 11 protein subunits and several cofactors. After the photoexcitation of the primary donor, a dimer of chlorophyll a $(Chla)_2$, P700, an electron is transferred via a chlorophyll (A_0) to a phylloquinone (A_1) and then to the iron sulfur clusters, F_x, F_A and F_B (Fig. 11.9) [52].

An improved electron density map of Photosystem I yielded the location of an additional ten beta-carotenes as well as five chlorophylls and several loop regions, and eventually revealed the locations of and interactions among 17 protein subunits and 193 non-covalently bound photochemical cofactors.

Mechanism of the electron transfer and the spin dynamic effects in were discussed in [53]. The energy transfer and charge separation kinetics in core Photosystem I (PSI) particles of *Chlamydomonas reinhardtii* has been studied using ultrafast transient absorption in the femtosecond-to-nanosecond time range. For the first time, the transient difference spectrum for the excited reaction center state and the formation and decay of the primary radical pair and its intermediate spectrum directly from measurements on open PSI reaction centers were directly resolved. Several possibilities are discussed for the intermediate redox states and their sequence which involve oxidation of P700 in the first electron transfer step or only in the second electron transfer step, which would represent a fundamental change from the presently assumed mechanism. Three different redox intermediates on the timescale up to 100 ps were distinguished. Formation and decay of the primary radical pair and its intermediate spectrum were also observed. The data were rationalized by a likely inclusion of an additional redox state in the electron transfer pathway.

A new electron transfer mechanism for PS I where the accessory chlorophylls function as the primary electron donor(s) and the A0 Chl(s) are the primary electron acceptor(s) was proposed [54]. This new mechanism resolved in a straightforward manner the difficulty with the previous scheme, where an electron would have to overcome a distance of approximately 14 Å in <1 ps in a single step. The decay of the light-induced SCRP $[P_{700}^+A_1^-]$ and the associated ESEEM have been studied in thylakoid membranes, cellular membranes, and purified PS I prepared from the wild-type strains of *Synechocystis* sp. PCC 6803, *Chlamydomonas reinhardtii*, and *Spinaceae oleracea*. [55]. Complete reduction of the iron-sulfur electron acceptors F_A, F_B, and F_X and a partial photo-accumulation of the reduced quinone electron acceptor A_{1A} was observed in presence of a reductant at and $T < 220$ K. It was found that when all of the iron-sulfur centers are oxidized, only the echo modulation associated with the A branch $[P_{700}^+A_{1A}^-]$ radical pair is observed. In all of the systems investigated the time dependences of the out-of phase ESEEM can be interpreted in terms of different proportions of the signal associated with the $[P_{700}^+A_{1A}^-]$ and $[P_{700}^+A_{1B}^-]$ radical pairs, suggesting that bidirectionality of the electron transfer in PS I is a common feature of all species. An electron spin-polarized signal of the $P_{800}^+A_1(Q)^-$ state in the homodimeric reaction center core complex of *Heliobacterium modesticaldum* was detected in [56].

The first successful detection of quantum oscillations in photosynthetic RCs of plant PS I was achieved by combining transient EPR with pulsed laser excitation [57]. In this experiment the sample was irradiated with a short laser pulse and the time evolution of the transverse magnetization was monitored in the presence of a weak microwave magnetic field. The experiments were carried out using fully deuterated samples to improve signal-to-noise ratio required to reveal the quantum oscillations.

^{13}C MAS and photo-CIDNP ssNMR provided essential data for local electron spin densities of the ground state of a photoinduced radical pair formed in PS I particles (nuclear spins in natural abundance) in the dark state and under continuous illumination with white light [58]. The photo-CIDNP signals were assigned to a single Chl *a* molecule, which is probably the P2 cofactor of the primary donor P700. According to the photo-CIDNP data of PS I, all ^{13}C NMR signals are emissive (Fig. 11.10). These data were analyzed in the framework of two alternative mechanisms: (1) three-spin mixing (TSM) mechanism in which, the coupling between the two electron spins provides for enhanced polarization of the two constituent radical ions followed by a transfer of polarization to the nuclear spins by an anisotropic hyperfine coupling; and (2) the differential decay (DD) mechanism, when, due to different *g*-factors of the radicals and hyperfine couplings of the two electron spins, the polarization of the constituent radical ions can arise from evolution to a superposition of the singlet and triplet state of the pair and subsequent decay of the pairs in a triplet state. It was concluded that the TSM dominates over

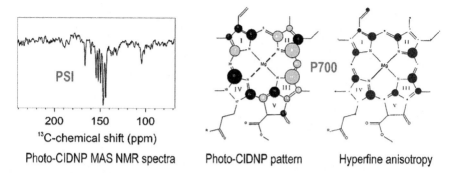

Fig. 11.10 Photo-CIDNP spectra and patterns of Chl a molecules observed in PS I. The size of the circles is semiquantitatively related to the signal intensity. All the observed photo-CIDNP enhanced NMR signals are negative (emissive) [58]

the DD mechanism and the former is responsible for the remarkable strength of the photo-CIDNP in PS I.

HF EPR/PELDOR/RIDME methodology that led to establishing three-dimensional structure of the transient radical pairs $P_{865}^{\bullet+}Q_A^{\bullet-}$ in frozen-solution RCs derived from Rb. Sphaeroides has been extended to the transient radical pair pairs $P_{700}^{\bullet+}A_1^{\bullet-}$ [59], where P_{700} is the primary electron donor and A_1 is phylloqui-none electron acceptor in the fully deuterated PS I from the cyanobacterium *Synechocystis* sp. PCC 6803 (Fig. 11.11). By combining and analyzing data of these three different magnetic resonance experiments, the interspin distance and

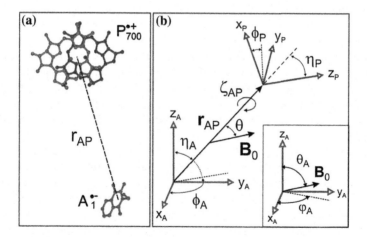

Fig. 11.11 a Arrangement of the cofactor radicals pairs $P_{700}^{\bullet+}$ and pairs $A_1^{\bullet-}$ within PS I. **b** Geometrical representation of the respective g-tensor frames of $A(x_A, y_A, z_A)$ and $P(x_P, y_P, z_P)$ in terms of the polar angles η_A, φ_A and η_P, φ_P and the dipolar vector $\mathbf{r}_{AP}$. The inset shows the g-value selection in the A-frame of reference by the external magnetic field [59]

relative orientation of the radical pair components have been established. Details of angular arrangement of these cofactor radicals within PS I and graphical solution for angles can been seen in Fig. 11.10.

It should be noted that this HF EPR/PELDOR/RIDME study established geometry of SCRP without the need of any a *priori* data on the pair structure (such as, for example x-ray structure) and only relied on approximate locations of the molecules forming the radical pair.

11.3.3   Reaction Centers of Photosystem II

PS II is the largest and the most complex photosynthetic machinery when compared with bacterial RC and PS I. PS II can be found in cyanobacteria and plants and consists of about 20 different protein subunits and 14 integrally bound lipids [3, 60–63]. Specifically, PS II includes six redox cofactors that are capable of trapping trap electrons (or holes) in minima of Gibbs energy These cofactors are the oxygen-evolving complex, the amino acid residue tyrosine (Tyr), the reaction centre chlorophyll, pheophytin the plastoquinone molecules, Q_A and Q_B All these cofactors except Q_B are bonded to a twisted pair of hydrophobic proteins known as D1 and D2. The D1 and D2 proteins form the scaffolding of the PS II complex.

A schematic view of the photosystem II (PSII) complex in the thylakoid membrane is shown in Fig. 11.11 [64].

The reaction centers of PS II of higher plants and cyanobacteria to a large degree resemble that of the purple photosynthetic bacteria (bRCs) [62]. Similar to bacterial RC and PS I, magnetic resonance techniques, and EPR in particular, were found to be especially useful in investigating structure and electron transfer processes in PS II. One example involves establishing the role Mn^{2+} ions are playing in photosynthesis and PS II. Investigation of effects of Mn^{2+} on saturation curves of EPR spectra of pheophytin anion radical and P_{680}^{+} from 15 to 200 K and photoinduced changes of the fluorescence yield of chlorophyll in samples with different contents of Mn led to a conclusion that PS II contains a cluster of four Mn atoms, two of which can be replaced by Mg^{2+} or any other divalent metal [66]. The distances between Mn and pheophytin as well as between Mn and P_{680} were estimated from these EPR data (ibid).

The other fruitful avenue of investigation of PSII by EPR includes studies of SCRPs [67]. For example, a light-induced spin-polarized triplet state has been detected in a purified PS II by EPR at liquid helium temperature. The electron spin polarization pattern was attributed to the triplet originating from radical pair recombination between the oxidized primary donor chlorophyll, P-680^{+} and the reduced intermediate pheophytin, I^{-}. A stable light-induced EPR signal was also reported in PS II particles and in chloroplasts at 5 K. It was concluded that the signal is formed when the reaction center is in the state $D^{+}P_{680}Ph^{-}$ (D^{+} is an oxidized donor to P_{680}). The signal has been simulated by assuming the presence of at least two distinct radical pairs that differ slightly in the distance separating the radicals of the pairs. The distance between the radicals of the pair was calculated to be 6–7 Å (ibid) (Fig. 11.12).

Other types of PSII reaction-center core, such as those isolated from pea and the green alga *Scenedesmus*, were examined by EPR [68]. These studies revealed two types of triplet spectrum in addition to the spin-polarized reaction-center triplet: (1) a triplet formed upon continuous illumination at 4.2 K and attributed to a monomeric phaeophytin molecule; and (2) a triplet that was stable in the dark at 4.2 K following an illumination. The latter was assigned to the radical pair Donor $+I^-$. This study provided evidence that an electron donor to chlorophyll P680 is present in the polypeptide D1-polypeptide D2-cytochrome b-559 core complex.

Spin-polarized transient EPR spectra of $^3P_{680}$ in $D_1D_2cytb_{559}$-complexes of PS II were studied as a function of temperature [69]. The spin polarization as well as the rise time of the EPR signals, which are characteristic of A triplet formation via recombination from the primary radical pair $P_{680}^{+\cdot}Pheo^{-\cdot}$, was characterized via the spin polarization and the rise time of the EPR signals. Below 100 K the EPR spectrum of $^3P_{680}$ revealed characteristic zero-field splittings (zfs) and was associated with the triplet state of monomeric chlorophyll-a (Chl). The data indicated a transition between the two triplet states of different electronic origin. A narrowing of the spectrum at high temperature was interpreted as thermally activated delocalization of the triplet excitation over at least two Chl units with different orientations with respect to the PS II reaction center.

Another example was an EPR study of the triplet state of chlorophyll generated by radical pair recombination in PS II reaction center [70]. The spin state of the non-haem Fe^{2+} was varied using the CN–binding method. Polarization transfer from the radical pair to Q_A-in PSII, where the Fe^{2+} was low spin, was detected. It was suggested that both the short triplet lifetime in the presence of Q_A^- and the lack of polarised Q_A-might be explained in terms of the electron transfer mechanism for triplet quenching involving the semiquinone. It was proposed that this mechanism may occur in PS II due the triplet-bearing chlorophyll being adjacent to the pheophytin at low temperature as suggested from structural studies.

Particular emphasis of several studies has been placed on the structure and location of the cation radical and the triplet state at the donor side of PS II and on consequences for the initial charge separation process. The triplet states of PS II core particles from spinach at different reduction states of the iron–quinone complex of the reaction center primary electron acceptor were studied by time-resolved CW EPR [71]. With doubly reduced primary acceptor, PS II triplet state was detected and characterized by zero-field splitting parameters $|D| = 0.0286$ cm^{-1}, $|E| = 0.0044$ cm^{-1}. When the primary acceptor was singly reduced either chemically or photochemically, a triplet state of a different spectral shape was observed, bearing the same D and E values and with characteristic spin polarization pattern arising from RC radical pair recombination. A sequence of electron-transfer events in the reaction centers was proposed that explains the dependence of the triplet state properties on the reduction state of the iron–quinone primary acceptor complex.

Photo-CIDNP MAS ssNMR is another useful technique that allowed for comparative analysis of spin effects and electronic structure of ^{15}N-labeled RCs from

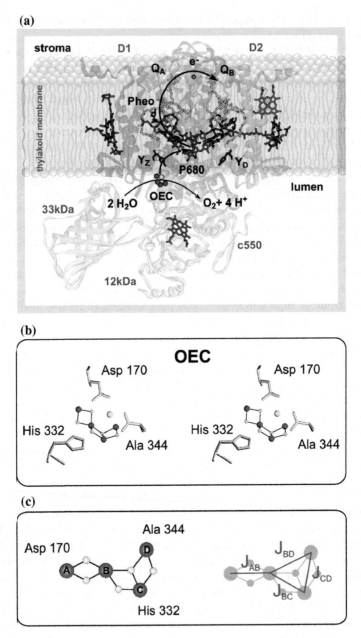

Fig. 11.12 *Panel a* shows a schematic view of the photosystem II (PSII) complex in the thylakoid membrane that is based on the 3.0 Å crystal structure of Loll et al. [65]. The cofactors of the D1, D2, cyt c550, and cyt b559 proteins are placed in color on *top* of the proteins. *Panel b* presents a stereoview of the Mn_4O_xCa cluster together with selected ligands. *Panel c* displays a schematic top view (approximately along the membrane normal) of the Mn_4O_5 core of the Mn_4O_xCa cluster [63]

spinach PS I and PS II [72]. Under illumination of PS I and PS II enhanced absorptive (positive) and emissive (negative) signals were observed. The two emissive as well as the enhanced absorptive set of signals were attributed to Chl cofactors based on the assignment obtained for Chls in solution. Results showed that the electron spin distribution in PS I, apart from its known delocalization over 2 chlorophyll molecules, reveals no marked disturbance, whereas the pattern of electron spin density distribution in PS II was inverted in the oxidized radical state. Assuming the signals arise from a Type-1 histidine having a deprotonated p-position, it was suggested that the donor would be a negatively charged [Chl-His]$^-$ complex in the ground state, and a neutral radical in the photo-oxidized state. Unifying those aspects, a hinge-type model for the donor of PS II was proposed.

11.4 Oxygen Evolving Complex

11.4.1 Natural Photosynthesis

The most essential for the formation of the biomass on earth is the biochemical process of splitting of two water molecules into molecular oxygen with the formation of energy reached P_{680} in the reduced state and release of four protons. In nature this energy-demanding process is carried out by the oxygen evolving complex (OEC), which is an oxo-bridged cluster of four manganese and one calcium ions (Mn_4O_5Ca), embedded in a protein matrix [73–75].

The photocycle involves for intermediate states: $S_0 \rightarrow S_1 \rightarrow S_2 \rightarrow S_3 \rightarrow S_4$ that are illustrated in Fig. 11.13 [64]. The oxidized cluster (Mn_4O_5Ca) is reduced by P_{680} and P_{680}^+, in turn, is reduced a redox-active tyrosine (Tyr[sup161]) to generate a neutral tyrosine radical which acts as an oxidant for water splitting.

Fig. 11.13 The Kok cycle includes a sequence of proton and electron transfer events at each S-state transition, along with the most probable oxidation states of the cluster. Probable oxidation states for the S_3 and S_4 states of the Kok cycle are currently being debated and have not been indicated here [64]

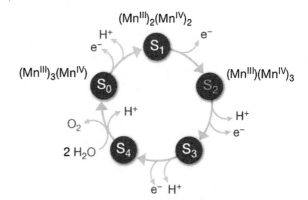

With wealth of experimental and theoretical work devoted to the OEC, here we restrict the discussion to the most recent data on specific spin effects in the unique (Mn_4O_5Ca) cluster.

Structural models of OEC have been examined theoretically with the emphasis on evaluating effects of different protonation states of ligated water molecules and the oxo bridges [74]. As illustrated by Fig. 11.14, deprotonation is the absolute requirement of the water splitting process. Therefore, protonation states of water ligands and oxo bridges are directly involved in tuning the electronic structures and oxidation potentials of OEC. The magnetic coupling in the S_2 state of the OEC, the total spin state of the cluster, the individual site spin values for each ion, and hyperfine couplings were calculated using the broken-symmetry DFT methodology (BS-DFT). Overall, four classes of models protonation patterns were considered (Fig. 11.15). The first class includes four water molecules, two coordinated to Mn_A and two to the Ca^{2+} cofactor. The second class investigated the possibility of protonation of the μ-oxo bridge, $O(5)$, between Mn_A and Mn_B.

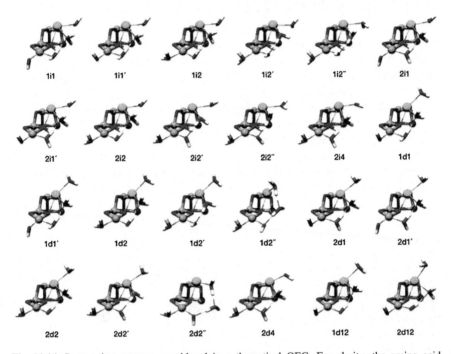

1i1   1i1′   1i2   1i2′   1i2″   2i1

2i1′   2i2   2i2′   2i2″   2i4   1d1

1d1′   1d2   1d2′   1d2″   2d1   2d1′

2d2   2d2′   2d2″   2d4   1d12   2d12

Fig. 11.14 Protonation patterns considered in a theoretical OEC. For clarity, the amino acid ligands are not shown. In labels "*min*" and "*mdn*" *m* indicates the parent model from which the model was constructed *n* indicates the specific water deprotonation site; and *i* or *d* indicates proton-shift isomers and deprotonated derivatives of the parent models, respectively [64]

Fig. 11.15 Model **1d2'**, one of the structurally, energetically, and spectroscopically consistent models for the S$_2$ state of the OEC. An interactive model of the molecule is available as a web-enhanced object in the HTML version [64]

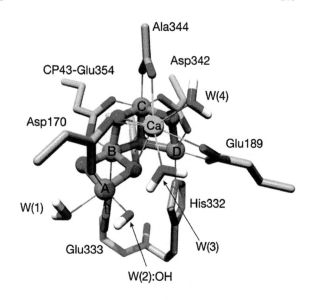

Comparison of these possible protonation models with available experimental data indicated that models **1d2'** and **1d2"** (Fig. 11.15) are the most consistent.

Structural and spin parameters of the PS II Mn$_4$O$_5$Ca cluster, specifically, valences and exchange coupling parameters of S$_0$ and S$_2$ states and preferred spin-coupling schemes for the S$_0$ and S$_2$ states, were established in the pulse [55]Mn-ENDOR and EPR experiments [75]. For the analysis of the EPR and ENDOR data of the Mn$_4$O$_x$Ca cluster, the spin-Hamiltonian of a system with n coupled Mn ions was suggested:

$$H = \sum_i \beta B_0 \cdot g_i \cdot S_i + \sum_i S_i \cdot a_i \cdot I_i + \sum_i S_i \cdot D_i \cdot S_i - \sum_{i<k} J_{ik}(S_i S_k) \quad (11.1)$$

where the indices i and k run from A to D (the four Mn ions), β is the Bohr magneton, g_i is the g-matrix for the electron spin of the i-th Mn ion, S_i and I_i are the operators of the electron spin and nuclear spin of the i-th Mn ion, respectively, a_i is the tensor of hyperfine interaction (HFI) for the i-th Mn ion, D_i is the zero-field splitting (ZFS) tensor for the electron spin of the i-th Mn ion, J_{ik} is the exchange between the i-th and k-th Mn ions, and · denotes matrix multiplication.

These data (summarized in Table 11.2 and Figs. 11.16 and 11.17) provided the basis for proposing the following molecular model for the S$_0$ → S$_1$ and S$_1$ → S$_2$ transitions. In the S$_0$ → S$_1$ transition, central Mn$_B$(III) is oxidized to Mn$_B$(IV) followed by deprotonation of the μ_2-hydroxo bridge between Mn$_A$ and Mn$_B$ and the contraction of this distance from 2.85 to 2.72 Å. In the S$_1$ → S$_2$ transition, Mn$_D$ is oxidized from Mn$_D$(III) to Mn$_D$(IV) and water molecule and either is not deprotonated or transfers the proton within an H-bridge to a nearby residue.

Table 11.2 Ranges for the strength of the exchange coupling constants of the individual Mn pairs in the Mn_4O_xCa cluster derived from the detailed search, i.e., after application of the T2(S_2) and T2(S_0) structure filters [75]

| | Mn oxidation state | | | | Exchange coupling constants (cm^{-1}) | | | |
	A	B	C	D	J_{AB}	J_{BC}	J_{CD}	J_{BD}
$S_2(A)$	III	IV	IV	IV	-180 ± 60	-60 ± 25	-40 ± 10	-5 ± 15
$S_2(C)$	IV	IV	III	IV	-130 ± 15	-45 ± 15	-32 ± 5	-10 ± 8
$S_0(A)$	IV	III	III	III	-25 ± 10	-50 ± 25	-60 ± 40	-10 ± 10
$S_0(B)$	III	IV	III	III	-80 ± 60	-40 ± 15	-40 ± 15	-10 ± 10
$S_0(C)$	III	III	IV	III	-35 ± 15	-40 ± 20	-69 ± 10	-5 ± 15
$S_0(D)$	III	III	III	IV	-25 ± 10^a	-50 ± 30	-40 ± 15	-5 ± 15

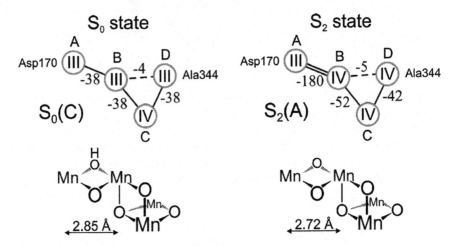

Fig. 11.16 Top: Preferred spin-coupling schemes for the S_0 and S_2 states of the Mn_4O_xCa cluster in PS II that specify the individual Mn oxidation states (*roman numbers*) and the exchange couplings, J_{ik}, (*arabic numbers*; in cm^{-1}) between the Mn ions. The relative coupling strength is represented by the type of connecting lines between Mn ions [75]

Fig. 11.17 Molecular interpretation of the $S_0 \rightarrow S_1$ and $S_1 \rightarrow S_2$ transitions. The oxygen atoms of the slow (W_s) and fast (W_f) exchanging substrate water molecules are indicated as *black dots*. W_s is identified here with the μ-OH/μ-O bridge between Mn_A and Mn_B (See details in [75])

Simulation of the HF EPR (W-band, 94-GHz) S_2 multiline signal of single-crystal PS II from cyanobacterium *Thermosynechococcus elongates* yielded the following principal components of the *g*-matrix: 1.997, 1.970, and 1.965 [76]. Evidence for delocalization of the triplet state 3P680 in the D1D2cytb559-complex of photosystem II was provided by detecting spin polarized transient ESR spectra of 3P680 in D1D2cytb559-complexes of Photosystem II [77]. The spin polarization and the rise time of the ESR signals indicated triplet formation via recombination from the primary radical pair $P^{+\cdot}$680Pheo dark-stable radical state Y$\sqrt{D}$ of the redox-active tyrosine.

Establishing tyrosine as an electron carrier between chlorophyll aII and the water-oxidizing manganese complex was an important step in understanding the mechanism of water splitting (reviewed in [78]). This state was attributed to a neutral radical of Tyr-160, which is an intermediate acceptor for the Mn-custer oxidation. The tyrosyl radical and plastoquinone, an anion radical, Q_{AP}^-, were shown to have different vibrational lines in the difference FTIR spectrum, a positive 1478 cm^{-1} line and positive (1482 and 1469 cm^{-1}) for the former and latter species, respectively [79]. The flash-induced Fourier transform IR (FTIR) difference spectrum of the oxygen-evolving Mn cluster upon S_1-to-S_2 transition indicated that in PS II from *Synechocystis* 6803 a tyrosine residue specifically labeled with ^{13}C at the ring-4 position, is coupled to the Mn cluster, and the vibrational modes of this tyrosine are affected upon S_2 formation [80]. It was suggested that Mn cluster and the tyrosine are linked via chemical and/or hydrogen bonds and the structural changes of the Mn cluster are transmitted to the tyrosine through these bonds. HF EPR at 94 GHz (W-band) was used to study the dark-stable tyrosine radical $\Psi\Delta$ in single crystals of PS II isolated from the thermophilic cyanobacterium *Synechococcus elongates* [81]. Magnitude and orientation of the g-matrix of YD$^\cdot$ and related information on several proton hyperfine tensors was deduced from analysis of angular-dependent EPR spectra.

Further details on evaluation of structural models for the photosynthetic water-oxidizing complex derived from spectroscopic and x-ray diffraction signatures can be found in several publication [37–39, 82–96].

As a recent example, using quantum mechanics/molecular mechanics (QM/MM) calculations structures with each cation in both the resting state (S_1) and in a series of reduced states (S_0, S_{-1}, and S_{-2}) was optimized (Fig. 11.18) [92]. It was determined that the X-ray crystal structures with either Ca^{2+} or Sr^{2+} are most consistent with the S_{-2} state (i.e., Mn$_4$[III, III, III, II] with O4 and O5 protonated).

11.4.2 Water Splitting in Artificial Clusters

Artificial water splitting in manganese clusters has been studying extensively and recently reviewed [93–97]. Overall, over 100 structures of tetra-Mn-oxo complexes can be found in the CCDC database containing either cube-like $M_4(OX)_4$ cores (cubans), unsymmetrical cubelike cores, or the planar butterfly cores $[Mn_4O_2]^{n+}$.

Photosystem II

Oxygen-Evolving
Complex

(with Ca^{2+} or Sr^{2+})

X-ray
reduction
⟶

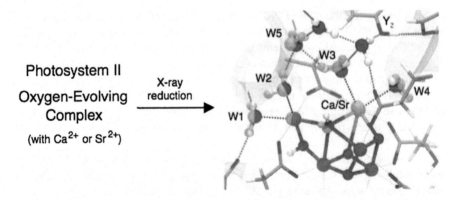

Fig. 11.18 QM/MM-optimized structure of the Ca^{2+}-OEC in the S$_{-2}$ state with both O4 and O5 protonated. Atoms in *light red* show atoms whose positions are different when Sr^{2+} is the heterocation. Bonds to OEC ligands (from subunit D1 unless otherwise noted) and terminal waters (W1–W4) are shown as *gray* and *black dashed lines*, respectively. Hydrogen bonds are shown as *red dashed lines*. For clarity, His332, CP43-Arg357, and other waters are not shown [92]

In comprehensive review [94], data on catalytic water oxidation by synthetic manganese-oxo complexes, particularly those with a cubical Mn$_4$O$_4$ modeling the PSII-WOC were summarized and discussed. On the base of the EPR, ENDOR, NMR, ESI-MS, Fourier-transform infrared (FTIR), electrochemical and X-ray absorption near edge structure (XANES) experiments the cubanes Mn oxidation states were established. Cubanes [Mn$_4$O$_4$]$^{6+}$ were found to be strong oxidizing agents [94]. By suspending the oxidized cubane, [Mn$_4$O$_4$L$_6$]$^{+}$, into a proton-conducting membrane (Nafion) pre-adsorbed onto a conducting electrode and electroxidizing the photoreduced butterfly complexes by the application of an external bias, catalytic evolution of O$_2$ and protons from water exceeding 1000 turnovers was shown. achieved. Replacing the external electric bias with redox coupling to a photoanode incorporating a Ru(bipyridyl) dye allowed to use sunlight as the only source of energy. These results appears to be a remarkable achievement.

Suggested general scheme of the O$_2$ evolution during WOS on manganese cluster is presented in Fig. 11.19. Important role of the Proton-Coupled Electron Transfer (PCET) was stressed.

The rates of water oxidation/O$_2$ evolution catalyzed by six cobalt-oxo clusters including the Co$_4$O$_4$ cubanes, Co$_4$O$_4$(OAc)$_4$(py)$_4$ and [Co$_4$O$_4$(OAc)$_2$(bpy)$_4$]$^{2+}$, using the common Ru(bpy)$_3$$^{2+}$/S$_2O_8$$^{2-}$ photo-oxidant assay were compared.[97] Kinetic results reveal a deprotonation step occurs on this pathway and that two electrons are removed before O$_2$ evolution occurs. The Co$_4$O$_4$ cubane core is shown to be the smallest catalytic unit for the intramolecular water oxidation pathway below 1.5 V (Fig. 11.20). The authors concluded that the origin of catalytic activity by Co$_4$O$_4$ cubanes illustrates three key features for water oxidation (1) four one-electron redox metals, (2) efficient charge delocalization of the first oxidn. step across the Co$_4$O$_4$ cluster, allowing for stabilization of higher oxidizing equiv., and (3) terminal coordination site for substrate aquo/oxo formation.

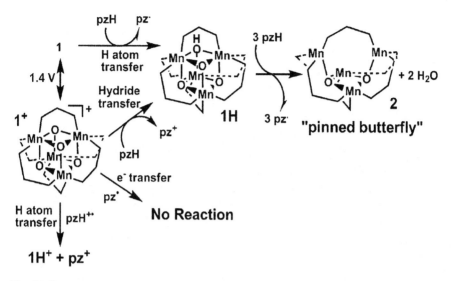

Fig. 11.19 Proton-Coupled Electron Transfer (PCET) reactions of the cubane model complexes $Mn_4O_4L_6$, 1a and 1a$^+$ (L$^-$ = (C$_6$H$_5$)$_2$PO$_2$$^-$) [94]

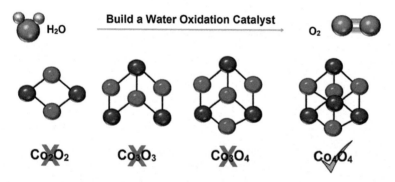

Fig. 11.20 A scheme illustrating that only Co_4O_4 cubane core is the smallest catalytic unit for the intramolecular water oxidation pathway [97]

In conclusion, light energy conversion in the photosynthetic reaction centers is characterized by the high energetic efficiency and the quantum yield close to 100 %. Such remarkable efficiency was achieved by the fulfillment of several principal conditions, which also should be realized in artificial systems of high efficiency (reviewed in [29, 30])

1. Direct electron transfer (ET) should be sufficiently fast to yield formation of the pair D$^+$A$^-$ *and to prevent the fast pair recombination (CR)*. This condition requires the donor and acceptor groups to be disposed at an optimum distance of about 7–8 Å relative to each other.

2. The process should be thermodynamically favorable along the electron transfer pathway: the electron-transfer driving force (ΔG_0) and redox potentials of the donor and acceptor groups in each stage should be optimal.
3. A molecular mobility of the immediate surroundings of the donor and acceptor groups (most probably in the nanosecond time scale) should be provided.
4. Charge-separated pair D^+A^-, while keeping its strong chemical reactivity, should be isolated from any side reactions.

For water splitting into O_2 and H_2 by four-electro oxidation mechanism to be effective, the manganese or other transition metal clusters of high redox potential should be used. This potential should be collected by four step by step one electron transfer and protons release and the process accomplish by one step four electron oxidation of deprotonated two water molecules in a corresponding cluster.

References

1. J.F. Allen, E. Gantt, J. Golbeck, B. Osmond (eds.), *Photosynthesis. Energy from the Sun* (Springer, Dordrecht, 2008)
2. A.F. Collings, C. Critchley, *Artificial Photosynthesis: From Basic Biology to Industrial Application* (Wiley-VCH, Weinheim, 2005)
3. R.E. Blankenship, *Molecular Mechanisms of Photosynthesis* (Blackwell, Oxford, 2001)
4. A. Marchanka, A. Savitsky, W. Lubitz, K. Mo"bius, M. van Gastel, J. Phys. Chem. B **114**, 14364 (2010)
5. Shusuke Katagiri, Yasuhiro Kobori, App. Magn. Reson. **37**, 177 (2010)
6. O.G. Poluektov, L.M. Utschig, M.C. Thurnauer, G. Kothe, App. Magn. Reson. **31**, 123–143 (2007)
7. Y.E. Kandrashkin, A. van der Est, App. Magn Reson. **31**, 105–122 (2007)
8. K.M. Salikhov, Y.N. Pushkar, J.H. Golbeck, D. Stehlik, App. Magn. Reson. **24**, 467 (2003)
9. A. Schweiger, G. Jeschke, *Principles of Pulse Electron Paramagnetic Resonance* (Oxford University Press, 2001)
10. W. Lubitz, *EPR in Photosynthesis*, eds. by B.C. Gilbert, M.J. Davies. *Electron Paramagnetic Resonance. Specialist Periodical Reports* (Royal Society of Chemistry, Cambridge, UK, 2004), vol. 19, p. 174
11. A.J. Hoff, Q. Rev. Biophys. **17**, 153 (1984)
12. G. Zwanenburg, P.J. Hore, Chem. Phys. Lett. **203**, 65 (1993)
13. E. Daviso, A. Alia, S. Prakash, A. Diller, P. Gast, J. Lugtenburg, J. Matysik, G. Jeschke, Phys. Chem. C **113**, 10269–10278 (2009)
14. J.J. Warren, N. Herrera, M.G. Hill, J.R. Winkler, H.B Gray, J. Am. Chem. Soc. **135**, 11151 (2013)
15. A.T. Taguchi, P.J. O'Malley, C.A. Wraight, S.A. Dikanov, J. Phys. Chem. B **118**, 9225 (2014)
16. S.K. Misra, K. Mobius, A. Savitsky, *Multifrequency Electron Paramagnetic Resonance*, ed. by S.K. Misra (2011), p. 875
17. M.R. Hartings, I.V. Kurnikov, A.R. Dunn, J.R. Winkler, H.B. Gray, M.A. Ratner, Coord. Chem. Rev. **254**, 248 (2010)
18. H.B. Gray, J.R. Winkler, Electron flow through metalloproteins. Biochim. Biophys. Acta Bioenergetics **1797**, 1563–1572 (2010)
19. C.A. Whited, W. Belliston-Bittner, A.R. Dunn, J.R. Winkler, H.B. Gray, J. Inorg. Biochem. **103**, 906–991 (2009)

20. D. Alvarez-Paggi, W. Meister, U.W.I. Kuhlmann, K. Tenger, L. Zimanyi, G. Rakhely, P. Hildebrandt, D.H. Murgida, J. Phys. Chem. B **117**,6061 (2013)
21. Stuchebrukhov A.A.R.A. Marcus, J. Phys. Chem. **99**, 7581 (1995)
22. D.V. Matyushov, J. Phys. Chem. **139**, 025102/1-025102/12 (2013)
23. A.N. Volkov, N.A.J. Nuland, PLoS Comput. Biol. **8**, 1002807 (2012)
24. P.C.P. de Andrade, Probability current in protein electron transfer: Lowdin population analysis. Int. J. Quant. Chem. **112**, 3325 (2012)
25. G.I. Likhtenshtein, L.A. Syrtsova, V. DSamuilov, E.N. Frolov, A.U. Borisov, Bogatyrenko, V. R. Photosynthetic reaction center, *XII International Botanic Congress: Theses of Reports* (Nauka, Leningrad, 1975), p. 429
26. G.I. Likhtenshtein, A.I. Kotel'nikov, A.V. Kulikov, L.A. Syrtsova, V.R. Bogatyrenko, A.I. Mel'nikov, E.N. Frolov, A.J. Berg, Int. J. Quant. Chem. **16**, 419 (1979)
27. G.I. Likhtenshtein, A.V. Kulikov, A.I. Kotelnikov, V.R. Bogatyrenko, Photobiochem. Photobiol. **3**, 178 (1982)
28. G.I. Likhtenshtein, Role of orbital and dynamic factors in electron transfer in reaction centres of photosynthetic systems. J. Photochem Photobiol A: Chem **96**, 79–92 (1996)
29. G.I. Likhtenshtein *New Trends In Enzyme Catalysis and Mimicking Chemical Reactions* (Kluwer Academic/Plenum Publishers, Dordrecht, 2003)
30. G.I. Likhtenshtein, *Solar Energy Conversion. Chemical Aspects* (WILEY-VCH, Weinhem, 2012)
31. G.I. Likhtenshtein, *Chemical Physics of Redox Metalloenzymes* (Springer-Verlag, Heidelberg, 1988)
32. R. Huber, EMBO J. **8**, 2125 (1989)
33. G.I. Likhtenshtein, A.E. Shilov, Zhurnal Fiz. Khem. (Russian J. Phys. Chem.) **44**, 849 (1970)
34. N.N. Semenov, A.E. Shilov, G.I. Likhtenshtein, Dokl. Akad. Nauk SSSR **221**, 1374 (1975)
35. B. Kok, B. Forbush, M.M. McGloin, Photochem. Photobiol. **11**, 457 (1970)
36. H. Dau, M. Haumann, Coord. Chem. Rev. **252**, 273 (2008)
37. J. Yano, J. Kern, V.K. Yachandra, H. Nilsson, S. Koroidov, J. Messinger, Chapter 2, Section 2 *Geometric and Electronic Structure of the* Mn_4CaO_5 *Cluster*, eds. by P.M.H. Kroneck, M.E. Sosa Torres. Sustaining Life on Planet Earth: Metalloenzymes Mastering Dioxygen and Other Chewy Gases. Metal Ions in Life Sciences (Springer, 2015), vol. 15, pp. 13–43
38. K.M. Davis, Y.N. Pushkar, J. Phys. Chem. B **119**, 3492 (2015)
39. F. Rappaport, M. Guergova-Kuras, P.J. Nixon, B.A. Diner, J. Lavergne, Biochemistry **41**, 8518 (2002)
40. C. Lancaster, D. Roy, *Comprehensive Series in Photochemical & Photobiological Sciences*, vol. 9 (Primary Processes of Photosynthesis, Part 2) (2008), p. 5
41. P.K. Fyfe, J.A. Potter, J. Cheng J.C. M. Williams, A.J. Watson, M.R., Jones, Biochemistry **46**, 10461 (2007)
42. E. Daviso, S. Prakash, A. Alia, P Gast, J. Neugebauer, G. Jeschke, Matysik, Proc. Natl. Acad. Sci. USA **106**, 22281 (2009)
43. R.A. Goldstein, S.G. Boxer, Biophys. J. **51**, 937 (1987)
44. J.S. van den Brink, T.E.P. Hermolle, P. Gast, P.J. Hore, A.J. Hoff, **100**, 2430 (1996)
45. S. Prakash, A. Alia, P. Gast, de Groot, H. J.M. Matysik, G. Jeschke, J. Am. Chem. Soc. **128**, 12794 (2006)
46. K.M. Salikhov, C.H. Bock, D. Stehlik, Appl. Magn. Reson. **1**, 195 (1990)
47. S.A. Dzuba, M.K. Bosch, A.J. Hoff, Chem. Phys. Lett. **248**, 427 (1996)
48. G. Kothe, M.C. Thurnauer, Photosynth. Res. **10**, 349 (2009)
49. G. Jeschke, J. Chem. Phys. **106**, 10072 (1997)
50. A. Savitsky, A.A. Dubinskii, M. Flores, W. Lubitz, K. Möbius, J. Phys. Chem. B **111**, 6245 (2007)
51. M. Haumann, P. Liebisch, C. Muller, M. Barra, M. Grabolle, H. Dau, Science **310**, 1019 (2005)
52. A. Amunts, H. Toporik, A. Borovikova, N. Nelson, J. Biol. Chem. **285**, 3478 (2010)
53. M.G. Müller, J. Niklas, W. Lubitz, A.R. Holzwarth, Biophys. J. **85**, 3899 (2003)

54. A.R. Holzwarth, M.G. Muller, J. Niklas, W. Lubitz, Biophys. J. **90**, 552 (2006)
55. S. Santabarbara, I. Kuprov, P.J. Hore, A. Casal, P. Heathcote, M.C. Evans, Biochemistry **45**, 7389 (2006)
56. Ryo Miyamoto, Hiroyuki Mino, Toru Kondo, Shigeru Itoh, Hirozo Oh-oka, Biochemistry **47**, 4386 (2008)
57. G. Kothe, S. Weber, R. Bittl, E. Ohmes, M.C. Thurnauer, J.R. Norris, Chem. Phys. Lett. **186**, 474–480 (1991)
58. A. Alia, E. Roy, P. Gast, H.J. van Gorkom, H.J.M. de Groot, G. Jeschke, J. Matysik, J. Am. Chem. Soc. **126**, 12819 (2004)
59. A. Savitsky, J. Niklas, J.H. Golbeck, K. Moebius, W. Lubitz, J. Phys. Chem. B (Ahead of Print)
60. T. Cardona, A. Sedoud, N. Cox, A.W. Rutherford, Biochim. Biophys. Acta **1817**, 26 (2012)
61. C.D. van der Weij-de Wit, J.P. Dekker, R. van Grondelle, I.H.M. van Stokkum, J. Phys. Chem. **115**, 3947–3956 (2011)
62. S.P. Årsköld, V.M. Masters, B.J. Prince, P.J. Smith, R.J. Pace, E. Krausz, J. Amer. Chem. Soc. **125**, 13063 (2003)
63. L.V. Kulik, B. Epel, W. Lubitz, J. Messinger, J. Am. Chem. Soc. **129**, 13421 (2007)
64. W. Ames, D.A. Pantazis, V. Krewald, N. Cox, J. Messinger, W. Lubitz, F. Neese, J. Am. Chem. Soc. **133**, 19743 (2011)
65. B. Loll, J. Kern, W. Saenger, A. Zouni, and. J. Biesiadka Nature **438**, 1040 (2005)
66. A.V. Kulikov, V.R. Bogatyrenko, G.I. Likhtenstein, S.I Allakhverdiev, V.V. Klimov, V.A. Shuvalov, A.A. Krasnovski, Biofizika **28**, 357–363 (1983)
67. A.W. Rutherford, M.C. Thurnauer, Proc. Natl. Acad. Sci. USA **79**, 7283 (1982)
68. C. Demetriou, J.H. Lockett, Biochem. J. **252**, 921 (1988)
69. A. Kamlowski, L. Frankemoeller, A. van der Est, D. Stehlik, A.R. Holzwarth, Berichte der Bunsen-Gesellschaft **100**, 2045 (1996)
70. Y. Deligiannakis, A.W. Rutherford, II. Biochim. Biophys. Acta **1365**, 354 (1998)
71. W.O. Feikema, P. Gast, I.B. Kleninab, I.I. Proskuryakov, Biochim. Biophys. Acta **1709**, 105 (2005)
72. A. Diller, E. Roy, P. Gast, H.J. van Gorkom, H.J.M. de Groot, C. Glaubitz, G. Jeschke, J. Matysik, A. Alia, Proc. Natl. Acad. Sci. **104**, 12767 (2007)
73. Y. Umena, K. Kawakami, Jian-Ren Shen Nobuo Kamiya, Crystal structure of oxygen-evolving photosystem II at a resolution of 1.9 Å. Nature **473**, 55–60 (2011)
74. K.N. Ferreira, T.M. Iverson, K. Maghlaoui, J. Barber, S. Iwata, Science **303**, 1831 (2004)
75. L.V. Kulik, B. Epel, W. Lubitz, J. Messinger, J. Am. Chem. Soc. **129**, 13421 (2007)
76. C. Teutloff, S. Kessen, J. Kern, A. Zouni, R. Bittl, FEBS Lett. **580**, 3605 (2006)
77. A. Kamlowski, L. Frankemoeller, A. van der Est, D. Stehlik, A.R. Holzwarth, Berichte der Bunsen-Gesellschaft **100**, 2045 (1996)
78. S. Gerken, K. Brettel, H.T. Witt, FEBS Lett. **237**, 69 (1988)
79. S. Kim, C. A Sacksteder, K.A. Bixby, B.A. Barry, Biochemistry **40**, 15384 (2001)
80. T. Noguchi, Y. Inoue, X.-S. Tang, Biochemistry **38**, 399 (1999)
81. W. Hofbauer, A. Zouni, R. Bittl, J. Kern, P. Orth, F. Lendzian, P. Fromme, H.T. Witt, W. Lubitz, Proc. Natl. Acad. Sci. USA **98**, 6623–6628 (2001)
82. M.M. Najafpour, A.N. Moghaddam, Y.N. Yang, E.-M. Aro, R. Carpentier, J.J. Eaton-Rye, C. H. Lee, S.I. Allakhverdiev, Photosynth. Res. **114**, 1–13 (2012)
83. G.C. Dismukes, R.E. Blankenship, *The Origin and Evolution of Photosynthetic Oxygen Production*, eds. by T. Wydrzynski, K. Satoh. Photosystem II. The Light-Driven Water: Plastoquinone Oxidoreductase. Advances in Photosynthesis and Respiration Springer (Dordrecht, 2005), vol. 22, p. 683
84. C. Teutloff, S. Pudollek, S. Kessen, M. Broser, A. Zouni, R. Bittl, Phys. Chem. Chem. Phys. **11**(31), 6715 (2009)
85. S. Gerken, K. Brettel, H.T. Witt, FEBS Lett. **237**, 69 (1988)
86. S. Kim, C.A. Sacksteder, K.A. Bixby, B.A. Barry, Biochemistry **40**, 15384 (2001)

87. J.E. Bartlett, S.V Baranov, G.M. Ananyev, G.C Dismukes, Philos. Trans. Royal Soc. B: Biol. Sci. **363**, 1494, 1253 (2008)
88. K.M. Davis, Y. Pushkar, J. Phys. Chem. B (Ahead of Print) (2015)
89. S. Mitsuo, I. Hiroshi, Y. Shusuke, S. Michihiro, A. Fusamichi, S. Jian-Ren, Y. Kizashi, Chem. Phys. Lett. (Ahead of Print) (2015)
90. A.G. Gabdulkhakov, V.G. Kljashtorny, M.V. Dontsova, Crystallography Reports (Ahead of Print) (2015)
91. V. Krewald, M. Retegan, N. Cox, J. Messinger, W. Lubitz. S. DeBeer, F. Neese, D.A. Pantazis. Chem. Sci. (Ahead of Print) (2015)
92. L. Vogt, M.Z. Ertem, P. Pal, Brudvig, V.S Batista, Biochemistry **54**, 820 (2015)
93. J.P. McEvoy, G.W. Brudvig, Chem. Rev. **106**, 4455 (2006)
94. G.C. Dismukes, R. Brimblecombe, G.A.N. Felton, R.S. Pryadun, J.E. Sheats, L. Spiccia, G.F. Swiegers, Acc. Chem. Res. **42**, 1935 (2009)
95. M.M. Najafpour, M.Z. Ghobadi, B. Haghighi, J.J. Eaton-Rye, T. Tomo, J.-R. Shen, S.I. Allakhverdiev Biochem. (Moscow) **79**, 324 (2014)
96. S.M. Lang, I. Fleischer, T.M. Bernhardt, R.N. Barnett, U. Landman, Nano Lett. **13**, 5549 (2013)
97. P.F. Smith, C. Kaplan, J.E. Sheats, D.M. Robinson, N.S. McCool, N. Mezle, G.G. Dismukes, Inorg. Chem. **53**, 2113–2121

Chapter 12
Spin Labeling Methods

Abstract This chapter is a brief review of the main results obtained by the methods of spin and triplet (phosphorescence) labels. Methods of physical labelling using unpaired electron spins such as nitroxide and other stable radicals, transition and lanthanide metal complexes, and phosphoresce chromophores in the triplet state have proven to be powerful tool for the study of structure and dynamics of a body of complex molecular objects, biological system in particular. Structure, molecular breathing, conformational transitions in labeled biological polymers and membranes in solid and liquid states have been investigated using the arsenal of traditional and recently developed physical methods. Interest to spin labeling is fueled by the many promising applications. Measurements of electric potential, redox status, oxygen concentration and acidity in various materials including those of biological relevance are amenable for the physical labeling techniques as well.

12.1 Introduction

A number of problems concerning the structure and dynamics of complex molecular systems including example from biology can be solved by a modification of a chosen portion of the object of interest by a physical label such as a stable radical, paramagnetic complex, luminophore, or Moessbauer atom followed by monitoring the label's properties by appropriate physical methods. The nitroxide spin labelling technique was pioneered by Ogawa and McConnell group [1]. Since then application of spin labeling in chemistry, biology and physics has received increasing attention. At present the SciFinder program offers 4100 references with the keywords "**nitroxide spin label and spin probe**".

Four types of spin-spin interactions provide opportunities for measuring spin-spin distances and the dynamics of the labelled system: (1) the dipole-dipole interaction arises from the magnetic dipole field of one paramagnetic center inducing a local magnetic field at the site of another paramagnetic group; (2) the exchange interaction is caused by overlap of the orbitals occupied by unpaired

© Springer International Publishing Switzerland 2016   289
G. Likhtenshtein, *Electron Spin Interactions in Chemistry and Biology*,
Biological and Medical Physics, Biomedical Engineering,
DOI 10.1007/978-3-319-33927-6_12

electrons as the particles approach each other (3) the electron-nuclear dipole (hyperfine) interaction which can affect the powder pattern and spectra arising from rotational motion, and (4) the contact hyperfine (spin-electron spin nucleus) interaction, which arises from a finite electron spin density at the nucleus. The investigation of spin electron effects in labelled molecular objects by traditional and modern physical methods, first of all for example electron spin resonance, can provide a wealth of information for establishing the system's structural and dynamic properties.

The theoretical and experimental basis of these approaches was described in Chaps. 1–3, 5–7.

12.2   Proteins

12.2.1   Distance Measurements. Double Spin Labeling

In light of the importance of investigation of molecular objects structure, the spin labeling technology has been using. The method of double spin labeling (DSL), invented pioneered by Likhtenshtein in 1968 [2] is based on specific modification of chosen groups in the object of interest by two or several spin labels, nitroxides or complexes of paramagnetic metal, followed by analysis of the effects of the spin-spin interactions on the label ESR spectra or spin relaxation times. Such an analysis allows one to measure distances between the spin labels and to characterize the distance distribution.

The first applications of dipole-dipole spin-spin interactions to the investigation of protein surface topography were based on changes in the lineshape of nitroxide radical ESR spectra arising from interactions with a second nitroxide radical or a paramagnetic ion [2–5]. This approach allows one to estimate distances between the paramagnetic centers up to 2.5 nm. Later, the higher sensitivity of power saturation curves of a radical to interactions between the radical and paramagnetic ions maximum in specific cases up to 15 nm was demonstrated [6–10]. In parallel, the effects of the spin exchange interaction on the ESR spectra and spin relaxation were studied and were used for establishment the structure of systems under investigation, such as nitrogenase and non-heme protein [11, 12]. Since then, double labeling methods and their modification have being widely employed in structural investigations in chemistry and biology [12–35] (and references therein).

In the past decades, the energetic development of advanced ESR pulse methods, as powerful tools for obtaining structural information in the nanometer range, has also driven progress in methods of spin labeling (Chap. 6). Discovery of site-directed spin labeling (SDSL) of proteins markedly expanded the spin labeling potential [16].

12.2.2 Nitroxide Labels

Site directed spin labeling (SDSL) and advanced pulse ESR methods opened a new era in applications of the double spin labeling technique. SDSL is the substitution of a selected amino acid with cystine via the site directed mutagenesis technique following by chemical modification with a sulfhydryl reactive nitroxide radical, for example, (1-Oxy-2,2,5,5-tetramethyl-3-pyrroline-3 methyl) methanethiosulfonate (MTSL) or 2,2,6,6-tetramethylpiperidine-1-oxyl-4-amino-4-carboxylic acid) [16].

As an example, in one study eight T4 lysozyme (T4L) mutants, 61/135, 65/135, 135/80, 61/86, 65/86, 61/80, 65/80, 65/76 were prepared and doubly labeled with methanethiosulfonate spin-label (MTSSL) (Fig. 12.1) [17].

The distances between the labels obtained by DQC-ESR at 9 and 17 GHz were found to range from 20 Å for the 65/76 mutant to 47 Å for the 61/135 mutant. The high quality of the dipolar spectra allowed the determination of the distance distributions and the shape of these distributions caused by the presence of multiple conformations of the spin-label.

For distance estimation between labeled chosen groups it is important to know the label conformations and motion. The two spin labeled 72R1 and 72R2 mutants of T4 lysozyme, which bear the spin labels derivatives of MTSSL (Fig. 12.1) at a solvent-exposed helix site were examined by X-band ESR spectroscopy [18]. A model of slow reorientations of the whole protein superimposed on fast chain conformational jumps and fluctuations in the minima of the chain torsional potential was used for the simulations employing the Stochastic Liouville equation (SLE) formalism It was shown that spin labels yield spectra typical of slightly mobile nitroxides related to conformers, whose mobility is limited to torsional fluctuations (Formula 12.1).

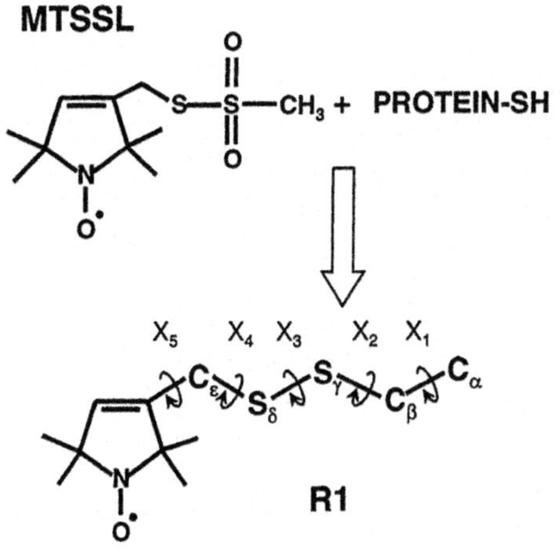

Fig. 12.1 Protein labeling with MTSSL spin-label used in this work. After labeling, the nitroxide moiety is tethered to the backbone via a side chain referred to as R1. Note that the C_α and C_β carbons are on the cysteine residue with the C_α position in the backbone. The dihedral angles X1−X5 are also shown [17]

Formula 12.1

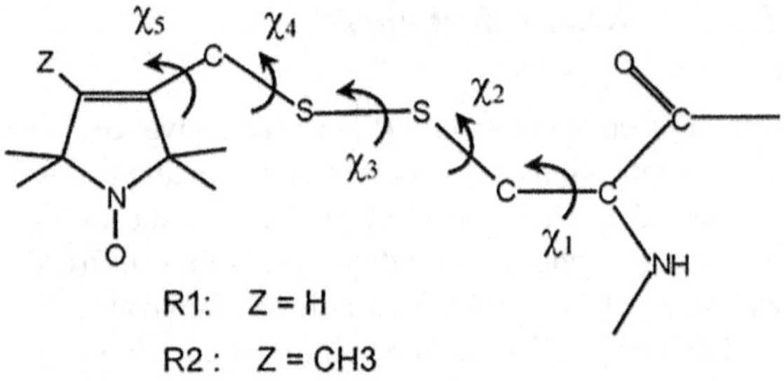

R1: Z = H

R2 : Z = CH3

Formula 12.2

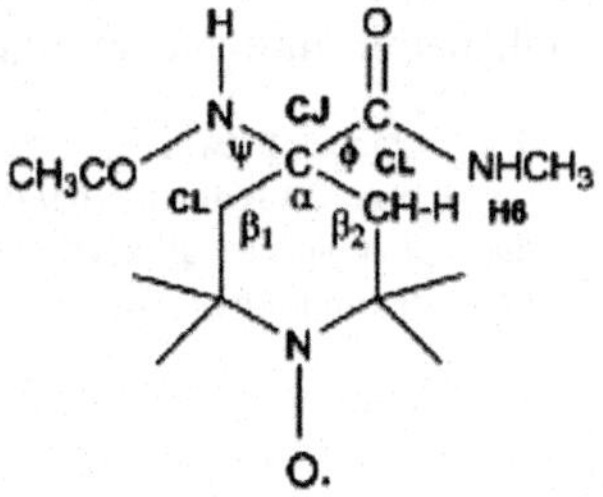

The conformationally constrained nitroxide amino acid TOAC first described in [19] possesses advantages over commonly used flexible spin labels and has attracted much interest in recent years is widely used. Results of measurement of distance between two TOAC nitroxide spin labels on peptides and other objects by continuous-wave-EPR spectroscopy were reviewed [20] (Formula 12.2).

Method for measuring nanometer-scale distances between specific sites in proteins based on the analysis of CW ESR saturation curves was designed and developed in [6–9]. Successful application of the method involves one slower-relaxing paramagnetic center with T_1 times that are long enough for the ESR line to be observed and for the CW ESR saturation curve to be obtained, and a second center characterized by short relaxation times, that is, $T_{1f}\ \mu_{ef}/r^3 \ll 1$ (Sects. 3.3.2, 3.30–3.33). The method was shown to be effective in solution structural problems, in particular for long distance estimates. Application of the method for measuring nanometer-scale distances between specific sites was first illustrated by the estimation of distances between the heme group of human hemoglobin and the nitroxide fragment of three spin labels attached covalently to the β-93 SH-group [7].

In thirteen single-cysteine variants of spin labeled myoglobin (Fig. 12.2), iron-nitroxyl interspin distances in the range of 17–30 Å were detected by saturation recovery measurements (SRM) of the enhancement of the nitroxyl spin lattice relaxation rates between ~30–140 K, and by spin-echo measurements (SEM) of the enhancement of spin-spin relaxation rates at 10–30 K [21]. The values of distances obtained by SRM and SEM were found to be in reasonable agreement with values calculated with the insight II program. The further development of the method relaxation-enhancement method based on—measurement of distance between spin labels was discussed in [22–24].

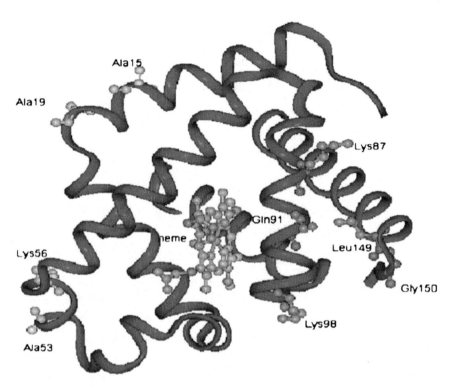

Fig. 12.2 Ribbon diagram of sperm-whale myoglobin, based on Brookhaven Protein Database pdb1mbw file, showing nine native side chains that were replaced by cysteine mutations for attachment of spin labels [21]

The majority of experiments on pulsed dipolar ESR spectroscopy, DEER and DQC, are performed in frozen samples. In work [25], the effects of the freezing rate and concentration of the conformation of biomacromolecules and/or spin-label was elucidated on doubly-spin-labeled T4 lysozyme (Fig. 12.3). Analysis of experimental results in glycerol, as a cryoprotectant, at concentrations of 10 and 30 % led to conclusions that freezing affects mainly the ensemble of spin-label rotamers, but the distance maxima remain essentially unchanged. Advantages of the use of the new five-pulse DEER sequence over the standard four-pulse DEER was demonstrated on T4 lysozyme MTSL spin labeled at position 8, 44, 65, 128, 135 [26]. DEER-5 was recorded on a 12 μs time scale, which allowed the determination of distances up to 8 nm using matrix deuteration [26].

In excellent work [27], nitroxide spin-labels were inserted at 37 different positions in T4 lysozyme for a total of 51 different distance pairs for DEER experiments (Fig. 12.4). In the analysis of these experiments the rotamer population of the attached spin-labels was established using a novel multiple-copy restrained-ensemble (MCRE) simulation method. The MCRE simulations were found to match

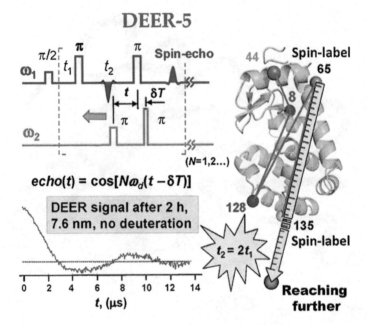

Fig. 12.3 Scheme illustrating the DEER-5 experiment and position of spin labels on T4-lysozyme [26]

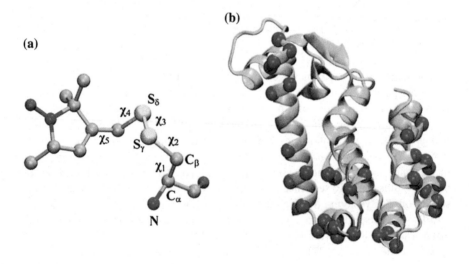

Fig. 12.4 Position of spin labeled attached to T4 lysozyme [27]

Fig. 12.5 View of the cytoplasmic face of inactive rhodopsin (PDB entry 1GZM) showing modeled R1 side chains for all sites studied here. The *dotted lines* connect pairs where the interspin distance distribution was determined by using DEER. Each distribution was measured before and after light-activation. The R1 side chains are modeled from known conformations [28]

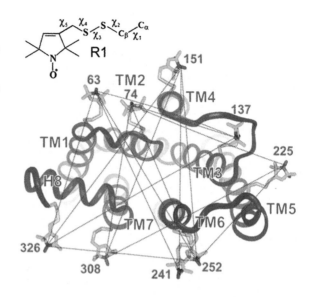

the experimental distance histograms obtained from ESR/DEER measurements of 51 distance distributions for 37 nitroxide spin-labeled pairs and were are in good agreement with available distance determinations from X-ray crystallographic structures.

16 pairs of the nitroxide side chain R1 (Fig. 12.5) were introduced into rhodopsin and interspin distances of the pairs located on the solvent exposed surface were measured in the protein inactive (**R**) and light-activated state (**R***), using DEER [28].

PELDOR (cf. Chap. 6) was used to measure a distance between two 2,2,6,6-tetramethylpiperidine-1-oxyl-4-amino-4-carboxylic acid spin labels incorporated at positions 1 and 16 in a linear peptide antibiotic Alamethicin comprised of 19 aminoacids (Fig. 12.6) in hydrated egg-yolk phosphatidylcholine (ePC) vesicles at 77 K [29].

Question of a great interest is measurement between transition metal center and nitroxide spin label. In this direction two methods, pulsed electron–electron double resonance (PELDOR) and relaxation-induced dipolar modulation enhancement (RIDME) were compared on the heme-containing and spin-labeled cytochrome P450cam [14]. The five-pulse RIDME experiment has been shown to be more suitable than PELDOR for the distance measurements between the Fe^{3+} ion and nitroxide spin label, that is, (1) the RIDME experiment allows obtaining the whole dipolar spectrum (Pake doublet) and requires shorter measurement time as compared to the PELDOR time trace.

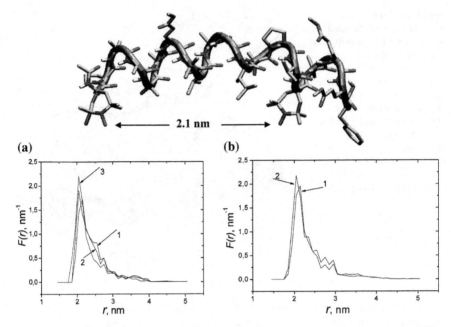

Fig. 12.6 Distance distribution functions F(r) between the two spin labels for frozen Alm[1,16] at 77 K. In the *upper part* of the figure, a model of Alm[1,16], built from the crystal structure of Alm[16] (molecule A), (27) is shown (see details in [29])

12.2.3   Miscellaneous Labels

Studies using copper complexes were the first reported double spin label experiments which provided the means to measure spin-spin distances up to 2 nm [2]. The reliability of measuring copper nitroxide separations by the DEER time domain method was demonstrated on a rigid double labeled nitroxide-copper complex [30].

A new type of spin labeling for nanometer-range distance measurements in proteins involving the use of Gd^{3+} complexes was introduced in [31, 32]. Two proteins, p75ICD and $\tau_C 14$, were labeled at selected cysteine thiol groups with either two nitroxides or two Gd^{3+} spin labels, followed by spin-spin distance measurements using a pulse X-band (9.5 GHz, 0.35 T) spectrometer for the nitroxide-labeled proteins and a W-band (95 GHz, 3.5 T) spectrometer for the Gd^{3+}-labeled proteins. The "zero" dead-time four-pulse DEER sequence and two-pulse sequence techniques provided the means to measure nanometer-scale distances. The X-band measurements gave a nitroxide−nitroxide distance distribution with a maximum at 2.5 nm, and the W-band measurements gave a Gd^{3+}−Gd^{3+} distance distribution with a maximum at 3.4 nm. Figures 12.7 and 12.8 illustrates W-band DEER results of doubly nitroxide-labeled p75ICD and the positions of the two Gd^{3+} labels separated by distance 29 nm.

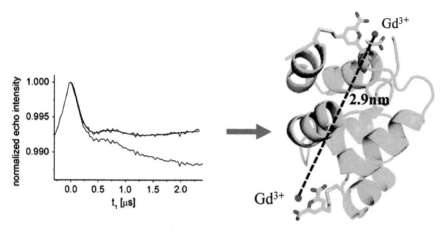

Fig. 12.7 W-band DEER results (40 K) for doubly nitroxide-labeled p75ICD [32]

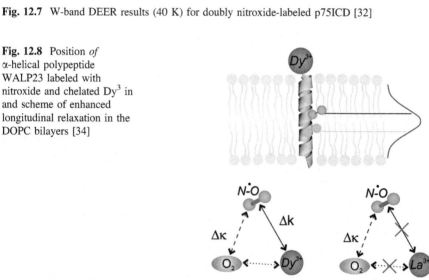

Fig. 12.8 Position *of* α-helical polypeptide WALP23 labeled with nitroxide and chelated Dy^3 in and scheme of enhanced longitudinal relaxation in the DOPC bilayers [34]

The other fruitful avenues of spin labeling structural investigation was using of nitroxides in combination with lanthanide ions [33, 34]. In a system of three different types of paramagnetic species incorporated in DOPC bilayers, α-helical polypeptide WALP23 labeled with nitroxide, chelated (DOTA) Dy^{3+} ions on the bilayers surface, and membrane-dissolved O_2 (Fig. 12.7), the longitudinal relaxation of nitroxide spin-labels was measured [34]. In some experiments paramagnetic Dy^{3+} ions were replaced with diamagnetic La^{3+}. Measurements of the longitudinal relaxation of nitroxide radicals were performed with the inversion recovery pulse sequence at the X band at temperatures between 20 and 100 K.

A combination of paramagnetic labels, consisting of nitroxide radical and the terpyridine complex of Gd^{3+} was reported [36]. In the model double labeled

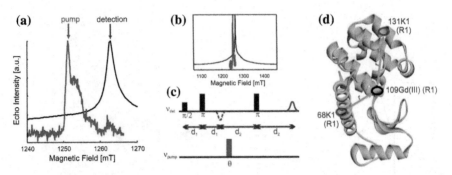

Fig. 12.9 Selective measurements of field swept ED EPR spectra, pulse sequence, and investigated mutants in Gd(III)–nitroxide DEER measurements in T4-lysozyme. **a** Selected measurements of ED EPR spectra of [Gd(DOTA)] and nitroxide radicals. *Arrows* mark positions of pump and detection pulses. **b** Corresponding ED EPR spectra in a broader field range. **c** Sequence of microwave pulses used in the DEER experiment to measure Gd(III)–nitroxide distances. **d** Ribbon model of wild-type T4L (PDB ID code 1L63) where the spin-labeled sites have been marked with ellipses. See details in [33]

molecule, the spin-spin distance 2.43 nm was measured by the DEER time domain and ED EPR techniques. The new method has been successfully applied to distance measurement in proteins and membranes. As an example in two double mutants of bacteriophage T4-lysozyme with a genetically encoded unnatural amino acid *p*-acetyl-L-phenylalanine (*p*-AcPhe) either at site 68 or at site 131 and a cysteine at position 109, spin-spin distances up to 5.0 nm were measured by DEER [33]. Selective measurements of field swept ED EPR spectra, the pulse sequence, and the mutants investigated by Gd(III)–nitroxide DEER measurements in T4-lysozyme are presented in Fig. 12.9.

The further development of the relaxation-enhancement method based on— measurement of distance between nitroxide and lanthanide labels with an upper measurable distance limit of about 5–6 nm was tested on a water-soluble protein (T4-lysozyme) [35].

To overcome the common requirement for pulsed ESR techniques to be performed at cryogenic temperatures to reduce the rapid electron spin relaxation rate and to prevent averaging of electron-electron dipolar interaction due to the rapid molecular tumbling, T4-lysozyme, labeled at positions 65 and 80 or 65 and 76 with a triarylmethyl (TAM)-based label, was immobilized on Sepharose [37]. Under these conditions, the relaxation time T_m of the attached protein was 0.7 μs. The TAM spin label is characterized by high resolution ESR. Distances between the TAM labels detected using double quantum coherence (DQC) showed a narrow distance distribution and were close to those expected from the protein models.

Despite the encouraging achievements on the experimental measurement of interspin distance in model relatively simple proteins, the application of spin labeling to important complicated objects of unknown structure remains to be a challenging task.

12.2.4   Protein Dynamics

The principles underlying the application of the nitroxide spin label method, as a tool for experimental investigation of protein molecular dynamics ("breating"), were formulated in the late 1960s by Likhtenstein [38]. In parallel, complementary research on protein dynamics were performed by a proposed Moessbauer atom labeling [39]. Transglobular conformational change in myoglobin and lysozyme as a consequence of a protein molecular dynamics was first demonstrated in [40].

A suggested approach for the investigation of protein dynamics relied on the well-known finding that the rotation of a molecule in a condensed phase is modulated to a great extent by the molecular dynamics of the surrounding molecules [13]. This phenomenon is caused by relatively tight packing of molecules of liquids and amorphous solids on the one hand and the existence of static and dynamic defects in these systems on the other hand. Therefore, the particulars of nitroxide motion report on characteristics of the dynamics of the surrounding medium. An important feature of nitroxides is their structural anisotropy which causes anisotropy of magnetic interactions, such as the hyperfine splitting and the g-tensor. In the immobilized state of radicals, ESR ("powder") spectra arise from contributions of individual spectra (spin packets) whose features depend on the molecular orientation relative of the applied magnetic field. Radical motion leads to averaging of the hyperfine interaction and the g-tensor which causes significant changes to the ESR spectrum.

The correlation time for isotropic and anisotropic rotation τ_c for a nitroxide in a homogeneous medium over the range of fast ($\tau_c = 10^{-9} \cdot 10^{-10}$ s) and slow ($\tau_c = 10^{-7} \cdot 10^{-8}$ s) rotational diffusion, can be determined via the theoretical framework of ESR spectra line shapes developed by Kivelsonn [41], Freed [42, 43]. The detailed behavior of the line shape in the fast motional limit gives information about spin spin (transverse) (T_{2s}) relaxation of radicals. The theory of ESR spectra in the slow motion region has been a long-standing research thrust of the Freed group [44].

The study of very slow motion ($\tau_c = 10^{-3}$–10^{-6}) is based on the fact that the electron spin lattice (longitudinal) relaxation time (T_{1s}) is typically longer than contributions arising from multiple relaxation pathways in, for example, nitroxide ESR spectra. Saturation recovery and analysis of spectra of the second harmonic of the dispersive component of the ESR absorption (V_2') are commonly used in the saturation transfer techniques [45–47]. The parameters of the saturation transfer depend on the electron spin lattice relaxation time (T_{1s}) and the nuclear spin lattice relaxation time (T_{1n}). Rotational motion competes with relaxation processes which 'quench' the relaxation pathway. A saturating 'pump' field is an example of pathway quenching. For instance, at ambient temperature and in nonpolar solvent, the competition can be take place only if the rate of motion is comparable to the electron spin-lattice relaxation rate ($1/\tau_c = 1.5$–3×10^5 s^{-1}). At low temperatures measurements of slower motions can be studied up to $1/\tau_c = 10^3$ s^{-1}.

In pulse ELDOR radical motion can be directly investigated [46]. The rotation rate can also be directly monitored by methods in which spin echo techniques are used in combination with magnetization transfer. A number of CW and pulse ESR techniques based on saturation by MW radiation of high power have been discussed in Chaps. 5 and 6. For systems which require detailed knowledge of mechanism and time scale of motion, practical application of these methods require developing an appropriate model with subsequent extensive spectral simulations.

The general features of molecular dynamics of proteins modified with nitroxide, fluorescence, phosphorescence, Moessbauer and radical pairs labels were described in [13–15, 48]. and references therein. The first direct experimental evidence of nanosecond intramolecular mobility of proteins at ambient temperatures was obtained using spin and Mössbauer labels and probes [13, 38, 39, 48, 49] and was confirmed in further publications [50–52]. For example, a hydrophobic aromatic derivative of a nitroxide radical was embedded in the human serum albumin binding site and the mobility of the spin probe was traced by ESR spectroscopy [49]. The apparent reciprocal correlation time ($1/\tau_c = 10^8\,\mathrm{s}^{-1}$) interpreted as a characteristic frequency of the immobilized probe was found to be independent of viscosity and, therefore, was attributed to the local mobility of the label. It was concluded from the apparent energy ($E_{app} = 2.5$ kcal/mole) and entropy ($\Delta S^{\#}app = 15.0$ e.u.) activation that in proteins the probe mobility follows the mobility of the flexible walls of the protein binding site with a similar frequency. Similar effects were observed in spin labeled cross-linked polymers. The nano second dynamics of the HSA binding site around the dansyl moiety of the dual fluorophore-nitroxide probe was monitored indirectly by the temperature dependent relaxation shift $\Delta\lambda^{fl}max(T)50^{55}$ and directly using picosecond fluorescent time-resolved technique [51].

The resolution of ESR spectrum and therefore efficiency of its application for study molecular dynamics was found to be markedly improved by enhancing the g-resolution of the spectrum by going to higher field. The practical demonstration of high field ESR advantages in molecular biological objects using 2-mm ESR spectroscopy was first demonstrated by Krinichnyi et al. [52]. Resolution may also be improved if higher frequencies are not available by using isotopically substituted spin labels, e.g., nitroxides enriched with ^{15}N and ^{2}H [53].

To investigate a protein molecular dynamics, thirty single Cys substitution mutants of phage T4 lysozyme were prepared and spin-labeled with a SH group-specific nitroxide reagent and mobility of the label nitroxide fragment were detected [54]. Experiments indicated that at interhelical loop sites, the side-chains had a high mobility, consistent with high crystallography thermal factors. The authors suggested that mobility the side-chains is determined by backbone dynamics. Experimental data on motion of spin labels 72R1 (methanethiosulfonate, MTSSL) and a methylated analog with hindered internal motion 72R2 located on solvent-exposed helix site of T4 lysozyme were analyzed using the Stochastic Liouville equation methodology [55]. X-band ESR spectraf R2 spin label were typical of slightly mobile nitroxides indicating that their mobility is limited to torsional fluctuations (damped oscillations), while the label R2 spectra showed presence of constrained and mobile chain conformers.

More detailed information about motion of nitroxide side chains *(R1) or (R2)* attached to solvent-exposed helical sites, 72 or 131 of T4 lysozyme was obtained using multifrequency ESR spectroscopy (9, 95, 170, and 240 GHz) in temperature range of 0.2°–32° [56]. It was shown that ESR at lower frequencies is more sensitive to the slower motions, whereas ESR at higher frequencies is more sensitive to the faster motions. The global tumbling of the protein and the internal motion consisting of backbone fluctuations and side chain isomerizations were considered in terms of the slowly relaxing local structure (SRLS) model.

Data on motions in proteins detected by changes in the average distance and/or the shape and width of the distribution using DEER (Fig. 12.10) have been reported in [57]. The following dynamic effects were described: movement of the K^+ Channel Gate and helices in photoactivated rhodopsin, distortion of BtuB (the outer membrane *E. coli* vitamin B_{12} (CNCbl) transporter) at osmolytes and crystal contacts, conformational equilibria Na^+-coupled leucine transporter, and large amplitude conformational changes in ATP-binding cassette transporters.

To kinetically trap the high-pressure conformational ensemble for subsequent DEER data collection over the range of 0–3 kbar a method was developed for rapidly freezing spin-labeled proteins under pressure [58]. Application of high

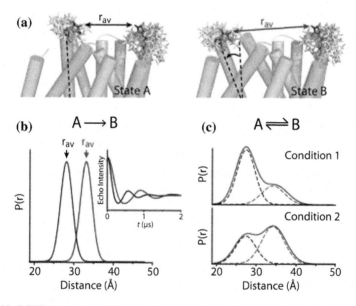

Fig. 12.10 DEER Detection of Triggered Conformational Changes. **a** Hypothetical motion of a transmembrane helix (*orange*) during the transition from state A to state B alters the average distance (r_{av}, *arrows*) between spin labels. **b** If states A and B are distinct conformers of different energies, the spin echo decay (*inset*). **c** A and B represent two conformations present in equilibrium [57]

hydrostatic pressure increases the population of excited states for study of the multiplicity of states in the ensemble and microsecond to millisecond exchange between them. The data on apomyoglobin directly revealed the amplitude of helical fluctuations.

12.3 Nucleic Acids

The general principles of nitroxide spin labeling for solving problems of structure and dynamics of DNA and RNA are similar to those for synthetic polymers and proteins. Four trends, relevant for both DNA and RNA research, may be pointed out: (1) new methods of nuclear acid labeling, primarily site directed technology, (2) distance measurements between a pair or pairs of sites (3) study the conformations and dynamics of the local environment of sites in the vicinity of a label position; and (4) conformational transitions arising from interactions with other species. The observed EPR spectra of attached spin labels can be influenced by three modes of motion: (i) tumbling of the entire molecule; (ii) torsional oscillations about bonds that connect the nitroxide moiety to the macromolecule;(iii) local macromolecular structural fluctuations at the labeling site and (iv) bending of the polymer chains.

Excellent experimental measurements of structure and dynamics of nuclear acids were carried out [59, 60]. Work on side directed chemical modification of DNA with nitroxide derivatives and ESR examination of structure and dynamics of labeled biopolymers have paved the way for further detailed investigations in this extremely important area of molecular biology [59–63]. Commonly, nucleic acid SDSL studies employ nitroxides covalently attached to a base, the 2-position of sugar, or phosphate of specific nucleotides of a target strand. Recently, studies involving nitroxides inserted noncovalently into specific sites of DNA duplexes has also been reported [60]. Solid-phase chemical synthesis [61] and enzymatic approaches [62] have been explored for site-specific incorporation of modified nucleotides. The side directed spin labeling technique (SDSL) pioneered by the Hubble group [16] for studying membrane proteins has proved to be a powerful tool for studying structures and dynamics of nucleic acids as well.

Site-Directed labeling in conjunction with ESR spectroscopy was utilized for studying the backbone dynamics in solution along the GCN4-58 bZip sequence of a protein both in the free state and bound to DNA [63]. Experiments revealed a mobility gradient, indicating that side chain motions reflect backbone motions. It was shown that the backbone motions are damped in the DNA-bound state, although a gradient of motion persists with residues at the DNA-binding site being the most highly ordered.

Nitroxide spin labels were attached to the 2'-aminocytidine sugar in the lower stem of the mini TAR DNA, in the upper loop, and near the bulge of the macromolecule(Fig. 12.11) and their tumbling were detected by high-frequency 236.6 GHz EPR [58]. Labels attached to the 2'-aminocytidine sugar in the mini c

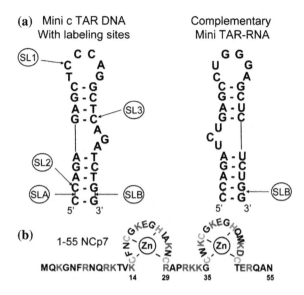

Fig. 12.11 a Secondary structure of mini c TAR DNA and mini TAR RNA and the position of spin-labels used in this study primarily of mini c TAR DNA. b Primary structure of 1–55 NCp7. The internal dynamics of mini c TAR DNA probed by Electron Paramagnetic Resonance of nitroxide spin-labels at the lower stem, the loop, and the bulge [58]

TAR DNA underwent fast, anisotropic, hindered local rotational motion of the spin probe, which was faster in the lower stem. More flexible labels attached to phosphorothioates at the end of the lower stem tumbled isotropically in mini c TAR DNA. Nanosecond motions of large segments of the oligonucleotide detected by 9.5 GHz EPR was enhanced by increasing the ratio of the nucleocapsid protein NCp7 to mini c TAR DNA.

A twisting motion of Tet repressor (TetR) and the human immunodeficiency virus type 1 reverse transcriptase (RT) on interaction with nucleic acid substrates or inhibitors in solution were revealed by measurement of distances between interacting nitroxides at positions 22/22' or 47/47' using X-band ESR [64]. In the core of the protein, spin-label side chains located near the tc-binding pocket or at position 202 indicated different conformations for the tc- and DNA-complexed repressor. The interspin distances obtained were found to agree with the relevant crystal data of the so-called open and closed conformations. PELDOR techniques were applied for studying the distance distributions between two spin labels attached to single-stranded DNAs and double-stranded DNAs of 12 nucleotides in experiments performed in frozen at 77 K glassy water/glycerol [65]. It was found that that duplexes possess a substantially narrower distribution, as compared to the single-stranded DNAs. Isertion of the nonnucleotide 2-hydroxymethyl-3-hydroxy-tetrahydrofuran led to a slight decrease of the mean distance between two spin labels, while insertion of the nonnucleotide insert bis(diethylene glycol)-phosphate results in a remarkable broadening of the distance distribution.

This pulse technique was also employed for measurement interspin distances up to 5.6 nm in DNA duplexes with spin probes inserted on the strand [(4:19) and (4:20)] [66]. It was shown that a conformational transition induced by a high concentration of trifluoroethanol in deuterated aqueous buffer is accompanied by a

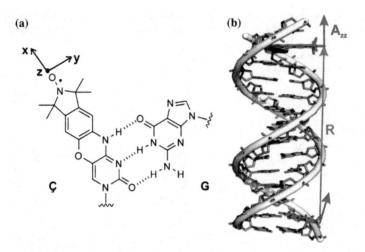

Fig. 12.12 **a** Rigid spin label Ç, a cytidine analogue, base-paired with guanine. The magnetic axis system of the nitroxide hyperfine tensor A and the G-tensor are coplanar to the molecular axis system (x, y, z). The paramagnetic center (unpaired spin) is located at the center of the N–O bond. **b** The duplex DNA(1, 12) containing the two spin labels Ç shown in *red*. The *numbers* in parentheses indicate labeled base pair positions. The interconnecting distance vector R and relative orientations of nitroxides can be determined by the PELDOR experiments [60]

0.8 nm distance change between labels attached to the duplex position (4:20). Incorporation of two rigid nitroxide spin labels, attached to a series of 20 base pair DNA duplexes through three hydrogen bonds (Fig. 12.12) followed by orientation-selective PELDOR (X-band (9 GHz/0.3 T) and G-band (180 GHz/6.4 T) experiments made it possible to establish the conformational flexibility of helical DNA [60, 68]. On the basis of spin label distance measurements in the range of 2–4 nm a dynamic model for ds-DNA molecules where stretching of the molecule leads to a slightly reduced radius of the helix induced by a cooperative twist–stretch coupling was suggested. The twist–stretch coupling is negative and accounts for an angular variation of $\pm 22°$ between the N–O axis of nitroxide spin labels positioned one helical turn above one another.

The interspin distance, label orientations and conformational dynamics of DNA duplexes and a DNA/protein complex noncovalently labeled were established by the PELDOR technique [67]. In the DNA duplexes, the distance distribution from the orientation-averaged PELDOR time trace was found to have a mean distance of 69.2 Å, while after complexation with the Lac repressor protein this distance diminished to 64.6 Å. This finding was attributed to deformation of the DNA.

Divalent metal (Mg^{2+}) ion-dependent folding of the hammerhead ribozyme from *Schistosoma mansoni* was monitored with pulse DEER by measuring nanometer-scale distances between paramagnetic spin-labels attached to 2′-amino-substituted $U_{1.6}$ and $C_{L2.1}$ positions in loops I and II of the extended HHRz, of RNA (Fig. 12.13) [68]. DEER measurements detected ribozyme folding detected by

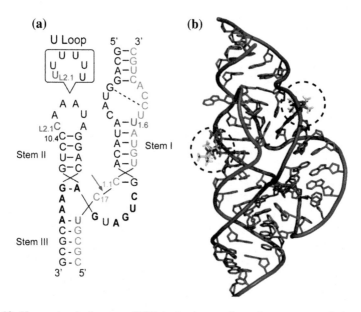

Fig. 12.13 Hammerhead ribozyme (HHRz) structure. **a** Secondary structure of the extended HHRz from *S. mansoni*. **b** Model of spin-labeled extended HHRz created from the crystal structure [68]

changes in spin–spin distance populations with the addition of Mg^{2+} ions, showing a strong, metal-dependent increase in the distance distribution at 2–3 nm.

Positions U4–U14, U4–U15, U14–U26, and U15–U26 of the 27 nucleotide of neomycin-responsive riboswitch were modified with spin labels and spin-spin distances were measured by PELDOR [69]. These distances were unchanged upon neomycin binding. which implies that the global stem-loop architecture is preserved in the absence and presence of the ligand. To overcome the problem of wide distribution of interspin distances and increase the precision of the distance measurements nitroxide spin labels in model DNA duplexes, 5′ were replaced with Triarylmethyl (trityl, TAM) based spin labels. The replacement allowed narrowing of the distance distributions by a factor of 3.72.

12.4 Biological Membranes

Biomembranes are formed from lipid molecules arranged, as a rule in bilayers. Molecular dynamics is a fundamental process directly related to stability and functional activity of biomembranes. Four main types of molecular motion of lipids modified by spin labels, commonly a nitroxide, are studied: rotational motion, lipid chain flexibility, lateral diffusion and jumps from one layer to another ("flip-flop")

[70–72]. Key examples of spin labeling in investigation of biomembrane structure and mobility are presented in works [73–75].

The phase diagram of 1,2-dipalmitoyl-*sn*-glycerophosphatidylcholin labeled (labeled at the end-chain 16-PC)−cholesterol binary mixtures versus temperature was studied using 2D-ELDOR at Ku band with the "full Sc−method" [73]. The experiments were performed at temperatures 25, 35, 48, and 57 °C, with six samples covering the range of 0–50 % cholesterol. showed that As a result of the experiments, phase diagram consists of liquid-ordered (L_0), liquid-disordered (L_d), and gel phases, and that there are two phase coexistence regions was revealed. Dynamic molecular structures of the membrane phases were and phase boundaries were determined by an analysis of the line shape changes and the homogeneous T_2's, extracted from the pure absorption spectra in the 2D-plus-mixing-time representation The dynamic behavior of spin-labels with nitroxides located in different positions on the aliphatic chains was first investigated with high-frequency (2-mm) ESR spectroscopy [74]. On the basis of temperature dependence of the spin probe spectra over the range 120–260 K (Fig. 12.14), it was concluded that in the temperature range 220–260 K the nitroxide rotation is essentially anisotropic with correlation time $\tau_c = 10^{-7}$–10^8 s and about 10^{-9} s at physiological temperatures.

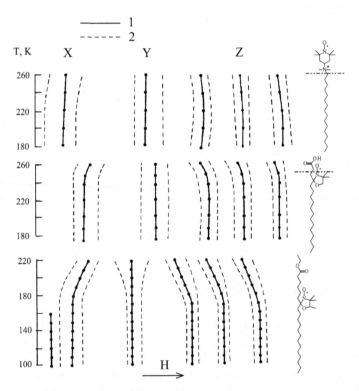

Fig. 12.14 Temperature dependences of the position (*1*) and line width (*2*) of the 2-mm ESR spectra of nitroxide spin probes (*right*) in lecitin lyposomes [74]

The first studies of lateral diffusion in membranes were performed by McConnell and McFarland [75]. A small drop of spin labeled lipid was inserted in oriented multilayer films and the diffusion process was monitored by transformation of the ESR spectra from exchange narrowed singlet to triplet from magnetically diluted probes. The coefficient of translational diffusion was found to be 10^{-8} cm^2 s^{-1}. In another work [76], detailed information on the translational diffusion of doxyl stearic spin probes bearing ^{15}N and ^{14}N in different parts of the lipid molecules was obtained by ELDOR [76]. The main result of that work was to determine the membrane dynamic profile, that is, the sequence of weakening of the exchange interaction between ^{15}N and ^{14}N fragments located at the following position of the doxyl stearic probes: 12:12 > 16:12 $\approx$ 12:16 $\gg$ 12:5 $\gg$ polar head.

Phosphocholine spin labels on the lipid headgroup and different positions on the acyl chain were used to detect perturbation of the HIV gp41 fusion peptide (FP) (Fig. 12.15), on lipid bilayers containing different cholesterol concentrations [77]. The following finding were reported: (1) the experiment showed a cooperative effect on the lipid order versus the gp41 FP/lipid ratio; (2) gp41 FP induces membrane ordering in all lipid compositions studied; (3) in high-cholesterol-containing lipid bilayers, where gp41 FP is in the β-aggregation conformation, its effect on the lipid ordering reaches deeper into the bilayer.

Data on rate constants for physical label encounters in model lipid membranes obtained by the cascade method.

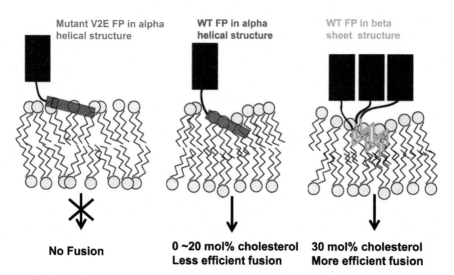

Fig. 12.15 Schematic representation of the modes of HIV gp41 FP insertion into lipid bilayers, and its effects on membrane order and fusion activities. The FP in β-sheet conformation is represented as a trimer simply because gp41 is a trimer [77]

12.5 Miscellaneous

12.5.1 Spin Label-Spin Probe Method (SLSPM)

Three versions of the SLSPM have been proposed and implemented by Likhtenshtein group: (1) Immobilized nitroxide and paramagnetic transition complex **(SLSPM 1) freely defused in solution,** (2) paramagnetic complex on a macromolecule and nitroxide in solution **(SLSPM 2)**, and (3) embedded radical and paramagnetic complex randomly distributed in a glassy solution **(SLSPM 3)**. The first variant **of SLSPM 1** proposed by Likhtenshtein et al. [78] and developed by the Hyde group [79] was based on the dynamic spin-spin exchange interaction of a stable radical, generally a nitroxide, attached to a molecular object of interest with a freely diffusing spin probe, which is chemically inert paramagnetic species capable of diffusing freely in solution. Because for a particular pair of paramagnetic species, the value of the dynamic exchange rate constant k_{ex} depends on microviscosity, steric hindrances, and the distribution of electrostatic charges, the method was intensively employed for investigating the microstructure of objects and systems of interest, proteins in particular, in the vicinity of the spin label. The theory of dynamic spin exchange is described in Sect. 1.2.3. A specific case of **(SLSPM 2)** is provided by an immobilized paramagnetic center and a freely diffusing nitroxide probe **(SLSPM 2)** as described in [7] and Sect. 12.5.3. Measurement of depth of immersion of spin labels nitroxide fragments in biomembranes was subject of investigation in the frame of **SLSPM 3** [7] and referenced therein.

After pioneering publications in [78, 79], principles of **SLSPM 1** were utilized in recent works [80–90]. In work [80], the accessibility parameter values of effect of fast-relaxing spin probe Ni(II) ethylenediaminediacetic acid (NiEDDA) obtained by the long pulse saturation recovery (EPR at W-band (94 GHz)) were found to be 0.30 and 0.28 μs^{-1} for the nitroxide spin labeled cortical membranes (PCD) 0.36 and 0.34 μs^{-1} for the CBDs of nuclear membranes, respectively. The saturation recovery was used also for monitoring effect of NiEDDA on spin-lattice relaxation rates of nitroxide spin labels along the 128–135 sequence in T4 lysozyme [81]. Data on ESR spectra and on gradient of accessibility of nitroxide segments of spin labels attached to the fourth transmembrane domain (TMDIV) to paramagnetic spin probes NiEDDA and O_2 were reported [82]. The experiments showed that the accessibilities to NiEDDA are out of phase with accessibilities to O_2, indicating that one face of the helix is immersed in the lipid bilayer while the other is solvated by an aqueous environment. The EPR spectral line shapes of the spin labels indicated considerable motional restriction.

12.5.2 Spin Oximetry

Molecular oxygen plays the role of a spin probe in the frame of the SLSPM 1. The electron spin transverse relaxation time of a nitroxide spin is short and, therefore, the radical ESR spectra are not sensitive to O_2 in ambient condition and spin oximetry can be realized only by direct or non-direct measurement of relatively longer longitudinal parameters [83–91]. Nevertheless, stable aromatic radicals (derivatives of triarylmethyl radical, Indian ink), phtalocyanine, and micrococrystals (fusinite and carbohydrate chars) show narrow singlet ESR lines sensitive to O_2 concentration in water and biological systems such as cells, organs, small animals and even parts of human body [88].

In pioneering work of James Hide group, a method for measuring the oxygen diffusion-concentration product was invented [83]. The method is based on the dependence of the spin-lattice relaxation time T_1 of the spin label on the bimolecular collision rate with oxygen. Both time-domain and continuous wave saturation methods showed strong Heisenberg exchange between spin label and oxygen which contributes directly to the T_1 of the spin label, while dipolar interactions are negligible. The efficiency of the method was illustrated on dimyristoylphosphatidylcholine liposomes modified with stearic acid and sterol-type labels [84]. The effect of O_2 on the spin lattice relaxation time T_1, derived from saturation curves of the CW ESR spectra of nitroxide radicals, was examined for radicals incorporated into phosphatidylcholine (PC) liposomes and attached to lysozyme and albumin [85]. O_2 increased the rates of spin-lattice and spin-spin relaxation by the factor $k_{ex}C$, where k_{ex} is a spin-exchange constant and C is the oxygen concentration. The experiments indicted that for the liposomes, $k_{ex}C$, increased as the spin-label approached the center of the bilayer, while for the tagged proteins, $k_{ex}C$ was very close to that of the radicals in aqueous solution.

Advantages in measurement of the oxygen transport parameter using saturation-recovery EPR at Q- and W-band as compared at X-band have been demonstrated in saturated (DMPC) and unsaturated (POPC) lipid bilayer membranes with the use of stearic acid (n-SASL) and phosphatidylcholine (n-PC) spin-labels [86]. The contribution of dissolved molecular oxygen to the relaxation rate was is expressed in terms of the oxygen transport parameter $W = T_1^{-1}(\text{Air}) - T_1^{-1}(N_2)$, which was found to be a function of both concentration and translational diffusion of oxygen in the local environment of a spin-label. Benefits of SR EPR spin-label oximetry at Q- and W-band such as an ability for studying samples of small volume, ~ 30 nL, a higher resonator efficiency parameter and a new technique for canceling free induction decay signals, were demonstrated.

Figure 12.16 illustrates a profile of the oxygen transport parameter across the POPC membrane obtained by X-, Q- and W-band spectroscopy.

Various aspects of the spin oximetry applications have been reviewed elsewhere [87–91].

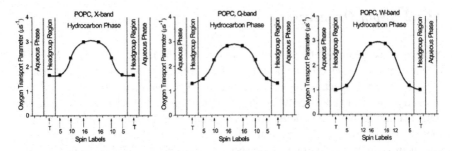

Fig. 12.16 A profile of Oxygen transport (accessibility) obtained at *X*-, *Q*-, and *W*-band. Measurements and equilibrations with gas for *X*- and *Q*-band were performed at 29–30 °C. Measurements and equilibrations with gas for *W*-band were performed at room temperature [86]

12.5.3   Estimation of Local Charges by Spin Label-Spin Probe Method 2 (SLSPM 2)

SLSPM 2 invented by Likhtenshtein with coworkers [92] is based on measurement of local charge Z_X in the vicinity of a paramagnetic particle (such as the active site of metalloprotein or a spin label) colliding with a nitroxide with known charge Z_p. Unknown values of Z_X were calculated with the use of the classic Debye equation.

$$\frac{k^+}{k^0} \text{ or } \frac{k^-}{k^0} = \frac{Z_p Z_X \alpha}{\exp(Z_p Z_X \alpha) - 1} \qquad (12.1)$$

where k^+, k^- and k^0 are the rate constants of encounters for positively charged, negatively charged, and neutral uncharged particles, respectively; and $\alpha = e^2/ k_B T \varepsilon_0 r$, where e is the charge of an electron; k_B is the Boltzmann constant; temperature T = 293 K, ε_0 is the dielectric constant of water, and r is the distance between the charges in the encounter complex. The data on values of (k^+/k^0) and (k^-/k^0), determined by measurement of the rate constants of spin exchange, were applied to two types of problems: (1) investigation of the electrostatic fields in the vicinity of a spin label or spin probe using a second paramagnetic species with a different charge, ferricyanide anion or diphenylchromium cation [7, 14, 92, 93] and (2) monitoring the effect of a paramagnetic species, such as a complex of param-agnetic ion with a protein or the active site of cytochrome c and myoglobin on the spin relaxation parameters of nitroxide spin probes of different charges freely dif-fusing in solution [7].

The second method was used for estimation of local charge in the vicinity of the heme proteins active center using spin probes of different charges [7], A mea-surement of the apparent spin-exchange rate-constants (k^+, k^-, and k_0) of the positively charged negatively charged and neutral nitroxide spin-probes in dynamic interaction with hemin, ferricyanide anion, and heme proteins in aqueous solutions allowed to determine the apparent local charges (Z_H) in the vicinity of the heme groups [7]. A similar approach was employed in [94] for determination of the electrostatic potential near the surface of calf thymus DNA. The dynamic exchange interaction between a spin label, ^{14}N-nitroxide derivative of 9-aminoacridine attached to DNA, and spin probes, free ^{15}N-labeled nitroxides of different charges, was monitored by electron-electron double resonance (ELDOR).

12.5.4   Measurement of Spin Labels Immersion Death by SLSPM 3

In solving problems of enzyme catalysis, molecular biophysics of proteins, biomembranes and molecular biology, one must also know the depth of immersion of paramagnetic centers in a biological matrix, i.e. the availability of enzyme sites to substrates, distance of electron tunneling between a donor and an acceptor group, position of a spin-label in a membrane and in a protein globule, distribution of the electrostatic field around the PC, etc.

Methods for measurement of depth of immersions of radicals proposed in [7, 95, 96]. are based on the determination of the distance of closest approach of an immersed spin label (nitroxide radical, for example) and a spin probe, paramagnetic complex distributed uniformly in a vitrified matrix, by examining the effect of the dipole-dipole interactions between them using the CW ESR saturation technique A similar approach examining the kinetics of spin-lattice relaxation for paramagnetics in solids was developed in [10].

For an estimation of a radical depth of immersion (R_0) of a radical in a matrix special algorithms were used [95, 96]. Two cases were theoretically analyzed: (1) at a low concentration of paramagnetic ions, when the spin-spin average distance ($R_{av} > R_0$) and (2) at a high concentration of paramagnetic ions ($R_0 > R_{av}$). The second approach examined positions of the nitroxide segments of spin-labeled probes IV and V in different biomembranes [96]. The paramagnetic ion, Co^{2+} or $K_3Fe(CN)$, were introduced together with an ion antagonist, Pb^{2+} or Al^{3+}. These diamagnetic ions prevent adsorption of the paramagnetic ions on the membrane surface and the paramagnetic ions distribute only in the vitrified solvent phase. The investigation revealed a variety of spin-labeled lipid conformations in different biological membranes. For example the R_0 values for spin probes **IV-V** in *M. capsulatus* membranes were consistent with more extended conformations

Formula 12.3

IV   V

than those from the chromatophore, chloroplast and microsome membranes (Formula 12.3).

The proposed methods has been employed for estimating the depth of immersion of paramagnetic metals in active centers of metalloenzymes [7] and references therein.

12.5.5 Proton Spin-Lattice Relaxation Affected by Paramagnetic Ions and Spin Probes

In this section, we have provided brief description of method of estimation of local electrostatic charge (or potential) around specific protons in molecules of interest. According to [97–99], the spin-lattice relaxation rate ($1/T_{1p}$), via the dipole-dipole mechanism, of a species bearing proton nuclei during an encounter with a free paramagnetic (spin probe, R), can be described by the value of the apparent spin-lattice relaxation rate constants $k = (1/T_{1p})/[R]$, which characterizes the frequency of encounter between the species. Nitroxide radicals of different electrostatic charge were used as spin probes [7, 100, 101]. The ratio of the apparent

Formula 12.4

spin-lattice relaxation rate constants k^{+-}/k^0 where k^{+-}/k^0 related to charge and uncharged molecule, respectively, reflects the effect of electrostatics upon the encounter interaction. The feasibility of this approach for the measurement of the electrostatic potential around protons was tested on simple organic molecules and amino acids As an illustration, the experimental data of a number of small charged (ethylamine, propionate anion and imidazole) and uncharged (ethanol) molecules were found to be in reasonable agreement with the results of a molecular modeling calculation of the local electrostatic potential [100]. The feasibility of this approach for the measurement of the electrostatic potential around protons was also tested on simple organic molecules, amino acids and monosaccharides (Formula 12.4).

12.6 Spin PH Probes

In recent years, thanks to the spin pH probes techniques the measurement of local acidity in biological objects including organs and even living organisms received increasing attention. As was demonstrated in the pioneering work [102], EPR spectra of stable nitroxides of the imidazoline and imidazolidine types have been shown to be sensitive to pH and serve as spin pH probes [102]. Data on the synthesis of a wide set of pH-sensitive nitroxides of different sensitivity, stability to reduction, lipophilicity, covalent binding properties to macromolecules, and applications have been reported [103–106] and references therein. As an example, the EPR spectra of a series of 4-dialkylamino-2,5-dihydroimidazole nitroxides with pyridine-4-yl, 4-dimethylaminophenyl or 4-hydroxyphenyl groups in position 2 of the imidazole ring were shown to be pH-sensitive due to consecutive protonation of the amidino moiety and the basic group(s) at position 2 of the imidazole ring over a pH range from 2 to 6.5 [105] (Formula 12.5).

The original concept of dual function pH and oxygen paramagnetic probes based on the incorporation of ionizable groups into the structure of persistent triarylmethyl radicals,TAMs, was recently realized using an asymmetric monophosphonated TAM probe with the simplest doublet hfs pattern (Fig. 12.17) [104]. The high efficiency of the proposed probe to oxygen, leading to an accuracy of pO_2

Formula 12.5

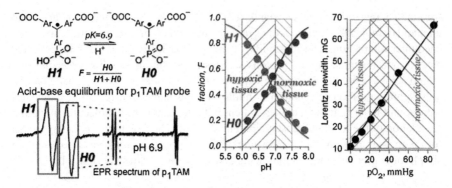

Fig. 12.17 Scheme illustrating the use triarylmethyl radicals, for independent measurement pH and oxygen. [Dhimitruka I, Bobko AA, Eubank TD, Komarov DA, Khramtsov VV. Phosphonated trityl probes for concurrent in vivo tissue oxygen and pH monitoring using electron paramagnetic resonance-based techniques [104]

measurements of ≈ 1 mmHg, originates from an extraordinarily low line width of the synthesized deuterated derivative, p_1TAM-D ($\Delta H_{pp} \leq 50$ mG, Lorentz line width, ≤ 20 mG). The presence of a phosphono group in the p_1TAM-D structure provides pH sensitivity to its EPR spectra in the physiological range of pH from 5.9 to 8.2. A significant difference in extracellular pH and pO_2 between tumor and normal mammary gland tissues and the effect of animal breathing with 100 % O_2 on tissue oxygenation were demonstrated in L-band EPR experiments performed in breast tumor-bearing mice.

For quantifying changes and spatial distribution of pH in vivo and to monitor spatial distribution in pathological conditions, specially developed pH phantom tubes have been designed [106]. High precision measurements of the hyperfine coupling constant resulting in mean pH values accurate to less than 0.1 pH was demonstrated. Ricently, results of biomedically relevant research on direct measurement of local pH in water-in-oil ointments, proteins and proteinaceous matrix, biodegradable polymers and phospholipid membranes as well as in animals in vivo and potentially in human subjects obtained by employing the nitroxide redox probes have been reviewed [105].

The direct measurements of acidity inside micropores of inorganic (hydrogels based on TiO_2, SiO_2, zeolites and kaolin) sorbents and organic (synthetic ion-exchange resins) using pH-sensitive nitroxide radicals (NR) as probes and ESR technique were carried out [107]. It was found that local pH values at the site of label location in the sorbents differ from pH values of the solutions by 1–2 units. The spin pH-metry was applied for measurement of local pH (pH^{loc}) in SiO_2 gel and aluminium oxides modified by incorporating F^- and SO_4^{2-}. The values (pH^{loc}) inside sorbents pores (excluding γ-Al_2O_3) were found to differ from the pH-values of external solution by 0.5–1.3 units.

12.6.1 Dual Fluorophore-Nitroxides (FNRO·) as Redox Sensors and Spin Traps

The spin redox probe technique utilizes the ability of nitroxides to be reduced via a reducing agent to the corresponding hydroxyl amine. CW ESR spectroscopy of nitroxide radicals (NRO·) are widely used for quantitative characterization of redox processes and protection from radical damage [13–15, 108–111] and references therein. Nevertheless, the sensitivity of the CW ESR technique is not sufficient for real time monitoring of picomole redox processes in biological systems. To overcome the problem a method dual fluorophore-nitroxide redox probe was invented and implemented.

The use of dual fluorophore-nitroxide compounds (FNRO·) as redox probes and spin traps is based on observations first made in the Likhtenshtein group [112]. It was shown that in a dual molecule, photochemical reduction of the nitroxide fragment results in decay of the ESR signal and enhancement of the fluorescence. The nitroxide segment acts as a quencher of the fluorescence of the chromophore fragment. and the redox process was monitored using two independent methods, ESR and the significantly more sensitive fluorescence technique. A similar effect was observed after chemical reduction of dual fluorophore-nitroxide compounds by ascorbic acids and other antioxidants, as well as nitric oxide and superoxide [112–120]. Trapping of free radicals by FNRO· can be also detected following ESR signal decay and fluorescence [114].

On the basis of available data, two mechanisms of fluorescence quenching in the dual compounds were proposed [51, 115, 116] the major mechanism, intersystem crossing (ISC); and the minor mechanism, irreversible intramolecular electron transfer from the excited singlet state of the fluorophore (donor) to nitroxide (acceptor), followed by fluorophore segment regeneration and hydroxylamine formation. The latter mechanism is also responsible for photoreduction. Since 1986 a number of applications of the dual nitroxide supper molecules for study redox processes was reported.

Using the abovementioned unique properties of the dual molecules, a method for the quantitative analysis of vitamin C in biological and chemical liquids has been proposed [117]. Pseudo-first-order reactions between the dual molecule (Formula 12.6) and ascorbic acid were monitored. An increase in the fluorescent intensity and decay of the EPR signal of the dual probe DT occurred with the same rate constant k_{red}. The calibration curves of k_{red}—ascorbic acid concentration for vitamin C analysis were used for performing vitamin C analysis in human blood plasma and commercial jucis. A collection of dual probes of different redox potential of the nitroxide segment and the fluorophore spectral property were synthesized, that made it possible to select reducing agents appropriate for different media (Fig. 2.15, Sect. 10) [116, 118].

For example, the dual PN probe possessing a high redox potential was employed for analyzing such antioxidants as quercetin and galangin down to a submicroconcentration scale [118] (Formula 12.7).

Formula 12.6

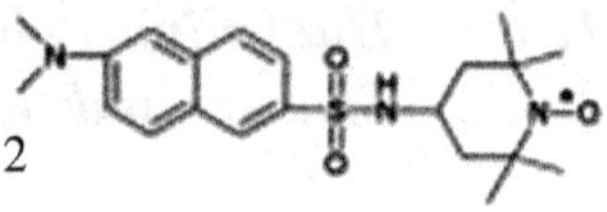

Formula 12.7

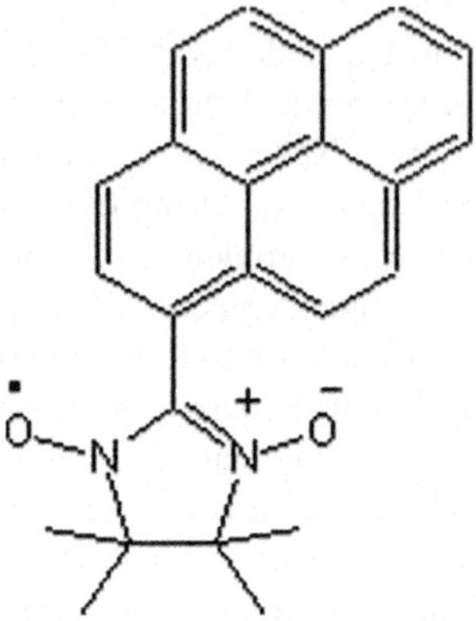

PN

A new method for NO analysis on the nano-concentration scale and real-time monitoring of the NO outflux from tissues using of pyren-nitronyl PN has been developed [119]. A reaction of PN with NO yields a pyrene–imino nitroxide radical (PI) and NO_2 and is accompanied by changes in the EPR spectrum from a five-line pattern (two equivalent N nuclei) into a seven-line pattern (two nonequivalent N nuclei) and by a drastic increase in the fluorescence intensity. The EPR methods were employed for the calibration of NO in the micromole concentration scale while the much more sensitive fluorescent technique allowed for the calibration of NO in the nanomolar concentration scale. This method was applied to real time monitoring of basal flux NO in tissue from pig trachea epithelia of 1.2 nmol/g × min.

A new rapid and highly sensitive method for real time superoxide analysis has been developed based on a reaction of fluorophore–nitronyl with O_2^- [118]. The reaction of PN with superoxide radicals, as a reduced agent, generated by the xanthine/xanthine oxidase system, yielded drastic increase (about 2,000 times) in the fluorescent intensity and a decrease of the EPR signal. In conditions, when the rate of superoxide production is slow ($\omega i < 2 \times 10^{-7}$ M/min), the dependence of rate of fluorescent increase (dI/dt) on the rate of superoxide production was used as a calibration curve for real time monitoring O_2^- in biological and chemical systems.

A bifunctional stilbene-nitroxide label (BFL1) immobilized on the surface of a quartz plate was applied for the parallel determination of the antioxidant status and measurement of micro- and macro-viscosity of the media [120]. This method allowed a measurement of the concentration of ascorbic acid in solution in a range of $(1-9) \times 10^{-4}$ M and the viscosity of a medium in the range 1–500 cP (Formula 12.8).

Formula 12.8

BFL1

Dual nitron-fluorescent probes (NFP) for monitoring redox processes were designed and used for monitoring redox processes [121]. Reactions of NFP with reactive radicals are accompanied by the appearance of an ESR signal from the nitroxide moiety and quenching of the fluorescence. These supermolecules are effectively used as fluorescence spin trap of high sensitivity. Three nitron-fluorescent probes, *p*-nitrostilbene-*tert*-butyl-nitrone, coumaryl-styryl-*tert*-butyl-nitron and a third compound based on 4-pyrrolidine-1,8-naphthylimido-methylbenzene as fluorophore were employed for study of a system containing inhibitors of components of the respiratory chain and the F_1-F_0-ATP synthase under a variety of physiological conditions.

It is important to note that in complex systems, biological ones in particular, not the antioxidant concentration but *the value of the rate constant of the nitroxide reduction* that can be taken as a quantitative characteristics of the antioxidant status of system of interest.

Thus, the application of dual fluorophore-nitroxide methodology can afford insights into biochemical redox processes and, therefore, can create a basis for the development of new methods for biomedical research and medical diagnostics.

12.6.2 Triplet Labeling

Triplet triplet energy transfer (TTET) and triple triplet annihilation are fundamental photophysical processes in chemistry and biology (Sect. 2.2). For instance, phototherapies based on the generation of reactive singlet oxygen during the process of triplet-triplet energy transfer from an excited phosphorescent chromophore to triplet O_2 are widely employed in medicine [122, 123].

As was first shown in the Likhtenshtein group, [14, 124–130], paramagnetism and the relative long life time of chromophores in the excited triplet state provide the possibility for these compounds to be used as phosphorescence (triplet) labels for solving a number of structural and dynamics problems. The theoretical basis of such applications, based on the spin exchange mechanism, was described in Chap. 2. In the limited frame of the present book, here we concentrate on only two specific problems, namely, protein conductivity probed by spin exchange and rear encounters in biomembranes, which can be solved utilizing the method of triplet labels.

Proteins spin exchange conductivity

The effect of the heme group on the lifetime of the excited triplet state of eosin isothiocyanate label attached to terminal amino group of different Hbs and myoglobin was measured by studying the kinetics of phosphorescence decay [127]. The effect of nitroxyl spin labels attached to His-15 on the phosphorescence spectra of lysozyme was also studied. Experimental data on the exchange reactions between centers localized on proteins with known structure (Hb, myoglobin, lysozyme, carboanhydrase, bacterial ferredoxin) indicated that the exchange interaction between the donor and acceptor centers decays exponentially and approximately relates to the corresponding dependence in non-conductive model systems. The equation proposed in [127] allows to estimate the distances between the centers.

An approach developed by Mekler et al. [128] is based on the following scheme

$$
\begin{array}{c}
h\upsilon_1 \\
D \rightarrow D^T \\
D^T + A^{S0} \rightarrow D + A^T \\
A^T + A^T \rightarrow A^{S0} + A^{S1} \\
A^{S1} \rightarrow A^{S0} + h\upsilon_2
\end{array}
$$

where D^T is a donor in the exited triplet state, A^{S0} is an acceptor in a ground singlet state, A^T is the acceptor in the exited triplet state generated during the encounter of D^T and A^{S0}, and A^{S1} is the acceptor in the excited singlet state. The main advantage of such an approach is that one may monitor relatively slow triplet quenching and triplet-triplet annihilation using very sensitive fluorescence techniques and measuring rate constants of these processes.

In the sarcoplasmic reticulum membranes [129], the annihilated delayed fluorescence (ADF) measurements were performed for a study of the self-interaction of triplet-excited probes (perylene, (I) 7,12-dimethylbenzanthracene, and 4-(2-anthryl)butanoic acid (II) and the triplet-excited probes quenching by ferrocene in sarcoplasmic reticulum membranes (SR) and in dimyristoylphosphatidylcholine liposomes. The dynamic processes in the system were characterized by the following parameters: the average ADF time decay of I and II in SR was 4.2×10^{-5} and 2.3×10^{-5} s, respectively and the triplet-state constant quenching of I by ferrocene was 2.1×107 (mM/g of lipid s^{-1}) in SR and 4.5×107 (mM/g of lipid s^{-1}). Triplet-triplet annihilation of perylen, benzantracene and pentacene which produced delayed annihilation fluorescence upon encounters occurs with a characteristic time about 1 ms, revealed free diffusion of the probes for distance of not less than 30 nm. Therefore, in SRM lipids "lakes" are not isolated by the membrane proteins.

Measurements of the kinetics of erythrosine (I) phosphorescence and ubiquone-10 quenched delayed fluorescence of 4-(2-anthryl)butanoic acid (II), and 12-(2-anthryl)dodecanoic acid (III) were carried out [130]. The experiments indicated that the quinone ring of ubiquinone was an effective quencher for the triplet states of probes II and III, whose chromophores are localized in the hydrophobic region of the lipid bilayer, and of probe I, localized in the polar region of the liposomes. Furthermore, the quenching rate constant of III is 2-fold higher than that of II. The conclusion was that the ubiquinone Q-10 molecule can effectively reach the membrane surface to play its role as an electron carrier.

12.6.3   Triplet-Photochrome Method. Spin Cascade

The triplet-photochrome labeling method (TPLM), invented by Mekler and Likhtenshtein [131], is based on ability of a photochrome (stilbene for example) to undergo sensitized photoisomerization through the triplet-triplet energy transfer from a donor molecule excited to its triplet state. Starting from *cis*-stilbene, which is not fluorescent at steady-state conditions, and measuring the rate of increase of emitted fluorescence, it has been possible to monitor the process of the sensitized *cis–trans* photoisomerization. Two unique properties of TPLM, namely, long lifetime of the sensitizer triplet state (up to 1 s) and the possibility of integrating data on the stilbene photoisomerization allow one to investigate very slow diffusion processes using highly sensitive fluorescence techniques. Quantitative measurement of encounters of proteins in viscous solution and membranes using very low concentrations of both the triplet and photochrome probes appears to be available by TPLM techniques.

Realization of the triplet-photochrome labeling method was demonstrated in a study of rare encounters in a system containing the erythrosinisothiocyanate (ERITC) sensitizer and 4-acetamido-4′-isothiocyanato-2,2′-stilbenedisulfonic acid (SITC) photochrome probe, both covalently bound to α-chymotrypsin [131]. TPLM was first applied to follow the protein–protein dynamic contacts in biomembranes [132] SITS and ERITC were bound covalently to a membrane preparation of Na^+, K^+ ATPase. Triplet–triplet energy transfer from the light-excited triplet ERITC to cis-SITS initiated the *cis–trans* photoisomerization of cis-SITS which was followed by a commercial spectrofluorimeter. The apparent rate constant of triplet–triplet energy transfer from ERITC to cis-SITS was found to be $k_{app} = 0.4 \times 10^3 M^{-1} s^{-1}$ (at 25 °C), while for unbound ERITC and SITS in solution $k_{app} = 7 \times 10^7 M^{-1} s^{-1}$. The drop of k_{app} in the case of labels bound to ATPase was a result of the increased media viscosity and steric factors.

For quantitative investigation of slow dynamic processes, in biological membranes in particular, the spin cascade method (SCM) was invented and developed by the Likhtenshtein group [133–135]. A cascade of photochemical and photophysical reactions includes triplet–triplet energy transfer between a triplet sensitizer and a fluorescence photochrome, which undergoes *cis–trans* photoisomerization, and triplet excited state quenching by a stable radical. The spin cascade system studied in [134, 135]. consists of the triplet sensitizer (erythrosin B), the photochrome stilbene derivative probe (4-dimethylamino-4-aminostilbene), and nitroxide radicals (5-doxyl stearic acid) quenching the excited triplet state of the sensitizer (Fig. 12.18)

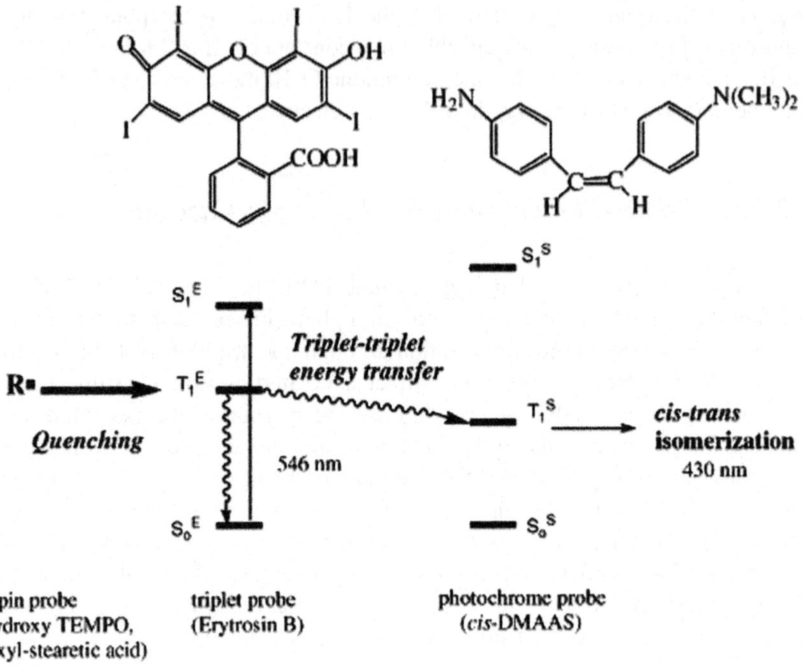

Fig. 12.18 Representation of energy levels of cascade reactants and competition between the $T_1^E \rightarrow T_1^S$ and $T_1^E \rightarrow S_0^S$ processes [135]

The experimental quenching rate constant of the cascade reaction kq and the rate constant of the triplet–triplet energy transfer k_T evaluated in 2D terms were obtained as kq = $(1.05 \times 10^{15}$ cm^2/(mol s) and k_T = $(1.26 \times 10^{12}$ cm^2/(mol s), respectively [135]. The values of diffusion rate constants in model lipid membranes together with similar data obtained from other methods, that cover characteristic times over eight orders of magnitude, were found to be in a reasonable agreement with the advanced theory of diffusion-controlled reactions in two dimensions (Fig. 12.19).

In addition, using the spin cascade method, the following dynamic parameters of the cascade system components can be experimentally measured: the spin label rotation correlation time and spin relaxation parameters, the fluorescence and phosphorescence polarization correlation times, the rate constants of the chromophore singlet and triplet state quenching by nitroxides, and the rate constant of photoisomerization. This set of parameters is a cumulative characteristic of the dynamic state of biomembranes in the wide range of the probes amplitude and characteristic time. Proficiency of the method can be expanded by a choice of the cascade components with the higher efficiency of triplet–triplet energy transfer, higher sensitizer lifetime, and by an increase of the time of integration of experimental data on a photochrome photoisomerization.

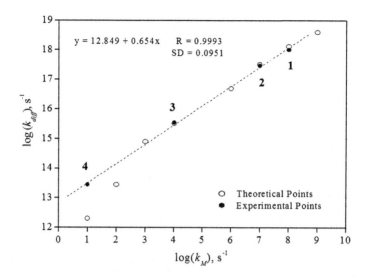

Fig. 12.19 Theoretical and experimental dependences of logarithms of the diffusion-controlled rate constants (k_{diff}) on logarithms of the unimolecular decay of the excited species (k_M) which characterizes the timescale of different methods [135]

Due to space constraints, data on physical labeling in physical chemistry, photo chemistry, biochemistry,biotechnology and biomedicine have not included in this Chapter. Basic information about these systems can be found in books [13–15, 109] and a recent review [136-139].

References

1. T.J. Stone, T. Buckman, P.L. Nordio, H.M. McConnell, Proc. Nat. Acad. Sci. USA **54**, 1010 (1965)
2. G.I. Likhtenshtein, Mol. Biol. (Moscow) **2**, 234 (1968)
3. J.C. Taylor, J.S. Leigh, M. Cohn, Proc. Natl. Acad. Sci. USA **64**, 219–206 (1969)
4. A.I. Kulikov, G.I. Likhtenshtein, E.G. Rozantzev, V. Suskina, A. Shapiro, Biofizika **17**, 42 (1972)
5. A.I. Kokorin, K.I. Zamaraev, G.L. Grigoryan, V.P. Ivanov, E.G. Rozantsev. Biofizika **17**, 34 (1972)
6. A.I. Kulikov, G.I. Likhtenstein, Biofisika **19**, 420 (1974)
7. Likhtenshtein, *Depth of immersion of paramagnetic centers*, eds. by L. Berliner, S. Eaton, G. Eaton. Magnetic Resonance in Biology (Kluwer Academic Publishers, Dordrecht, 2000), p. 309
8. G.D. Case, J.S. Jr, Leigh, Intramitochondrial position of cytochrome haem groups determined by dipolar interaction with paramagnetic cations. Biochem. J. **160**, 769 (1976)
9. A.I. Kulikov, G.I. Likhtenstein, Adv. Molecu.l Relax. Proc. **10**, 47 (1977)
10. M.K. Bowman, J.R Norris, Cross relaxation of free radicals in partially ordered solids. J. Phys. Chem. **86**, 3385 (1982)

11. L.A. Syrtsova, L.A. Levchenko, E.N. Frolov, G.I. Likhtenshtein, T.N. Pisarscaya, L.V. Vorob'ev, V.A Gromoglasova, Mol. Biol. (Moscow) **5**, 726 (1971)
12. D. Abdullin, F. Duthie, A. Meyer, E.S. Muller, G. Hagelueken, O. Schiemann, J. Phys. Chem. B (2015)
13. G.I. Likhtenshtein, *Spin Labeling Method in Molecular Biology* (Wiley Interscience, New York, 1976)
14. G.I. Likhtenstein, *Biophysical Labeling Methods in Molecular Biology* (Cambridge University Press, New York, Cambridge, 1993)
15. G.I. Likhtenshtein, J. Yamauchi, S. Nakatsuji, A. Smirnov, R. Tamura, *Nitroxides: Application in Chemistry, Biomedicine, and Materials Science* (WILEY-VCH, Weinhem, 2008)
16. W.L. Hubbell, C.J. Lopez, C. Altenbach, Z. Yang, Curr. Opin. Struct. Biol. **23**, 725 (2013)
17. P.P. Borbat, H. Mchaourab, J.H. Freed, J. Am. Chem. Soc. **124**, 5304 (2002)
18. F. Ferrarini, J.H. Freed, J. Phys. Chem. B **110**, 26260 (2006)
19. A. Rassat, P. Ray, Bull. Soc. Chim. Fr. 815 (1967)
20. J.C. McNulty, G.L. Millhauser, TOAS. *The rigid nitroxide side chain*, eds. by L. Berliner, S. Eaton, G. Eaton. Magnetic resonance in biology. (Kluwer Academic Publishers. Dordrecht, 2000), vol. 19, p. 277
21. D. Ulyanov, B.E. Bowler, G.R. Eaton, S.S. Eaton, Bioph. J. **95**, 5306 (2008)
22. S. Razzaghi, E.K. Brooks, E. Bordignon, W.L. Hubbell, M. Yulikov M, G. Jeschke, ChemBioChem **14**, 1883 (2013)
23. P. Lueders, S. Razzaghi, H. Jager, R. Tschaggelar, M.A Hemminga, M. Yulikov, G. Jeschke, Mol. Phys. Ahead of Print
24. G. Jeschke, DEER distance measurements on proteins. Ann. Rev. Phys. Chem. **63**, 419–446 (2012)
25. E.R. Georgieva, A.S. Roy, V.M. Grigoryants, P.P. Borbat, K.A. Earle, C.P. Scholes, C. P. Freed, J. Magn. Reson. **216**, 69 (2012)
26. P.P. Borbat, E.R. Georgieva, J.H. Freed, J. Phys. Chem. Lett. **4**, 170 (2013)
27. S.M. Islam, R.A. Stein, H.S. Mchaourab, B. Roux, J. Phys. Chem. B **117**, 4754 (2013)
28. C. Altenbach, A.K. Kusnetzow, O.P. Ernst, K.P. Hofmann, W.L. Hubbell, Proc. Nat. Acad. Sci. USA **105**, 7439 (2008)
29. A.D. Milov, R.I. Samoilova, Y.D. Tsvetkov, M. De Zotti, C. Toniolo, J. Raap, and. J. Phys. Chem. B **112**, 13469 (2008)
30. E. Narr, A. Godt, G. Jeschke, Angewan. Chemi. Inter. Ed. Eng. **41**, 3907 (2002)
31. A.M. Raitsimring, C. Gunanathan, A. Potapov, I. Efremenko, J.M.L. Martin, D. Milstein, D. Goldfarb, J. Am. Chem. Soc. **129**, 14138 (2007)
32. A. Potapov, H. Yagi, T. Huber, S. Jergic, N.E. Dixon, G. Otting, D. Goldfarb, J. Am. Chem. Soc. 132, 9040–9048 (2010)
33. L. Garbuio, E. Bordignon, E.K. Brooks, W.L. Hubbell, G. Jeschke, M. Yulikov, J. Phys. Chem. B **117**, 3145 (2013)
34. P. Lueders, H. Jager, M.A. Hemminga, G. Jeschke, M. Yulikov, J. Phys. Chem. Lett. **3,** 1336 (2013)
35. S. Razzaghi, E.K. Brooks, E. Bordignon, E.K. Hubbell W.L.M. Yulikov, G. Jeschke, ChemBioChem **14**, 1883 (2013)
36. P. Lueders, G. Jeschke, M. Yulikov, J. Phys. Chem. Lett. **2**, 604 (2011)
37. Z. Yang, Y. Liu, P. Borbat, J.L. Zweier, J.H. Freed, W.L. Hubbell, J. Am. Chem. Soc. **134**, 9950 (2012)
38. G.I. Likhtenshtein, A.P. Pivovarov, PKh Bobodzhanov, E.G. Rozantsev, N.B. Smolina, Biofizika **13**, 396 (1968)
39. E.N. Frolov, A.D. Mokrushin, G.I. Likhtenshtein, V.A. Trukhtanov, V.I. Gol'danskii, Doklady Akademii Nauk SSSR **212**, 165 (1973)
40. G.I. Likhtenshtein. Y.D. Akhmedov, L.V.L.A. Ivanov, L.A. Krinitskaya, Y.V. Kokhanov, Mol. Biol. (Moscow) **8**, 40 (1974)
41. D. Kivelson, J. Chem Phys. **33**, 1094 (1960)

42. J.H. Freed, *Theory of the ESR spectra of nitroxides*, eds. by L. Berliner. Spin Labeling. Theory and Applications (Academic Press, New York, 1976), vol 1
43. D. Kruk, J. Kowalewski, S. Tipikin, J.H. Freed, M. Moscicki, A, Mielczarek, M. Port, J Chem. Phys. **134**, 024508/1 (2011)
44. S.A. Dzuba, A.G. Mar'yasov, K.M. Salikhov, YuD Tsvetkov, J. Magn. Reson. **58**, 95 (1984)
45. J.S. Hyde, D.D. Thomas, Ann. N. Y. Acad. Sci. **222**, 680 (1973)
46. W. Froncisz, T.G., Camenisch, J.J. Ratke, J.S. Hyde, Rev. Sci. Instrum. **72**, 1837 (2001)
47. A.H. Beth, E.J. Biol. Magn. Reson. (Biomedical EPR, Part B) **24**, 369–407 (2005)
48. G.I. Likhtenshtein, *New trends in enzyme catalysis and mimicking chemical reactions* (Kluwer Academic/Plenum Publishers, New York, 2003)
49. E.N. Frolov, N.V. Kharakhoncheva, G.I. Likhtenshtein, Molec. Biol. (Moscow) **8**, 886 (1974)
50. V.R. Fogel, E.T Rubtsova, G.I. Likhtenshtein, K. Hideg, J. Photochem. Photobiol. A Chem. **83**, 229 (1994)
51. G.I. Likhtenshtein, D. Pines, E. Pines, V, Khutorsky, Appl. Magn. Reson. **35**, 459 (2009)
52. V.I. Krinichnyi, O.Y. Grinberg, V.R. Bogatirenko, G.I. Likhtenshtein, Y.S. Lebedev, Biofizika **30**, 216
53. M.T. Lerch, Z. Yang, E.K. Brooks, W.L. Hubbell, Proc. Nat. Acad. Sci. **111**, E1201 (2014)
54. Mchaourab S, M. A Lietzow, K. Hideg, and W. L. Hubbell, ,()
55. F. Tombolato, A. Ferrarini, J.H. Freed, J. Phys. Chem. B **110**, 26260 (2006)
56. Z. Zhang, M.R. Fleissner, D.S. Tipikin, Z. Liang, J.K. Moscicki, K.A. Earle, W.L. Hubbell, J.H. Freed, J. Phys. Chem. B **114**, 5503 (2010)
57. H.S. Mchaourab, P.R. Steed, K. Kazmier, Structure **19**, 1549 (2011)
58. Y. Sun, Z. Zhang, V.M. Grigoryants, W.K. Myers, F. Liu, K.A. Earle, J.H. Freed, C. P. Scholes, Biochemistry **51**, 8530 (2012)
59. B.H. Robinson, C. Mailer, G. Drobny, Ann. Rev. Biophys. Biomol. Struct. **26**, 629 (1997)
60. A. Marko, V. Denysenkov, D. Margraf, P. Cekan, O. Schiemann, STh Sigurdsson, T.F. Prisner, J. Am. Chem. Soc. **133**, 13375 (2011)
61. U. Jakobsen, S.A. Shelke, S. Vogel, S.T. Sigurdsson, J. Am. Chem. Soc. **132**, 10424 (2010)
62. S. Obeid, M. Yulikov, G. Jeschke, A. Marx, Angewen. Chem. Int. Ed. **47**, 6782 (2008)
63. L. Columbus, W.L. Hubbell, Biochemistry **43**, 7273–7287
64. H.J. Steinhoff, B. Suess, Methods **29**, 188 (2003)
65. N.A. Kuznetsov, A.D. Milov, V.V. Koval, R.I. Samoilova, Y.A. Grishin, D.G. Knorre, Y.D. Tsvetkov, O.S. Fedorova, S.A. Dzuba, Phys. Chem. Chem. Phys. **11**, 6826 (2009)
66. G. Sicoli, G. Mathis, O. Delalande, Y. Boulard, D. Gasparutto, S. Gambarelli, Angew. Chem. Int. Ed. **47**, 735 (2008)
67. G.W. Reginsson, S.A. Shelke, C. Rouillon, M.F. White, S.T. Sigurdsson, O. Schiemann, Protein-induced changes in DNA structure and dynamics observed with noncovalent site-directed spin labeling and PELDOR. Nucleic Acids Res **41**, e11 (2013)
68. N.K. Kim, M.K. Bowman, V.J. DeRose, J. Am. Chem. Soc. **13**, 8882 (2010)
69. I. Krstic, O. Frolow, D. Sezer, B. Endeward, J.E. Weigand, B. Suess, J.W. Engels, T.F. Prisner, J. Am. Chem. Soc. **132**, 1454 (2010)
70. W.K. Subczynski, J. Widomska, J.B. Feix, Rad. Biol. Med. **46**, 707 (2009)
71. S. Kuntz, F. Leinisch, K. Hideg, W.E. Appl, Magn. Reson. **37**, 455 (2009)
72. B. Dzikovski, J.H. Freed, *Membrane fluidity*, ed. by. T.P. Begley. Wiley encyclopedia of chemical biology, vol. **2**, pp. 728 (2009)
73. Y.-W. Chiang, A.J. Costa-Filho, J.H. Freed, J. Phys. Chem. B **111**, 11260 (2007)
74. V.I. Krinichnyi, O.Y. Grinberg, E.I. Judanova, M.L. Borin, Y.S. Lebedev, G.I. Likhtenshtein, Biofizika **32**, 59 (1988)
75. H.M. McConnell, B.J. McFarland, Quar. Rev. Biophys. **3**, 91 (1970)
76. J.S. Hyde, J.B. Feix, *Electron-Electron Double Resonance*, eds. by L.J. Berliner, J. Reubin. Biological Magnetic Resonance, Theory and Application (Plenum Press, New-York. 1989), vol. 8, p. 305
77. A.L. Lai, J.H. Freed, Biophys. J. **106**, 172 (2014)

78. G.I. Likhtenshtein, Y.B. Grebenshchikov, T.V. Avilova, Mol. Biol. (Moscow) **6**, 52 (1972)
79. J.S. Hyde, H.M. Swartz, W.E Antholine, *The spin probe–spin label method*, ed. by L. J. Berliner. Spin Labeling Theory and Application (Academic Press, NY, 1979), vol. 2, p. 72
80. L. Mainali, M. Raguz, T.G. Camenisch, J.S. Hyde, W.K. Subczynski, J. Magn. Reson. **212**, 86 (2011)
81. J. Pyka, J. Ilnicki, C. Altenbach, W.L. Hubbell, W. Froncisz, Biophys. J. **89**, 2059 (2005)
82. RS. Kaplan, J.A. Mayor, R. Kotaria, D.E. Walters, H.S. Mchaourab, Biochemistry **39**, 9157 (2000)
83. W.K. Subczynski, J.S. Hyde, Biochim. bioph. acta **643**, 283 (1981)
84. A. Kusumi, W.K. Subczynski, and. J. S. Proc. Natl. Acad. Sci. USA **79**, 1854 (1982)
85. E.I. Yudanova, A.V. Kulikov, Biofizika **29**, 925 (1984)
86. W.K. Subczynski, L. Mainali , T.G. Camenisch, W. Froncisz, J.S. Hyde, Spin-label oximetry at Q- and W-band. J. Magn. Reson. **209**, 142–148 (2011)
87. J.S. Hyde, Trends in EPR technology. Biol. Magn. Reson. 24, 409–428 (2005)
88. H.M. Swartz, N. Khan, Biol Magn Reson **23**, 197–228 (2005)
89. G. Ilangovan, J.L. Zweier, P. Kuppusamy, Methods in enzymology. Oxygen Sensing **381**, 747 (2004)
90. H. Hou, S.P. Mupparaju, J.P. Lariviere, S. Hodge, J. Gui, H.M. Swartz, N. Khan, Radiation Res **179**, 343 (2013)
91. E.M. Fisher, M. Khan, R. Salisbury, P. Kuppusamy, Cell Biochem. Biophys. **67**, 451 (2013)
92. G.I. Likhtenstein, Y.B. Grebentchikov, Y.V. Kokhanov, Mol. Biol. (Moscow) **4**, 782 (1970)
93. G.I. Likhtenshtein, *Chemical Physics of Redox Metalloenzymes* (Springer, Heidelberg, 1988)
94. J.L. Hecht, B. Honig, Y. Shin, W.L. Hubbell, J. Phys. Chem. **99**, 7782 (1995)
95. E.S. Kulikov, Cherepanova, and V. R. Bogatyrenko Theo. Expe Chem. **17**, 618 (1981)
96. E.S. Cherepanova, A.V. Kulikov, G.I. Likhtenstein, Biol. Membr. (Moscow) **77**, 51 (1990)
97. L.P. Hwang, J.H. Freed, J. Chem. Phys. **63**, 4017 (1975)
98. I.V. Alexandrov, *Theory of Magnetic Relaxation* (Nauka, Moscow, 1975)
99. B.M. Berdnikov, A.B., Doktorov, L.L. Makarshin, Theor. Eksp. Chem. (Kiev) **16**, 765 (1980)
100. G.I. Likhtenshtein, I. Vaisbuch, I. Adin, A. Shames, R. Glaser, Biophys. J. **77**, 443 (1997)
101. R. Glaser, A. Novoselsky, A. Shames, G.I Likhtenshtein, Isr. J. Chem. Special Lemieux Wolf Prize Laureate issue **40**, 263 (2000)
102. V.V. Khramtsov, L.M. Weiner, I.A. Grigor'ev, L.B. Volodarsky, Chem. Phys. Lett. **91**, 69 (1982)
103. I.A. Kirilyuk, A.A. Bobko, V.V. Khramtsov, I.A. Grigor'ev, Org. Biomol. Chem. **3**, 1269 (2005)
104. I. Dhimitruka, A.A. Bobko, T.D. Eubank, D.A. Komarov, V.V. Khramtsov, J. Am. Chem. Soc. **135**, 5904–5910 (2013)
105. V.V. Khramtsov, *In vivo Spectroscopy and Imaging of Nitroxide Probes*, ed. by A.I. Kokorin. Nitroxides: Theory, Experiment and Applications, p. 317 (2012)
106. S. Koda, J. Goodwin, V.V. Khramtsov, H. Fujii, H. Hirata, Anal. Chem. **84**, 3833 (2012)
107. L.S. Molochnikov, E.G. Kovalyova, I.A. Grigor'ev, A.A. Zagorodni, J. Phys. Chem. B **108**,1302 (2004)
108. H.M. Swartz, N. Khan, V.V. Khramtsov, Antioxidants and redox signaling **9**, 1757–1771
109. N. Kocherginsky, H.M, Swarts, *Nitroxide Spin Labels. Reactions in Biology and Chemistry* (CRC Press, 1995)
110. P. Kuppusamy, M.C. Krishna, Current Topics Biophys. **26**, 29 (2002)
111. Rumiana Bakalova, Zhivko Zhelev, Ichio Aoki, Tsuneo Saga, Clin. Cancer Res. **19**, 2503 (2013)
112. I.M. Bystryak, G.I. Likhtenshtein, A.I. Kotelnikov, O.H. Hankovsky, K. Hideg, Russ. J. Phys. Chem O. H. **60**, 1679 (1986)
113. G.I. Likhtenshtein, App. Bioche. Biotechn. **152**, 135 (2009)
114. R. Braslau, F. Rivera, E. Lilie E, M. Cottman, J. Org. Chem. **78**, 238–245 (2013)
115. Likhtenstein, K. Ishii, S. Nakatsuji, Photochem. Photobiol. **83**, 871 (2007)

116. G.I. Likhtenshtein, Pure Appl. Chem. **80**, 2125 (2008)
117. E. Lozinsky, V.V. Martin, T.A. Berezina, A. Shames, A.L. Weis, G.I. Likhtenshtein, J. Biochem. Biophys Meth. **38**, 29 (1999)
118. N. Medvedeva, V.V. Martin, G.I. Likhtenshten, J. Photochem. Photobiol A: Chem. **163**, 45 (2004)
119. E.M. Lozinsky, L.V. Martina, A.I. Shames, N. Uzlaner, A. Masarwa, G.I. Likhtenshtein, D. Meyerstein, V.V. Martin, Z. Priel, Anal. Biochem. **326**, 139 (2004)
120. P. Parkhomyuk-BenArye, N. Strashnikova, G.I. Likhtenshtein, J. Biochem. Biophys. Meth. **51**, 1 (2004)
121. S. Hauck, Y. Lorat, F. Leinisch, C. Kop, J. Abrossinow, W.E. Trommer, *Nitroxides: Theory, Experiment and Applications*, ed. by A.I. Kokorin (2012), p. 347
122. J. Adauwiyah, H.H. Suraiya, Med. J. Malaysia **65**, 297 (2010)
123. S. Titus, J. Hodge, Am. Fam. Physician **86**, 734 (2012)
124. V.M. Mekler, A.I. Kotel'nikov, G.I. Likhtenshtein, M.A. Berkovich, Biofizika **27**, 641 (1982)
125. IKh Yusupov, G.I. Likhtenshtein, Biophysics (English Translation) **57**, 197 (2012)
126. V.R. Likhtenshtein, *Stilbenes: Application in Chemistry, Life Science and Material Science* (WILEY-VCH, Weinhem, 2009)
127. A.I. Kotel'nikov, V.R. Fogel, V.R. Likhtenshtein, G.B. Postnikova, E. A. Shlapnikova EA. Molekulyarnaya Biologiya (Moscow) **15**, 281 (1981)
128. V.M. Mekler, A.I. Kotel'nikov, Likhtenstein. Biofizika **28**, 503 (1983)
129. V.M. Mekler, A.I. Kotel'nikov, G.I. Likhtenshtein, R.A. Kotel'hikova, L.V. Tatyanenko, V. I. Shvets, A.P. Kaplun, Biofizika **29**, 809 (1984)
130. V.M. Mekler, A.I. Kotel'nikov, G.I. Likhtenshtein, A.P. Kaplun, V.I. Shvets, Biofizika **29**, 779 (1984)
131. V.M. Mekler, G.I. Likhtenshtein, Biofizika **31**, 568 (1986)
132. Mekler, F.T. Umarova, Biofizika **33**, 720 (1988)
133. V. Papper, G.I. Likhtenshtein, N. Medvedeva, D.V. Khoudyakov, J. Photochem. Photobiol. A: Chem. **122**, 79 (1999)
134. V. Papper, N. Medvedeva, I. Fishov, G.I. Likhtenshein, Appl. Biochem. Biotech. **89**, 231 (2000)
135. N. Medvedeva, V. Papper, G.I. Likhtenshten, Phys. Chem. Chem. Phys. **7**, 3368 (2005)
136. G.I. Likhtenshtein, Nitroxides: 170 Years of history in biology and biomedicine. Intern. Res. J. Pure Appl. Chem. **8**, 1 (2015)
137. J.H. Park, W.E. Trommer, *Biological Magnetic Resonance*, eds. by S.S. Eaton, G.R. Eaton, L. Berliner (Kluwer Academic/Plenum Publisher, 2005), Chap. 11
138. S.A. Rice, A.R. Dinner, Adv. Chem. Phys. **150**, 170 (2012)
139. G.Y. Shevelev, O.A. Krumkacheva, A.A. Lomzov, D. Kuzhelev, V. Trukhin, O.Y. Rogozhnikova, V.M. Tormyshev, D.V. Pyshnyi, M.V. Fedin, E.G. Bagryanskaya, J. Phys. Chem. B (2015)

Conclusions and Outlook

Some two centuries ago the German poet, writer and philosopher Johann Wolfgang von Goethe had noted that *"The drama she* [Nature] *plays is always new, because she is always bringing new spectators"* [1]. While we cannot be fully certain what were exactly on the Goethe's mind at that time, one thing for sure he was not reflecting on electronic spins and their intriguing quantum mechanical interactions. However, today we are entirely confident that energetically weak but fundamentally important spin phenomena would always find some curious spectators among the scientists attempting to comprehend the Nature's greatest play. Nowadays' researchers equipped with modern spectroscopic methods, creative chemistry and deep knowledge of biology, comprehensive quantum mechanical theories and virtually unlimited computational resources take on some of the most difficult aspects of electron interactions in systems of ever growing complexity. These coherent efforts across the disciplines propelled the recent progress in many basic and applied fields that span from spin chemistry to spintronics and biological spin labeling.

Spin is a solely quantum-mechanical concept that does not arise from classical mechanics and cannot be easily rationalized without relativistic quantum mechanical interpretation developed by Dirac. However, the electronic spin and its interactions are responsible for an exceptional diversity in structural and dynamical properties of molecular systems bearing an unpaired electronic spin. This diversity is owned by four fundamental phenomena. The first two phenomena—electronic spin-spin exchange interaction (SSEI) and electron spin-nuclear spin contact interaction (SNCI)—are strictly quantum mechanical in nature while the other two —electronic spin-spin dipole-dipole interaction (SSDDI) and electron spin–nuclear spin dipole-dipole interaction (SNDDI) allow for semi-classical interpretation.

As discussed in this volume, SSEI could be either static or dynamic. The former is caused by a direct overlap of electronic orbitals of individual paramagnetic species (direct SSEI) or an effective overlap via a rigid chemical bridge (super SSEI). The dynamic exchange is realized upon molecular encounters of spin bearing compounds diffusing in solution or connected by a flexible bridge. In all of the aforementioned cases, the efficiency of the exchange is determined by the magnitude of the exchange integral J and the values of the interacting spins. The

© Springer International Publishing Switzerland 2016   327
G. Likhtenshtein, *Electron Spin Interactions in Chemistry and Biology*,
Biological and Medical Physics, Biomedical Engineering,
DOI 10.1007/978-3-319-33927-6

electronic spin is always partially delocalized within a molecule leading to appearance of non-zero spin density on nucleuses.

Electron spin exchange interaction plays the pivotal role in electron transfer and the long distance electron transfer (LDET) in particular. The LDET is well described by the Marcus-Levich equation that depends upon two main parameters —the value of the resonance integral (coupling factor) V and the Frank-Condon factor (FCF). Both these factors are also responsible for processes involving excited triplet states of chromophores including intersystem crossing (IC) via spin orbital interaction (SOI), triplet-triplet energy transfer (TTET) and triplet-triplet annihilation (TTA). In the frame of FCF, nuclear translational, rotational and vibration dynamics are all sufficiently affect the electron transfer and, in a lesser extend, the processes in triplet systems. A correlation between the long distance spin exchange (LDSE) and LDET has been established.

Dipolar interactions lead to a plethora of physical phenomena that determine materials' properties and serve as the basis of many informative experimental techniques. In classical approximation, a localized magnetic dipole (either electronic or nuclear spin) produces a local magnetic field in the location of the second spin. This interaction depends on the magnetic moments of the individual spins, the interspin distance, and the angle between the interspin vector and the external magnetic field. Dipolar interaction could be modulated by dynamic phenomena affecting the interspin distance (translational diffusion) or the angle (rotational diffusion) or both.

The theories of electron spin–electron spin dipolar interactions and electron spin–nuclear spin dipolar and contact interactions form the basis for spectroscopic methods capable of providing atomic-resolution structures of molecular systems possessing unpaired electronic spin(s), detailed data on molecular motion, and establishing spin distribution in non-crystallized molecular systems in any phase state. The arsenal of the spectroscopic methods includes conventional continuous wave (CW) ESR, ESR with optical and electrical detection (OD ESR and ED ESR, respectively), Fourier Transform (FT) and transient (TR) ESR and several important double resonance techniques. Electron-electron double resonance (ELDOR) could be carried out in both CW and time-domain mode. The latter is typically termed DEER (double electron-electron resonance) or PELDOR (pulsed electron-electron double resonance). Interactions of electronic and nuclear spins are effectively probed by electron spin-echo envelope modulation (ESEEM) and hyperfine sublevel correlation (HYSCORE) spectroscopy and electron-nuclear double resonance (ENDOR) including its several advanced modifications such as ENDOR with circularly polarized radiofrequency fields (CP-ENDOR), electron-nuclear-nuclear resonance (double ENDOR) and electron-nuclear-nuclear triple resonance (TRIPLE). Other techniques include two-dimensional (2D) ESR and two-dimensional electron-electron double resonance (2D-ELDOR), reaction yield detection magnetic resonance (RYDMR) and magnetically affected reaction yield (MARY), proton-electron double resonance imaging (PEDRI) and several others.

Spin-selective processes involving electron and nuclear spins are the main areas of interest for spin chemistry. One of the key phenomena arising from interactions between an unpaired electron and nuclear spins is dynamic nuclear spin polarization (DNP) effect originally predicted by Overhauser for metals. While specific mechanisms of DNP depends upon spin systems, DNP generally refers to a transfer of the spin polarization from electrons to nuclei resulting in a tremendous enhancement in population difference of nuclear spins for a given temperature and static polarization field (up to the theoretical limit of 660-fold for protons) and yielding highly polarized nuclear spin states. Spin-selective physical and chemical processes could also yield highly polarized nuclear and/or electronic spin states. These processes have a common mechanism that involves spin-orbital conversion of the singlet state (S_1) to the triplet state, which, in the presence of magnetic field B_0, is split into three states T_0, T_+ and T_-. The electron spin polarization occurs when $T_0 - T_-$ and $T_0 - T_+$ transitions are induced by electron spin relaxation or electromagnetic (RF or microwave) field satisfying the resonance condition. Understanding this fundamental phenomenon led to development of a number of advanced spin-selective methods including chemically induced spin-correlated dynamic electron polarization (CIDEP), chemical-induced dynamic nuclear polarization (CIDNP), and rapid development of various polarization schemes for dynamic nuclear polarization (DNP) in application to NMR in recent years. These methods provide researchers with unique experimental tool for gaining some deep insights into radical and ion-radical processes by establishing structures of short living intermediates and monitoring its spin dynamics while DNP holds the promise to dramatically improve sensitivity of NMR methods over an exceptionally broad field of applications—from structure of proteins to analysis of small molecules at low concentrations.

Recent development of new advanced ESR time-domain methods has been timed favorably to the rapidly accelerating fields at the interface of materials science, chemistry, and structural biology. An incomplete list of new abbreviations in ESR methods that are coming to laboratories across the globe includes Fourier transform ESR (FT ESR) and its two-dimensional version—2D FT ESR, electron spin echo (ESE) and electron spin echo envelope modulation (ESEEM), hyperfine sublevel correlation spectroscopy (HYSCORE) and its multidimensional version—double nuclear coherence transfer hyperfine sublevel correlation spectroscopy (DONUT-HYSCORE), electron spin echo magnetization transfer (ESE MT), two-dimensional electron spin echo correlation spectroscopy (2D SECSY), pulsed versions of electron-nuclear double resonance (pulse ENDOR) and electron-electron double resonance (PELDOR), electron spin transient nutation (ESTN) and its two-dimensional version (2D ESTN), phase inverted echo-amplitude detected nutation (PEANUT), magnetic isotope effect (MEF), quantum beats effect (QBE), and double quantum coherence pulsed ESR (DQC ESR). While this list is still significantly shorter of the main experiments practiced by NMR, we can say with certainty that even more advanced multi-spin experiments involving electronic states will be introduced in the near future.

Complementary information on the spin states and spin dynamics can also be obtained also by optical, Mössbauer (gamma-resonance, GR), neutron scattering and muon spin spectroscopy and magnetic measurements and these techniques are expected to continue playing major roles in the future.

Specifically, quantitative parameters of singlet-triplet transitions, triplet-triplet transfer, singlet and triplet excited state intermolecular quenching, intersystem crossing, long-distance electron transfer, and spin-polarized processes can be determined by absorption, fluorescence, IR and Raman spectroscopy including time-domain and two-dimensional versions of these methods that are often more informative. Further data on dynamics of intermolecular processes in liquids could be obtained from quenching of the excited triplet states by molecular oxygen, stable radicals and chromophores. Measurements of triplet-triplet transfer in bridged donor-acceptor pairs allow researchers to characterize the superexchange processes in molecular bridges and, potentially, lead to design of effective molecular wires and novel molecular electronic and spintronic devices. Single molecule fluorescence correlation, hole burning optical spectroscopy and organic light-emitting diode (OLED) spectroscopy are other examples of new and rapidly developing methods in this area.

Analysis of the superfine magnetic structure of the GR spectra of paramagnetic compounds bearing Mössbauer-active atoms such as ^{57}Fe, for example, enables accurate measurements of the intensity of internal magnetic fields and the value of the electron spin magnetic moment, as well as provide estimates of the exchange integral in bi- and multi-spin systems.

Other useful experimental techniques include static and dynamic measurements of magnetic susceptibility—well-established materials characterization techniques that gained tremendously in sensitivity with the invention of superconducting quantum interference device (SQUID). While magnetic susceptibility—a parameter indicating a degree of magnetization of a bulk sample in response to an applied magnetic field—lacks spectral resolution and additional data provided by ESR, analysis of this parameter as a function of magnetic field and/or temperature allows for elucidating electron spin effects including magnetic dipole moment per unit volume, the energy in zero magnetic field (the first order Zeeman term), as well as an estimate of the exchange integral J. Spatial resolution of these methods could be gained in scanning SQUID microscopy where a miniaturized SQUID is used to image surface magnetic field strength with micrometer resolution. This scanning SQUID microscopy has been already used to characterized microscopically heterogeneous high-temperature superconductor materials based on YBCO and BSCCO compounds.

Inelastic neutron scattering (INS) observes a change in energy of a neutron as it scattered by a sample. The method has been proven to be useful for probing a wide variety of different physical phenomena including magnetic and quantum excitations and electronic transitions. Inelastic neutron scattering echo techniques (neutron spin echo, NSE, and neutron resonance spin echo) are based on measuring the quantum mechanical phase of the neutrons in addition to their amplitudes and can detect the loss of polarization (magnetization) due to dephasing of the spins. The latter capability is very useful for studying electronic properties of solids including spin effects.

Muon spin resonance (μSR) spectroscopy is based on an implantation of short-lived spin-polarized muons (muon is an elementary particle similar to the electron with the same spin of 1/2 and roughly the same negative elementary electric charge as an electron but much heavier and, thus, having significantly lower gyromagnetic ratio than the electron) in matter and then detecting the influence of atomic, molecular or crystalline surroundings on their spin motion similar to the related spin resonance techniques ESR and NMR.

All of the aforementioned techniques together with the cornerstone ESR and NMR methods have been proved to be exceptionally powerful in investigating spin effects in chemical and biological processes that are of interest for either fundamental or applied science reasons. The utility of this combined approach has been illustrated on examples of investigation of electron spin effects in a number of chemical and physical processes and systems.

Chemically induced electron spin polarization (CIDEP) of reactive radicals can be attributed to a variety of mechanisms involving triplet state, radical pairs, spin-correlated radical pairs, radical triplet pairs, and cross-relaxation. With help of time-resolved EPR and related methods electron spin polarization (ESP) has been observed at nanosecond and microsecond time scale for radicals, radical pairs, and radical-ion pairs excited triplet states. Photochemically induced CIDNP (photo-CIDNP) and time-resolved photo-CIDNP magic-angle spinning (MAS) NMR allowed for detection of subtle changes in the electronic structure in photosynthetic reaction centers. Field dependent CIDNP spectroscopy has been used to measure the exchange interaction, J, in photochemically generated radical ion pairs of rigid donor-bridge-acceptor molecules. Further development and combination of these methods will be very useful in uncovering mechanisms of photosynthetic reactions in natural and artificial systems.

Advanced experimental methods call for rigorous theoretical interpretation and provide for critical tests of the existing and new theories, and, in turn, allow for new interpretation of the existing experimental data. For example, a semi-empirical approach for quantitative estimations of the effect of bridging groups on the long-distance electron transfer LDET has been developed. Analogy between superexchange in electron transfer (ET) and electron spin exchange (SE) allowed for estimating the values of attenuation (decay) parameter for ET and SE from experimental data on triplet-triplet energy transfer (TTET). It was shown that the experimental dependence of the ET rate constant in photosynthetic and chemical donor-acceptor (DA) pairs on the D-A distance is similar to the correspondent dependence predicted for SE. This observation allowed for employing available data on exchange integral J in a series bridged nitroxide biradicals for calculating and tabulating decay parameters as a function of individual chemical groups.

Electromagnetic as well as static magnetic fields can be effective in altering the yield and/or product distribution of chemical and biochemical reactions that involve species possessing an unpaired electron spin (*i.e.*, radical or radical ion intermediates). These spin effects are based on fundamental properties of spin correlated radical pairs (SCRP), in which S ↔T interconversion is driven by both internal magnetic interactions and the externally applied magnetic and/or electromagnetic fields. Detailed understanding of the elementary processes in SCRP allowed for predicting the changes in the singlet yield and, consequently, the reaction yield, as a function of the static magnetic field strength. The spin exchange integral J can also be estimated from the magnitude of the observed magnetic field effects. The exchange interaction at short separation is strong and inhibits S $\leftrightarrow$ T interconversion. Significant magnetic field effects were detected for photoinduced electron transfer between phenazine and amines, electron transfer from ethyflferrocene (Fc) and oxazine, the hydrogen abstraction reaction of 4-methoxybenzophenone with thiophenol and in photosynthetic reaction centers.

One of the most fascinating manifestations of magnetic field effects on living organisms appears to be in ability of some birds, mammals, reptiles, amphibians, fish, crustaceans and insects to navigate in the Earth's magnetic field. What is the molecular mechanism beyond such molecular compass and does this mechanism invoke spins? Currently, two main hypotheses are discussed in the literature. One mechanism is reminiscent to the human compass invention and invokes deposits of the iron oxide particles of magnetite (Fe_3O_4).The second suggested mechanism involved light dependent radical-pair which originates from an anisotropy of the hyperfine or fine interactions in a weak magnetic field experienced by unpaired electron spins in a redox process. Cryptochromes, flavoproteins implicated in multiple blue light-dependent signaling pathways in plants or circadian clocks in animals, have been proposed to be responsible for the magnetic field "compass" sensitivity of birds. It was suggested that in the cryptochrome/photolyase, the redox-active cofactor FAD and triad of trytophans (Trp-triad) are participating in electron transfer initiated by blue light producing a flavosemiquinone radical, $FAD^{\cdot-}$ or $FADH^{\cdot}$, and a radical derived from the Trp-triad.

Electromagnetic fields may also affect physicochemical and biochemical processes involving electronic spin states to a rather large degree. For example, in spin-correlated radical pairs conversion of T_- and T_+ to T_0 state is affected by irradiation with microwave: such irradiation increases the population of the singlet state and, therefore, increases the singlet product yield. This fundamental physical phenomenon provides the ground for understanding effects of microwave radiation on a number of radical and ion radical processes.

Radio-frequency (rf) field can affect radical and ion radical reactions when it has the same direction of circular polarization as the electron spin precession in the magnetic field. This mechanism provides the theoretical basis for detailed study of spin dynamic effects by OD-ESR spectroscopy and related RYDMR methods. Investigation of photoinduced ET in donor-acceptor systems exposed to both static magnetic and rf fields suggested effective S$\leftrightarrow$T state interconversion in the cation-anion radical pair when the rf field is in resonance with hyperfine splitting(s). The influence of isotropic electron–nuclear hyperfine and Zeeman interactions of the electron spins with a linearly polarized rf field on evolution of a spin-correlated radical pair was observed for specific systems such as comprised of anthracene-d_{10} as a donor and 1,3-dicyanobenzene (DCB) as acceptor, and in the photoinduced electron-transfer from perdeuterated pyrene to 1,3-dicyanobenzene (1,3-DCB). Static and oscillating magnetic fields affect radical recombination reactions by the same mechanisms and have been the subject of intense investigations in recent years.

Spin phenomena also manifest in the magnetic isotope effect (MIE). MIE is based on the spin conservation rule, according to which reactions with participation of radicals, ion-radicals, paramagnetic metal ions and molecules are spin selective; *i.e.*, the reactions proceed along the spin-allowed channels. MIE can be enhanced by selective microwave pumping of the radical pairs carrying magnetic nuclei. Such pimping stimulates triplet–singlet spin conversion increasing recombination probability and, therefore, the yield of the product(s) with magnetic nuclei. This phenomenon has been analyzed theoretically including effects of frequency and magnitude of the rf field, hyperfine coupling constants, and other magnetic and chemical parameters of the radical pair. MIE was observed experimentally for many chemical processes (photochemical and thermochemical reductions, fractionating oxygen isotopes in the photo-oxidation of water by molecular oxygen and enzymatic ATP synthesis catalyzed by magnesium ions) and for isotopic pairs and triads (H-D, $^{12}C-^{13}C$, $^{16}O-^{17}O-^{18}O$, $^{28}Si-^{29}Si-^{30}Si$, $^{32}S-^{33}S-^{34}S$, $^{72}Ge-^{73}Ge-^{74}Ge$, $^{235}U-^{238}U$, $^{198,200}Hg-^{199,201}Hg$, $^{24}Mg-^{25}Mg-^{26}Mg$.

Studies of electron spin interactions have many implications for understanding spin states and molecular structures of a large number of systems ranging from transient and persistent radicals, transition metal complexes, polymers, surfaces, and biological systems such as proteins, enzymes, biomembranes, nucleic acids, cells, organs, animals and even humans.

Such studies have many implications for both fundamental and applied science and spurring new technologies. For example, a light-induced reversible transformation of chemical species between two organic isomers having different magnetic properties (photo-magnetochromism) as a mechanism for magnetic data storage has been proposed and the feasibility has been demonstrated using several molecular systems.

Molecular systems exhibiting long-range magnetic order and spontaneous magnetization (molecular magnets) have attracted the special interest and this filed is expected to grow even further. A great diversity of known molecular magnetic systems fall into the following classes: single molecular magnets, single chain magnets and bulk magnets (the latter are typically classified as paramagnetic, superparamagnetic, ferromagnetic, antiferromagnetic, and ferromagnetic sytems).

The spin polarization phenomenon may arise from different exchange interactions between the unpaired electron in the singly occupied molecular orbital (SOMO) with the spin "up" and "down" electrons in the fully occupied molecular orbitals (FOMO) and between the SOMO and the FOMO spins. In charge-transfer salts composed of chains of alternating electron-donor and electron-acceptor molecules, the exchange interactions have shown to play the pivotal role.

Recent advances in nanotechnology and nanofabrication methods and the society needs for new advanced materials are placing spotlight on the spin clusters exhibiting strong interspin interactions. Technologically important transition and/or rare earth metal nanoclusters in either bare form, imbedded in a diamagnetic matrix or protected by organic ligands have the length scale of 1–100 nm. Commonly the spin exchange interactions between magnetic ions inside of such a cluster ($J \sim 10^2$ K) are several order of magnitudes stronger than for molecules in a diluted crystal ($\sim 10^{-3}$ K). For many nanoclusters the exchange interaction has been found to be antiferromagnetic and, therefore, on molecular level the nanoclusters can be considered as molecular ferrimagnets. Currently, an arsenal of modern physical methods including the magnetic susceptibility, ESR, NMR, electron spin transient nutation, Moessbauer, x-ray and photoelectron spectroscopy, neutron and mions scattering, resonance absorption of phonons and electrochemistry is being applied for studying such nanosystems providing wealth of detailed data on their unusual electronic structure and spin properties.

It should be noted here that transition and rare earth ions are not the only choice for building blocks for new advanced magnetic materials. Several purely organic materials that are based on isolated molecular one-dimensional (1D) chains have been described to demonstrate superparamagnet-like properties. Notable examples include molecular chains composed from thiazyl, imidazole-3-oxide-1-oxyl, chiral nitroxide radicals, metallorganics, dopted and undoped conjugated polymers. In such radical chain systems the Heisenberg exchange is primarily responsible for magnetic anisotropy because dipolar interactions and spin-orbit coupling are comparably weak. Other potential materials for purely organic magnetic materials and magnetic devices include π-conjugated superparamagnetic organic compounds including high-spin polyradicals, polyarylmethyl tri-, penta-, hepta-, and hexadeca radicals. Some of these compounds are particularly attractive as they exhibit high stability at ambient temperature and/or higher magnetic ordering temperatures have been attracted attention. This area of research had flourished over the last two decades and the current and future technology needs are expected to fuel its further grows in years to come.

Nature's solar energy conversion and storage systems that is found in photosynthetic organisms, including plants, algae and a variety of types of bacteria, provide us with an astonishing example in energy efficiency with the quantum yield close to one. The primary photochemical and photophysical processes in donor-acceptor D-A pair of photosynthetic reaction centers RCs lead to charge photoseparation (D^+A^-), where the cation-radical D^+ is a strong oxidant and the anion-radical A^- is a strong reducing agent. The most challenging problems of photosynthetis—prevention of the fast recombination in the charge-separated pair —was solved brilliantly by the mother Nature building subsequent chemical reactions that store energy in stable chemical bonds. Spin electron effects in bacterial photosynthetic reaction center (BRC), plants photosystem I (PS I) and photosystem II (PS II), the role of polynuclear manganese systems in the photooxidation of water, oxygen evolution complex (OEC)—all have been studied and described in detail in the recent years. The overall picture of energy transduction in photosynthesis had emerged providing us with a remarkable example of how a clever combination of ordinary active groups into an ordered structure would yield qualitatively new and remarkable properties. While x-ray crystallography revealed the intricate organization of photosynthetic centers, the understanding of spin-effects that run this nature's energy capture machinery would not be possible without advanced EPR (particularly pulsed and high field/multifrequency methods), double resonance (ENDOR and PELDOR), CIDNP, and NMR methods. The same methods were found to be essential in solving the mechanism of water splitting that gives the life on Earth. These lessons from the Nature should be taken into consideration when solving the problem of artificial water photochemical splitting into O_2. The first important lesson is that manganese or other transition metal clusters of high redox potential should be used. The second essential lesson is that the potential should be collected by four one-electron steps and be terminated by one step four-electron mechanism.

Even though x-ray crystallography has been exceptionally successful in understanding molecular structure (and to lesser degree function) of biological macromolecules including proteins, the growing consensus is that not all the problems of structural biology could be solved by forcing intrinsically flexible molecules into just a few conformations that can be found in the crystal. Some of these problems, including structure and dynamics of flexible and/or partially disordered domains as well as exceptionally large protein complexes could be solved by magnetic resonance methods and particularly those involving electron spins. One can modify a small fraction of the molecular system by a molecular spin label (niroxide and other stable radicals, paramagnetic complex, luminophores in excited triplet state) and then follow the labeled site by monitoring its spin state and properties by corresponding physical methods. The following magnetic spin-spin interactions provide the means for measuring spin-spin distances and molecular dynamics of the spin compound bearing spin electron: dipole-dipole electron-electron and electron-nuclei interaction, and spin exchange. Continuous wave and pulse electron spin resonance methods have proved to be exceptionally sensitive and informative for establishing structural and dynamic properties of molecular systems, including biological macromolecules. Molecular structure, local dynamics of molecular chains, conformational transitions in labeled biological and chemical polymers, biological and model lipid bialyer membranes in solid and liquid states have all been investigated.

Measurements of electric potential, dielectric properties, redox status, concentration of dioxygen and acidity in various objects including living organisms are readily available for the physical labeling techniques as well. One of the authors of this monograph wrote in the first book on spin labeling [2, 3]: "It is thus our hope that spin labeling will continue to be an effective tool for solving various complicated problems in molecular biology". Now after 42 years, it is evident that present–day reality has surpassed all our optimistic expectations.

Theoretical and experimental material described in the book have formed theoretical and experimental basis to pursue further studies on the detailed mechanisms of numerous chemical and biological reactions. It is our hope that it will also stimulate further developments of novel methods for structural and dynamic studies, and inspire engineers for constructing new efficient devices. Organic and, in particular, metalloorganic compounds bearing spins appear to be promising materials for optical data storage, spintronics applications, optical sensing, spin valves, magnetic switching systems, optical information processing devices, conducting and semiconducting magnetic materials, etc. Superparamagnet-like isolated one-dimensional materials, with their intrinsic bistability, are promising for the development of devices for data storage, sensors of light and heat and switching units. Upconversion phenomena have already attracted much attention due to potential applications in photocatalysis, photovoltics (dye-sensitized solar cells), nonlinear optics, and molecular probes. Among these applications we would like to emphasize wavelength shifting for spectroscopy, sensitized photoreaction by low-energy photons, luminescent probes for bioimaging and improving the energy conversion efficiency of solar cells by harvesting photons below the energy threshold.

As we look into the future, there are all reasons to believe that the progress in both spin effects theory and experiment would continue for years to come. The field is vast and still awaits for rigorous exploration by curious minds. What one cannot ever predict are the new unexpectedly bright ideas and the paradigm-shifting discoveries. But what we can promise for sure is that new spectators will certainly come to appreciate one of the Nature's greatest play that certainly involves the spins.

References

1. B. Saunders, *The Maxims and Reflections of Goethe* (The Macmillan Company, New York, 1906)
2. G.I. Likhtenshtein, *Spin Labeling Method in Molecular Biology* (Nauka, Moscow, 1974)
3. G.I. Likhtenshtein, *Spin Labeling Method in Molecular Biology* (Wiley Interscience, New York, 1976)

Index

© Springer International Publishing Switzerland 2016
G. Likhtenshtein, *Electron Spin Interactions in Chemistry and Biology*,
Biological and Medical Physics, Biomedical Engineering,
DOI 10.1007/978-3-319-33927-6

CPSIA information can be obtained
at www.ICGtesting.com
Printed in the USA
LVOW05*1413300716

498414LV00002B/46/P